IET ENERGY ENGINEERING SERIES 124

Power Market Transformation

Other volumes in this series:

Power Market Transformation

Reducing emissions and empowering consumers

Barrie Murray

The Institution of Engineering and Technology

Published by The Institution of Engineering and Technology, London, United Kingdom

The Institution of Engineering and Technology is registered as a Charity in England & Wales (no. 211014) and Scotland (no. SC038698).

© The Institution of Engineering and Technology 2018

First published 2018

The Institution of Engineering and Technology
Michael Faraday House
Six Hills Way, Stevenage
Herts, SG1 2AY, United Kingdom

www.theiet.org

While the author and publisher believe that the information and guidance given in this work are correct, all parties must rely upon their own skill and judgement when making use of them. Neither the author nor publisher assumes any liability to anyone for any loss or damage caused by any error or omission in the work, whether such an error or omission is the result of negligence or any other cause. Any and all such liability is disclaimed.

The moral rights of the author to be identified as author of this work have been asserted by them in accordance with the Copyright, Designs and Patents Act 1988.

British Library Cataloguing in Publication Data
A catalogue record for this product is available from the British Library

ISBN 978-1-78561-481-1 (hardback)
ISBN 978-1-78561-482-8 (PDF)

Typeset in India by MPS Limited
Printed in the UK by CPI Group (UK) Ltd, Croydon

To my dear grandchildren
Ella, Harris, Henry and Scarlett

Contents

Foreword

This book is written by an internationally renowned consultant with a lifetime's experience in the energy sector. The work is based on his most recent experience providing services across the world to investors, banks, government agencies and utilities engaged in restructuring the industry and implementing power markets. It draws on the authors previous experience operating as a senior manager with both a generation and transmission company as well as with a distribution company and equipment supplier. The analysis is illustrated with worked examples providing insight balancing fundamental methodology and academic theory with practical information. It follows the theme of two previous successful books published 10 years apart that track the developments in the sector. The work will be of value to those organisations engaged in managing the changes taking place in the sector and new entrants. It also provides tutorial material suitable for senior graduates and post graduates interested in a career in the industry. The author has a PhD based on related work, is a Fellow of the UK Institution of Engineering and Technology and a Fellow of the Chartered Management Institute.

Preface

This book describes recent developments in the power and energy markets that are driving major changes in the generation mix, network requirements and customer engagement. The industry now faces new challenges associated with global warming and escalating demand from the developing countries such as China and India. The effect of government interventions to manage changes and to meet emission targets has resulted in an industry that is neither a free market nor centrally coordinated and lacks strategic direction. The uncertainty has resulted in a dearth of investment in generation and unprecedented low margins of spare capacity. This book provides an analysis of the changes and quantifies their impact. It reviews strategic decisions in the management of changes in the sector and aims to identify the optimal way forward to meet the triple objectives of security, affordability and sustainability with low emissions. It focuses on the economic aspects of initiatives to provide insight into their interaction. It will also provide insight into the opportunities in the sector to the many new potential players.

The book is based on the most recent developments in the energy sector and brings together all the initiatives that interact physically through the common network and financially through the energy market. It draws on extensive international experience in all sectors of the industry to bring clarity to the complex interactions. The book will be of value to all countries undergoing a transition. It will include worked examples and questions providing a basis for study and examination by academic institutions.

The sector is undergoing a major transformation to meet environmental targets and engage with a more active consumer base supported by smart networks and meters with local generation. The implications of these developments to the sector are not well understood. The development of distributed generation was forecast in my earlier 2009 publication to increase to meet 40% of the power energy needs by 2050 (p289). This book explains the implications of these developments to the existing market incumbents and a lot of new potential players. It also analyses the relative economics of central and distributed generation and associated network charges. It is presented in four parts.

Part 1 – How we arrived at where we are?

The first part of the book reviews the recent developments that are taking place in the international power sector across the world. It discusses the drivers of change and their impact on the market players and system operation. It discusses the impact of strategic decisions to accommodate renewable sources and their adverse affects on the development and performance of the sector. It discusses the reaction of the industry and investors and the developing engagement with customers.

Part 2 – Review of low carbon generation technology options

The second part analyses the changing global landscape of the sector to meet the objectives of low emissions and cost whilst maintaining security. It analyses the economic and technical aspects of the use of: embedded small scale distributed generation; renewable generation sources; large and small scale nuclear and fusion; carbon capture and storage and biomass generation. Their potential to meet the evolving needs of the industry is analysed. The implications of developments to network requirements and charges are discussed.

Part 3 – How are the changes being managed within a market environment?

The third part discusses the market mechanisms in place and needed to manage the emerging developments. It discusses the balancing and capacity markets and the impact of renewable and embedded generation on conventional large scale generators. It reviews the impact on network utilisation and charging mechanisms. It appraises the potential for cross-border trading, demand side management and exploiting interaction with gas, heat and transport systems.

Part 4 – How do we get to where we want to be?

The fourth part focuses on the analysis of options to meet the three concurrent objectives of security, cost and sustainability in curtailing emissions to identify an optimum strategy. A set of scenarios is developed to provide a framework for analysis based on the use of an overview model coupled with a detailed model embracing the technical and operational aspects of system management. An approach to manage future trading and evaluate contract risks is outlined. It reviews smart grid developments and their potential to support the sector transition. The work concludes with a review of the governance and market arrangements to identify the need for any changes to meet the future requirements resulting from the impending transformation.

Dr Barrie Murray
September 2017
barriemurray.ems@btinternet.com

Part I

How we arrived at where we are?

The first part of the book reviews the recent developments that are taking place in the international power sector across the world. It discusses the impact of strategic decisions and where they have adversely affected the development and performance of the sector. It discusses policy and market developments and the reaction of the industry and investors and the developing engagement with customers.

Chapter 1 – Market developments – This chapter discusses the drivers of change to realise reductions in emissions whilst maintaining security of supply and containing end-user bills. It describes how governments have sought to influence the developments of renewable energy with subsidies and reviews their impact on markets, conventional generators and end users. The market design options are reviewed to establish how best to manage the transition in the industry and maintain competition. The impact on the development of the optimum plant mix is discussed along with facilitating demand side engagement.

Chapter 2 – Market technical challenges – This chapter reviews the technical issues arising as a result of the drive to decarbonise electricity generation that will potentially constrain its development. The adverse impact of higher proportions of non-synchronous generation on system inertia and frequency control is discussed. The demand profile is much more variable and difficult to predict. It is more difficult for potential investors in generation to predict utilisation and potential revenues. There is an impact on the operating regime of generation and their cost implications altering the optimal plant mix. The need for wind output curtailment increases exponentially when the non-synchronous capacity reaches around 50%. The implication to the operation of the distribution and transmission networks is reviewed along with the changes in interconnector flows due to intermittent generation. It concludes that there is no competitive market operating for generation at the distribution level to support competition between local generation sources, suppliers and grid-connected generation.

Chapter 3 – Impact of developments on market players – This chapter discusses the impact of the transition on market players. It reviews the impact on generating companies of the reduced utilisation due to their plant being increasingly displaced by renewable output. This has led to a collapse in revenues with company valuations falling by up to 80%. The impact of the cost of CO_2 emissions and emission performance standards that is driving coal-fired generation out of the market is analysed. The impact of subsidies and the costs to end users is quantified. The shift in the use of the transmission and the role of distribution companies is discussed as their networks become more active. The potential contribution of intermittent generation to the capacity margin is analysed and why the security margins are falling.

Chapter 1

Market developments

1.1 Industry in transition

The energy industry is in a state of transition trying to meet environmental constraints on emissions whilst maintaining security and containing costs. Ambitious targets have been accepted by governments that require an increasing proportion of energy to be derived from renewable and other low-emission sources. To achieve the targets, governments have introduced

- Subsidies to encourage the development of renewable sources like wind and solar;
- A tax on emissions like CO_2 penalising generation from coal and to a lesser extent gas-fired generation;
- Established a requirement that renewable energy must be used when it is available or compensated for lost revenue;
- Promoted the involvement of consumers in the market through the application of smart meters to encourage more efficient use of energy.

These developments have distorted the operation of the normal wholesale market and the financial position and investment decisions of its participants.

- The utilisation of conventional generation is reduced with its output being displaced by renewable sources;
- The conventional generation becomes non-viable financially and may be closed or mothballed reducing the plant margin;
- The renewable generation is intermittent, and its output variations need to be balanced using conventional generators that incur extra operational costs in part loading, starting and ramping up and down;
- A significant proportion of the new generation is small in scale and connected to the distribution system rather than the high-voltage grid;
- The embedded generation has transformed the distribution networks from being radial transmitters to active networks requiring real-time control;
- Wind generation location is dictated by wind conditions and will often require grid extensions to remote sites;
- System operation is challenged by the impact of non-synchronous generation that may make a limited contribution to maintain system inertia and frequency

control particularly during low-load conditions coinciding with high levels of
renewable output;

- Customer awareness of the energy market has been promoted through
  government-sponsored efficiency improvement schemes coupled with the
  application of smart meters;
- All energy users have been exposed to rising prices in part due to subsidies for
  renewable generation with energy-intensive industries needing exemption to
  stay financially viable.

This book discusses and quantifies the impact of these drivers on the industry and
its participants and proposes how the industries organisation and regulation may
need to change to accommodate them. This chapter focuses on the implications to
free market operation.

1.2   Impact of developments

The utilisation of conventional generation has been drastically reduced with output
being displaced by renewable generation that takes priority in dispatch. European
Union targets are set to generate around 34% of electrical energy from renewable
sources. This equates to an even larger reduction in marginal coal generation virtually driving it out of the market. The closure of generating stations is leading to
unprecedented low plant margins that will impact security.

At the same time, conventional generation is still required to be available to
replace renewable generation output at times of low wind and solar output. It is also
required to track the variations in renewable output leading to higher operating
costs in part loading at reduced efficiency and incurring ramping costs and more
frequent starts and stops. This operating role further undermines the financial viability of conventional generation and adds to plant wear and tear.

Capacity markets have been introduced to try and arrest the fall in the plant
margin but an undesirable consequence has been the promotion of small embedded
generation connected to the distribution system. This generation does not pay
transmission costs, and its output helps to reduce the use of system charges to
suppliers. This situation distorts the competition between larger transmission-
connected generation and the embedded generation. The smaller scale generation
may be based on diesel generation with high levels of emissions.

The expansion of embedded generation and renewable generation connected at
distribution voltages is **transforming the distribution networks from being
essentially radial and passive to becoming active**. In some areas, they are
reaching the stage of saturation requiring reinforcement. They will need more
sophisticated monitoring and control facilities similar to those employed at transmission levels. These developments and the advent of electric vehicles will lead to
significant distribution network development costs that have not been fully factored
into current government policy.

In contrast, the transmission system is facing a fall in its utilisation with a shift
of some 15% of conventional generation to distribution voltages. This is in part due

to it being practical to connect the lower unit size of renewable generation like wind turbines to distribution voltages with lower costs. The reduction in loading on transmission circuits is resulting in high voltages at times of low demand and the need to switch some circuits out of service. The connection of remote wind farms can be put out to tender and is resulting in an increase in the number of transmission operators.

The operation of the power system is becoming more complex due to the intermittency of renewable generation and its priority dispatch. The net system demand is much less predictable compounded by the limited visibility of the output of embedded generation and the demand response. The renewable generation is essentially non-synchronous and does not contribute in the same way to system inertia and the support of frequency control. Voltage control is also a problem with the variations in the reactive output from renewable generation. The requirements for reserve to counter intermittency require continual management.

End users are faced with price rises to support the subsidies paid to renewable generators with bills rising by 13% or more. They are encouraged to participate in the market with the roll out of smart meters but have seen little benefit. Much has been said about demand side response through the use of price signals and local storage but its costs and viability have yet to be proven.

1.3 Emission targets

Ambitious targets for the reduction of emissions were set by the EU that has had far-reaching effects on the power industry. **The subsidies introduced for renewable generation distort market operation and introduce risks** discouraging new investment in conventional generation. It has impacted on the profitability of existing market players and increased prices for end users.

The Dutch Parliament decided in 2016 to cut the country's CO_2 emissions by 55% by 2030. This would mean that the Netherlands would have to close all its coal-fired power plants. The decision taken by the Parliament would bring the Netherlands undoubtedly in line with the Paris Climate Change pact, which has a long-term goal of cutting CO_2 levels by 80%–95% by 2050. There is a question regarding the replacement generation. Last year, the country saw its greenhouse gas emission increased by about 5% in comparison to 2014, while carbon monoxide levels also rose last year, being 2% higher than in 1990. The rise was due to a boost in coal-fired power production. By 2015, the Netherlands had already closed five coal-fired power stations, but there are five other plants, with three of them coming into operation as recent as 2015. Norway intends to ban petrol and diesel car sales by 2025. Norway is the largest market for electric cars in the world, with electric vehicles accounting for 24% of those on the road.

The United Kingdom accepted that by 2020 15% of energy for electricity generation, heat and transport should come from low-emission sources. Based on the typical energy production data, Table 1.1 illustrates what is required to meet this target in the United Kingdom.

Table 1.1   Target reductions

	Electricity	Transport	Heating	Totals
Total energy, TW h	350	600	700	1,650
Target reduction, %	34	10	10	15
Renewable, TW h	119	60	70	249

It can be seen that the largest target reduction in emissions is in the electricity generation sector and has driven major changes in the generation mix and its location. The carbon tax is paid by electricity generators for every tonne of carbon dioxide (CO_2) they release. In the United Kingdom, this was frozen in 2014 at 18 pounds per tonne until 2021. Most British electricity companies absorb the carbon tax by transferring it to clients via increased power bills. Companies with low carbon generation such as nuclear or renewables take advantage from the higher electricity prices. British electricity producers have called on the government to preserve the country's carbon tax until at least 2025, putting them at odds with industrial groups who want it reduced. Industrial conglomerates have demanded the government abandon the tax, declaring it has made power prices in Britain uncompetitive. The carbon tax adds approximately £6/MW h to gas generation prices and £14/MW h to coal raising British wholesale electricity prices, which in 2016 traded at approximately 40 pounds/MW h. The carbon tax is designed to encourage generators to invest in low carbon electricity generation and declared as important to the country's efforts to meet its climate change objectives. Britain has a legally binding goal to reduce its emissions by 80% from 1990 levels by 2050 and has embarked on electricity market reforms (EMRs) aimed at increasing investment in low carbon nuclear and clean power. Britain also intends to stop coal-fired power generation by 2025.

By 2016, there was approximately 14.5 TW h of renewable transport against the 2020 target of 60 TW h. At a historic rate of 1 TW h/year, the target is unlikely to be reached and would need to increase to 12 TW h/year. By 2016, renewable heat production was 35 TW h against the target of 70 TW h. The historic rate at that time was 2.5 TW h/year and needed to be increased to 9 TW h/year to meet the 2020 target of 70 TW h/year. **Taken together to meet the 10% targets, heat and transport have to convert 80 TW h of plant to low carbon energy sources by 2020.**

The retail sector is reacting to higher costs with more consumers switching supplier. In the United Kingdom, 5 million customers (around 16%) changed supplier in 2016 that is more than in other sector including mobile phones (14%) and broadband (9%). This is in part due to more suppliers entering the market with smaller and mid-size units capturing 20% of the market. This development promotes competition with aggregators supporting the identification of switching opportunities and facilitating the process.

1.4   Unbundling

The key action in creating a competitive electricity sector is 'unbundling' the vertically integrated state organisations to separate those entities that are natural monopolies, like transmission and distribution, from those where competition can be introduced. The monopolies are subject to price regulation administered by a government agency (the Regulator) that reviews costs and plans and seeks to encourage efficiency improvements using incentive mechanisms. The preferred arrangement is for the owner of the transmission network to be a separate entity from the system operator or at least ring-fenced financially (the so-called Independent System Operator).

New participants in the sector have to be licenced and in the United Kingdom. Policies have encouraged participation to promote competition and, **by 2017, there were some 400 licence holders including**

- 160 domestic and non-domestic supply companies
- 40 non-domestic supply companies
- 150 generators
- 16 interconnectors
- 18 transmission operators, 14 distribution operators and 9 independents.

This indicates the scale of the problem of realising coordinated development of the sector through a period of rapid change. This is particularly difficult when a lot of the new players have a limited knowledge of the way the system has to be operated.

The distribution companies are usually separated into ownership of the network and electricity supply i.e. the business of buying electricity wholesale and selling to the end users against a contract or tariff. Both the generators and suppliers pay for using the networks against regulated Use of System tariffs. The competitive business of generation and supply is facilitated through the introduction of a pool or power exchanges to support trading. This could be administered by a separate government-backed agency as in some countries in the Middle East or by a branch of the existing utility as in South Africa.

It is now over 20 years since the start of the development of electricity markets, and a variety of models have been applied around the world usually starting with a mandatory pool leading up to a fully open multi-market model with a liberalised supply side. The basic model types that have been implemented are shown in Table 1.2 but there are variations on these depending on what features are included.

The principal motivation for these developments was that competition would lead to improved efficiencies and drive down prices. In some respects, the economists promoting this concept were naive, and in practice, this ambition has not always been realised for a number of reasons:

- It is difficult to create a liquid energy market supported by open price discovery with an adequate number of participants operating at the margin to engender competition;

Table 1.2   Energy market options

State utility	Single generating and transmission company supplying distribution companies wholesale against a published tariff
Single-buyer model	Multiple generators selling to a single buyer acting on behalf of all users and providing wholesale to distributors and large users
Mandatory pool	All independent generators bid output into the pool with distributors/ suppliers buying from the pool at the system sell price
Bilateral market	Generators and suppliers are free to arrange bilateral contracts with a balancing market run by System Operator (SO) to match real-time generation to demand
Power boards	Multiple generators with their own supply business with short-term trading through exchanges and real-time balancing by the SO

- It has proved difficult to involve the demand side in the competitive process because of their inherent inflexibility;
- Prices on the day are largely constrained by previous investment decisions and prevailing fuel prices, and it is difficult to encourage the longer term optimum plant mix and margin of spare capacity through a short-term market;
- Vertical integration between generators and suppliers enables cross-subsidy and price manipulation that discourages new entry;
- Traditionally, the state utilities have charged on average prices rather than the higher marginal prices, and there has been a reluctance to let prices rise to new entry levels.

To encourage investment in new generation, the market structure and its governance has to appear open with opportunity to compete on a fair basis. New entrants are averse to organisations where the transmission business administers the market operation and suspects that they will treat the original state generation more favourably. There may also be concern that the transmission network business has interests in advocating developments that support its business revenues and are not entirely independent. They may also directly compete with generation in the supply of reactive energy. **Government subsidies to renewable generation and their priority dispatch has distorted the operation of the wholesale market and added risks that discourage investment.**

The network Use of System tariffs are designed to cover the fixed cost of the assets used and metering and system service charges as well as system losses and operating costs. The fixed costs are based on the registered maximum capacity and peak transfers, while the variable costs are based on the metered energy transfers. In some implementations, the network charges vary according to location on the system. The UK approach was to establish zonal use of system prices based on the proportion of the assets used by an injection at the connection node. The concept was to encourage generation to locate in areas close to demand centres to avoid overloading the transmission system with long distance bulk energy transfer. In practice, location was often determined by opportunities founded on local industrial synergy. The preferred arrangement is for the transmission business to operate as a service provider with regulated tariffs that reflect the user capacity requirements and their energy transfers that affect system losses. The system operator also manages the transfers in and out of the network using a short-term balancing market to match supply with demand in real time. **The distribution networks were essentially passive but are increasingly becoming active with embedded generation.** This is leading to the need for a closer interface between the Transmission System Operator (TSO) and Distribution System Operator (DSO).

Where generators are also allowed to act as suppliers with their own demand, then they effectively operate as self-contained power boards with trading at the margin with other power boards through exchanges. The internal trading arrangements limit the scope for price discovery and also provide the opportunity to manipulate the market wholesale price by cross-subsidies between the internal business divisions. The wholesale price can be artificially inflated by generators charging their suppliers higher prices which are absorbed without raising end-user prices. This would limit the ability of a new supplier to buy energy wholesale and undercut the power boards offers to end users. To counter this, regulators are requiring large generators to contract some of their capacity to new suppliers.

1.5 Implementing generation competition

The unbundling process involves breaking up the existing state generation stock into a number of clusters that can effectively compete whilst maintaining their financial viability. In practice, it is difficult to ensure that there is a sufficient range of generators operating at the margin at all times of the year. The most important consideration for cluster design is avoiding concentrating market power in a few clusters enabling them to control the market price and undermine any competition for dispatch. In the longer term, the market structure and pricing mechanism needs to facilitate new entry.

The pool arrangement process is similar to normal operation with generators providing data to the dispatcher and receiving instructions. Its implementation can be realised by adapting the existing scheduling algorithm to work on price submissions. A disadvantage is that it does not provide a basis for establishing Power Purchase Agreements (PPAs) to support new entry. These could only be established on a bilateral basis outside the pool and would not be an open process engendering competition.

With a single-buyer model, the preferred arrangement is for an independent government-backed organisation to act as the energy-purchasing agent on behalf of all consumers. This model has the advantage of providing a mechanism to manage the development of the plant mix and also the capacity margin. **It is most appropriate in managing periods of significant change where there is a need to attract new entry to bolster the plant margin but at the same time ensure fuel diversity and supply security.** It also affords the opportunity to retain tariffs on the basis of average rather than marginal costs to meet wider social objectives. Competition in this environment can be managed through auctions for new capacity as well as for additional tranches of energy at each annual, monthly and day ahead stage in competition with import options. New entry can be fostered by the sale of clusters backed by PPAs, for a limited period, pending wider market liberalisation. A large residual state generator can be retained to manage balancing and also provide make-up and spill options for smaller Independent Power Producers (IPPs) to balance their commitment to end users. A key consideration with cluster development is to create packages that are attractive to investors to establish new sources of finance to support other developments.

The Single Buyer (SB) model is not too difficult or costly a mechanism to establish and can be implemented relatively quickly. Subsequently, the market can be developed further by enabling some direct contracting between generators and end users. This could initially include supply to works on the same site or the definition of a percentage of output capacity that can be traded directly. The security that this direct trading option affords may appeal to large end users operating continuous processes. In this centrally controlled environment, cluster organisations would build expertise in managing their costs and trading pending wider liberalisation. The cluster design needs to be based on longer term plans as well as creating packages that will attract new entrants. Where a larger residual state generator is retained initially, to provide security and balancing, this could be fragmented and sold off as the market matures. A power exchange in this environment provides a mechanism to support the SB process in short-term trading.

The multi-market model enables any form of bilateral contracting and trading with only the balancing mechanism as a mandatory requirement. Any PPAs would have to be established on a bilateral basis between potential generators and a supply business or large end user and would not necessarily be through an open competitive process. This option relies on investors seeing current and prospective market prices that are above new entry costs. New investment could be realised through the sale of clusters but would have to be coupled with contracts for energy with the SB or the option to establish bilateral contracts with large users on favourable terms. There is a danger in this process that large intensive users would be 'cherry picked' leaving other clusters with a poorer load profile.

It is difficult and costly to establish a multi-market model with effective competition and liquidity. A relatively large number of clusters would realise an acceptable Hirschman–Herfindahl Index (HHI). The clusters would be established to avoid concentrations of market power able to set the marginal price. As well as trading expertise, the maintenance of secure system operation and balancing relies on there being a level of expertise in the market participants. A power exchange will be required to facilitate both generators and suppliers trading to adjust their short-term position and enable price discovery. The establishment of this type of market may follow from the operation of a successful pool or single-buyer market model when the necessary skills and processes are in place. Its development is driven by the market participants rather than by a central organisation.

The generators need to establish what will be required to operate successfully in the emerging market environment. The services currently derived centrally will need to be devolved or arrangements put in place to buy in the services. There will be a requirement to review the interfaces with the System and Market Operator to establish the required data flows. The internal processes necessary to provide the data will need to be defined and developed. There will also be a requirement to develop new skills in trading and risk management by new entrants. The organisation necessary to support the front, middle and back office functions will need to be established. The skills will take time to develop, and advantage should be taken of the relatively controlled environment afforded by a SB market model to build the business process model and form the organisation. The development of the process definition could be managed centrally initially with each group left to manage its own implementation and information Technology (IT) provision.

1.6 Rationale for market development

There is no universally correct choice of market structure, and the preferred arrangements at a particular time will be influenced by the stage of development of the system and its operating performance. For developing countries, the emphasis will be on creating an environment with options to facilitate new entry with state involvement to help secure investment. For the more developed countries, the emphasis may be placed on realising improvements in operating efficiency through competition. A number of countries including South Africa and Ireland chose to operate with a mandatory pool to ensure engagement in the competitive process. But it is unclear how the optimum level or mix of generation investment can be fostered in a long-term capital intensive industry like electricity generation with only a day-ahead market. The expectation promoted by leading economists was that the market would solve all. In the Middle East and parts of Africa, the single-buyer model has been the preferred model.

Across Europe, the move has been to an open bilateral market with generators and suppliers free to make bilateral deals with just a balancing market centrally administered by the SO to balance supply and demand in the event. The 'gate closure' defined when contracted transfers had to be declared to the SO and was progressively reduced to 1 hr ahead of the event. Generally, the bilateral deals that suppliers were prepared to make were for 6 months to a year or so ahead with power exchanges enabling price discovery with the same short horizon. With construction time scales of several years and life expectancy of large-scale generation of 30 years or more investors usually require more income certainty. To realise this, generators built up parallel supply businesses that provide an outlet for their capacity. They effectively became smaller scale vertically integrated utilities. It was also alleged that they could manipulate market prices. The supply business is often initially based on area franchises but is progressively opened up to competitive supply. Typically, this starts with large consumers above 1 MW, through to customers above 100 kW and eventually down to the domestic market. In the United Kingdom, bilateral trading between generators and suppliers was formalised by NETA (new electricity trading arrangements) in 2001.

Given the freedom to choose generation, a 'dash for gas' has occurred in a number of countries with a number of players choosing to build combined cycle gas turbine generation due to its lower capital cost and higher efficiency when compared to coal. Generators are able to dispatch their own generation to meet their contracted demand with any errors cashed out in the balancing market. Usually, like NETA, there are no capacity/availability payments, and generators have to recover all their costs through their contracted energy sales. But there was often more than adequate capacity, largely due to the entry of gas-fired generation that was available at competitive prices from the world market for equipment supply. This situation changed as generators faced with low utilisation were driven out of the market. As the plant margin has fallen, in a functioning market, price spikes should occur but these have been suppressed.

The operation of these markets has been distorted by the introduction of an ever-increasing tranche of subsidised renewable generation that is given priority access to run when available. The introduction of payments for the provision of capacity has been an attempt to manage capacity provision but has resulted in highlighting distortions in network charging between transmission and distribution-connected generation. The perceived risks in the market from the interventions of governments and regulators further help to discourage investment. **It is arguable that in a situation of rapid change the single-buyer market model provides more options to manage the impact of changes.**

1.7   Establishing competition

In operation, a key factor affecting the level of competition and market liquidity is establishing a sufficient number of competing participants. A common approach used to assess the market concentration is based on the HHI, calculated from the number of players in the market where

$$\text{HHI} = s1^2 + s2^2 + s3^2 + \ldots + sn^2$$

where sn is the market share of the ith firm.

The index provides a measure of competition and the likelihood of collusion within the market. The lower this value, the more robust the likely level of competition in the market; this does however go hand in hand with an increased level of counterparty risk and risk to business sustainability. The US Department of Justice considers a market with a result of less than 1,000 to be a competitive marketplace; a result of 1,000–1,800 to be a moderately concentrated marketplace and a result of 1,800 or greater to be a highly concentrated marketplace. As a general rule, mergers that increase the HHI by more than 100 points in concentrated markets raise anti-trust concerns.

As an example in the UK market, there are six big players, and if we assume an equal market share, this would result in an HHI of 1,665 and is not considered very competitive, and there are proposals that require the generators to make capacity available to new suppliers. A further complication is that the mixture of plant in each competing company needs to ensure that there is always competition at the margin where prices are set. If one company has plant operating at the margin for a large period of the year, then it is able to set prices. Also, where companies are able to vertically integrate, then they are able to cross-subsidise between the generation and supply business divisions to manipulate prices and discourage new entry.

According to the UK Competition and Markets Authority (CMA) report of 2016, millions of people were paying more than they need to for their energy. As such, it proposed making a database of customers on high tariffs available to rival suppliers who could then offer cheaper deals. The CMA also proposed a 'social tariff' to protect prepayment meter customers, until smart meters are universal. The workings of this capped tariff were unclear, but no one can be forced to supply electricity and gas at a loss.

European energy markets are badly distorted by the costs of green energy support measures, which the CMA said will account for 37% of typical UK bills by 2020. Smaller suppliers are exempt from much of these, allowing them to undercut the big players. So whilst the 'Big 6' might not have been served any mortal blows by the CMA, their trading environment is still difficult, effectively playing against state-sponsored new entrants.

1.8 Facilitating demand side involvement

Another approach to foster competition in the short term is to involve the demand side in the bidding process. In general, this has not been successful with the principle exception of Nordpool. This model, as applied in Scandinavia, has proved enduring and fully embraces the demand side. The model enables active demand side participation with generation and demand offers/bids matched to establish the market price for each period and each zone. All energy is then traded at that clearing price unless covered by bilateral contracts. Consumers bid into a day ahead market blocks of demand together with the price at which they are prepared to reduce demand. These prices are plotted as an ascending curve and compared with the curve of generator bid prices to establish the intersection point when the price consumers are prepared to pay matches the price that generators are prepared to sell at as shown in Figure 1.1. If the expected system balance price exceeds their bid, then their demand is not met. If network constraints become active, then the market is split, and each zone establishes its own balance point. In practice, it is only larger consumers with demand management facilities or embedded generation that can compete.

A typical bid structure is illustrated in Table 1.3 for a supplier with three blocks of in-house standby generation or interruptible demand available at the prices shown. The total supplier demand for an hour is shown to the right of the table together with a reserve provision of 824 MW. The block bid prices and associated MW values are shown where the price is that, which the supplier is prepared to buy at or sell in increasing price order. The results of four different market-clearing prices are shown to illustrate the principles. If the market-clearing price is €55/MW h, then the supplier would take all 18,660 MW from the market.

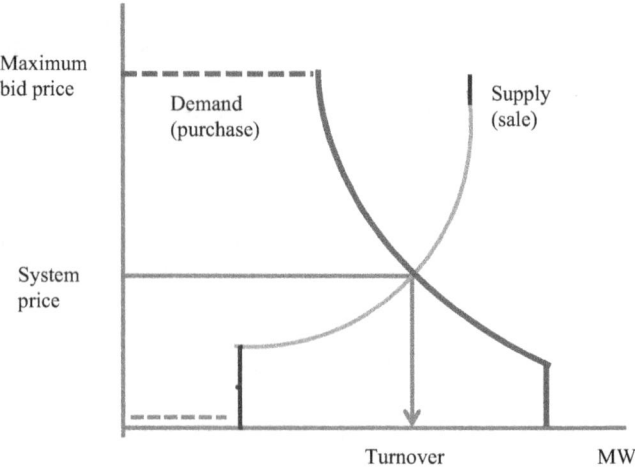

Figure 1.1   Nordpool model

Table 1.3   Nordpool bid structure

Generator block bid	1	2	3			
Available, MW	220	120	80		Reserve	824
Price, €/MW h	56	60	72			
Bid price	0	57	61	73		
Bid, MW	18,660	18,440	18,320	18,240	Local	Total
Hour	System price, €/MW h	Own generation			Import	Demand
1	55	0	0	0	18,660	17,836	18,660
2	57	220	0	0	18,440	17,836	18,660
3	61	220	120	0	18,320	17,836	18,660
4	73	220	120	80	18,240	17,836	18,660

If the price is €57/MW h, then it is cheaper to supply 220 MW from internal generator block 1 at 56/MW h. It can be seen through this process how the demand participates in setting the marginal price in competition exploiting the option of using internal generation or interrupting demand blocks.

The current power exchange arrangements principally cater for participants having generation and demand and therefore the flexibility to adjust their take from the system. Suppliers without generation have the option of taking supply from the market at the price offered or reducing their demand through management but cannot directly influence the clearing price.

1.9   Establishing the optimum level of generation capacity

The other major difficulty in establishing an effective competitive market is encouraging the right level of new entry and type of generation to meet longer term needs. The economic principal is that generation shortages will drive up marginal prices until they reach the new entry price level. In practice, large portfolio players may see sustaining high prices as better for their overall profit as all their energy can be sold at the higher marginal price. This contrasts with the arrangements in state utilities that generally operate at close to average regulated prices that will be lower than marginal prices. Their scale means they only require a modest tariff increase to finance new generation.

A more direct approach to establishing the optimum level of capacity is through an increment to the market price based on an estimate of the Loss of Load Probability. This was tried in the original UK pool and was designed to encourage generation availability at times of system stress but it did not provide a sustained source of revenue to finance new entry.

Another approach is to use direct capacity payments to generators to contribute to their fixed costs and reduce their risk. This is being introduced in the UK market through an auction process for capacity provision several years ahead. **The need to introduce these payments is largely due to the use of marginal generation being displaced by an increasing proportion of wind generation.** This reduces their revenue making them commercially non-viable. The Irish market has a high proportion of wind generation. Its operation is based on a pool but includes capacity payments based on best open cycle gas turbine capital costs, and this is often used as a benchmark. The problem with this arrangement is that the capacity payments are made to all generators and not just the new entrant. Economists claim that the extra costs will be recovered by reduced market energy prices as generators have less cost to recover. In practice, there is a risk that this will not occur with marginal generators costs influenced by the need to operate intermittently at part load with reduced efficiency. This is particularly the case with large tranches of wind generation, where gas turbines have to operate open cycle at efficiencies 30% less than their closed cycle performance to provide the flexibility.

Contracting for capacity requires an estimate of the level required to meet security criteria. Figure 1.2 shows a typical demand probability function for a power system together with the probability of generation being available based on outage statistics. It can be seen that there is an overlap when at high demand there may be a shortfall in capacity to meet it. This probability of generation shortfall and resulting loss of supply can be calculated. The optimum is where the cost of new entry equals the resulting reduction in the value of lost load to the consumers.

The result of a typical analysis is shown in Figure 1.3 and includes estimates of the number of hours in a year when some supplies may be interrupted. The graph shows the value of the lost energy and its rate of change plotted against the number of installed generation units. In this example, the target average lost load/year is set at 8 h and is realised with above 145 notional 500-MW

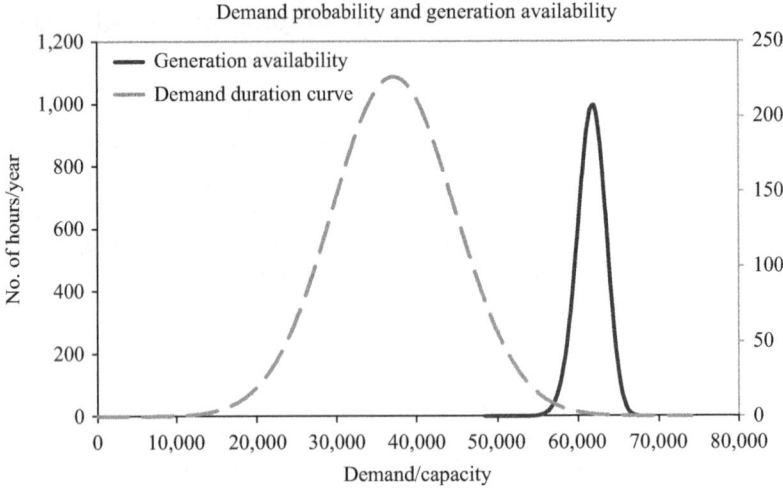

Figure 1.2   Demand/load function

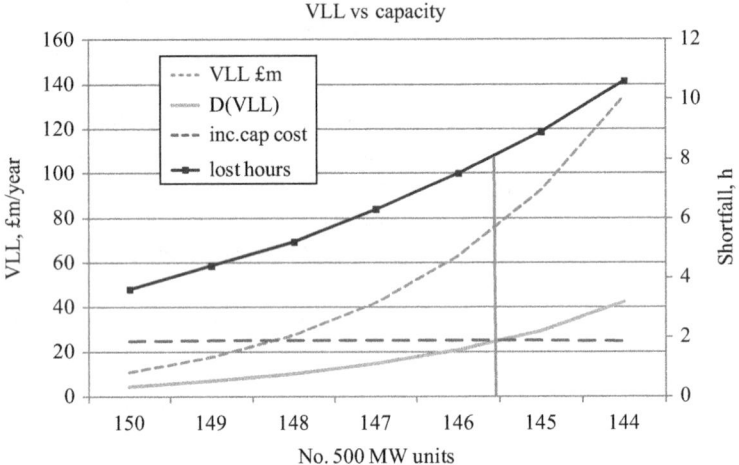

Figure 1.3   Optimum level of generation capacity

generators for this system. A complication in the current environment is the capacity contribution that can be expected from intermittent renewable sources like wind generation, and its analysis requires an extension of the probabilistic formulation.

The introduction of capacity payments and feed-in tariffs for renewable and nuclear generation undermines the competitive market approach, and the model begins to look more like a single-buyer model. These interventions also distort the development of the optimal plant mix. It is also assumed that capacity payments will be offset by reduced wholesale prices. In practice, the marginal generation setting prices may not receive capacity payments.

1.10 Plant Mix

A distinction needs to be drawn between optimising investment decisions with long-term horizons and encouraging operational efficiency through a short-term liquid market. The shape of the demand curve determines the optimal mix of plant. Nuclear generation, with a high capital cost but low operating cost, is most suitable for base load continuous operation. In contrast, open cycle gas turbines, with low capital cost but higher operating costs, are more suitable for peak lopping. In a developing situation, it is difficult for individual players to identify their long-term role or get long-term power purchase agreements. To some extent, risk can be managed by a generator owning a parallel supplier business. The other option is to establish a single-buyer model where a single entity contracts for long-term supply as well as short term. In this situation, an auction can be used to affect competition in the provision of new plant.

A particular problem is the advent of a large tranche of renewable generation operating on a take or pay basis with subsidies set by government. This seriously reduces the utilisation of other marginal generation and its competitive position. The result has been a number of generation closures with a dearth of new investment leading to unprecedented low plant margins. The energy market is undergoing unprecedented changes, behind the scenes globally, as nations seek to transform their power industries to be cleaner and greener, creating a whole host of challenges for the existing players.

1.11 Regulatory response

The European Energy Regulators reported that in 2015, the subsidy to renewable generation averaged €110/MW h in addition to wholesale prices and that this was adding an average 13% to end-user bills up from 6% in 2012. This was funded by feed-in tariffs in 21 countries and green certificates in 7. The intermittent renewable generation had no responsibility for balancing in half of the European countries with consumers bearing the costs. An average 16% of energy produced came from green sources receiving subsidies varying from 1% in Norway, rich in hydro, to 62% in Denmark. The predominant recipients were hydro, photovoltaic and bioenergy. Recognising the adverse impact on consumers, the European Regulators proposed changes including

- For the priority dispatch of renewable intermittent generation to be phased out;
- Avoid non-market approach to redispatch and Renewable Energy Source (RES) curtailment;
- Avoid net metering and ensure fair cost allocation.

Renewable generation should be integrated into the normal market. The priority dispatch of RES should be removed from new and phased out for existing plant with the removal of Article 11 of the Electricity Regulation. The cheapest generation should run in merit order, irrespective of type, to minimise consumer bills. Also, RES generation export should not restrict cross-border trading opportunities that may restrict overall costs. Where feasible, redispatch should also be based on the merit order, and only RES curtailed by transmission should be compensated based on market prices. **The use of net metering as a basis for network charging for consumers that also generate is considered unfair.** Embedded generators should pay based on their capacity, as other generators, to fully reflect their potential exploitation of the network. The view has also been expressed that intermittent RES should accept some responsibility for balancing.

1.12   Electricity market reform the United Kingdom

The developments in the industry created the need for changes in the design and operation of the market. In the United Kingdom, a review of the electricity market was launched by a white paper in the spring of 2011 proposing changes that were implemented during the first half of the decade. The EMR attempted a fundamental overhaul of the market arrangements to help to promote investment in energy infrastructure, especially low carbon generation. It included four main proposals:

1.   A capacity-based market
2.   An emissions performance standard
3.   A carbon price floor
4.   Revision of the renewable obligation and the feed-in tariff system

The proposed capacity-based market recognised that the revenues of conventional generators would be undermined by their utilisation being replaced by renewable generation. Table 1.4 shows a summary of the capital costs for gas and coal plant. It includes a calculation of the annual cost of servicing the capital that can be compared with the payments from the first round of the capacity market. It was expected to be around £50/kW/year but the outturn of £19.4/kW/year was influenced by a lot of small, often embedded generators. For the 400-MW generators shown in the table, the annual capacity payment would be around £7.7m/year:

$$400 * 19.4 * 1,000/1,000,000 = £7.7m/year$$

This can be compared against costs for gas of £23.78m/year and coal of £50.7m/year. It is not surprising that this would not encourage new entry and that plant margins have fallen. At the expected £50/kW, the capacity payment would have been £20m/year, for a 400-MW generator, and close to covering the capital cost of gas turbine generation costing £23.8m.

Table 1.4 Generation capital costs

Capital costs		Gas	Coal
Site and dev.	£m	9.7	27.5
EPC contract	£m	235.0	500.0
Electrical conn	£m	0.73	0.73
Gas connection	£m	2.47	
Spares	£m	14.3	31.9
Interest during const.	£m	11.3	23.2
Total investment cost	£m	273.5	583.3
Capital cost	£/kW	696.0	1,465.5
Capital cost pa	£/m	23.8	50.7

1.13 Irish I-SEM

The Single Electricity Market in Ireland was based on a stand-alone mandatory pool with regulated bidding and a capacity market until 2017. In line with 19 other European countries, it planned to move to a day-ahead market based on a pricing algorithm called 'EUPHEMIA'. The program simultaneously determines the market prices and is designed to facilitate convergence across Europe. An intraday market is also required to ensure that participants can accommodate changes in generation/demand positions between the time of the day-ahead market closure and the real-time event. This is becoming essential to accommodate the variations in the output of intermittent generation like wind and solar. This is intended to operate as a joint integrated cross-border market with other European countries and is based on a software system called XBID. Following 'gate closure', the TSOs will manage real-time operation through the use of a balancing market and dispatch process. They may, for system security reasons, change the position of participants in the day-ahead or intraday markets.

A complication for Ireland is that it is only connected to Europe through the United Kingdom. Any trades would have to flow across the UK network, and transfer capacity would have to be reserved.

1.14 Lessons learnt

What has been the outcome of these developments? Has it all been worthwhile?

- Energy prices have reduced in real terms but this was as much to do with the freedom to introduce cheaper gas-fired generation and the option to import cheaper coal;
- Attempts to engage the demand side in the market dynamics have largely failed;

- It is generally the view that the markets are illiquid and not competitive and are often dominated by incumbent players. Any assessment based on the HHI would concur with this view;
- Attempts have been made to weaken the market power of large incumbent players by requiring them to make part of their generation output capacity available to new market entrants so that they are able to buy wholesale and sell retail. But the incumbents, that also have supply companies, have been able to undermine new entry by adjusting their prices using internal company transfers.
- The short-term markets do not necessarily encourage the development of the optimum mix of generation with each new player making assumptions about their operating regime.
- Network charging was established on a zonal basis to encourage new generation location where it was most easily connected and close to load centres. This has often not resulted in the desired effect with other considerations related to local industry dominating location decisions.
- The different approaches used to set use of system charges at transmission and distribution levels have led to inconsistencies that distort the balance of choice in location of generation in favour of distribution.

1.15   Conclusions

In the current environment, governments are faced with a spectrum of objectives including securing future supplies, building up renewable energy to meet emission targets whilst maintaining fuel diversity and containing end-user bills (the so-called trilemma). In attempting to meet these objectives, governments have intervened in the market process by using subsidies to promote development of renewable technology and setting strike prices for energy contracts with clean generators. This has resulted in a market that is neither free nor centrally coordinated and has created a number of problems:

- Subsidies to renewable generators, and priority dispatch has undermined the operation of the wholesale market;
- Conventional generators have suffered reduced utilisation coupled with an increase in regulating duty to track variations in wind output undermining their financial viability;
- There is a shift in the operating regimes of conventional generation that will impact on the optimal plant mix;
- Investors see the regulatory bias in favour of renewables as introducing risks that undermine the business case to build new plants leading to unprecedented low margins;
- The costs of subsidies to renewable generation, and the costs of extending transmission and distribution to accommodate it, has led to high increases in end-user costs;

- System operation has become more complicated due to the need to accept on the system an increasing amount of intermittent renewable generation output, to predict demand and to manage dispatch and the interface with DSOs;
- Use of system charging arrangements on the distribution and transmission networks has distorted competition between grid-connected and embedded generation leading to an expansion of distribution-connected generation;
- The distribution networks are becoming more active rather than passive stretching the capacity of the network and leading to the need for reinforcement and the establishment of a distribution system operator;
- Policies to promote competition have resulted in hundreds of new players in the sector that undermines the ability to establish a holistic approach to system development.

Given the scale of the problem with the decarbonisation of heating and transport, still pending actions need to be initiated now to manage the increasing impact on the electricity sector. Regulators need to review the mechanisms used to encourage renewable generation and review the need for subsidies and priority dispatch. The distribution companies need to take up the role of a distribution system operator to manage their active networks. The DSO and TSO need to interact to manage the more active nature of the distribution network and exchange data to support demand prediction and manage security. Mechanisms need to be established to coordinate the interaction of an increasing number of market players.

Question 1.1 Given data shown on the costs of the capital and operating costs of gas and coal-fired generation establish the total generation costs if the load factor is reduced from 85% to 35% due to priority dispatch being given to renewable generation.

		Units	Gas	Coal
Plant	Capacity	MW	393	398
	Capacity net	MW	377	366
	Load factor	%	85	85
	Annual output	TW h	2.81	2.73
	Construction time	Years	2	4
	Project life	Years	20	20
Fuel	Efficiency	%	55	45
	Coal cost delivered	£/t		40.27972
	Coal calorific value	GJ/t		24
	Natural gas cost	£/mmBTU	3.842105	
	Natural gas cost delivered	pence/therm	38.42105	
	N gas calorific value	kW h/therm	29.32	
	N gas grossCV/netCV		1.11	
Finance	Cost of dept.	%	4.7	4.7
	Cost of equity	%	7	7

(Continues)

(*Continued*)

		Units	Gas	Coal
	Inflation	%	2.4	2.4
	Debt/equity split		0.8	0.8
	WACC (*D/E* = 0.8)	%	6.0	6.0
Cap costs	Site and dev.	£m	9.733	27.46
	EPC contract	£m	235	500
	Electrical conn	£m	0.733	0.733
	Gas connection	£m	2.466	
	Spares	£m	14.33	31.86
	Interest during const.	£m	11.26	23.20
	Total investment cost	£m	273.5	583.3
	Capital cost	£/kW	696	1,465
	Capital cost pa	£/m	23.78	50.71
O&M	Consumables	pence/kW h	0.02	0.113
	Labour rate	£/h	16.66	16.66
	No. of operators		6	14
	Operators labour*1.28	£M	1.12	2.62
	Maintenance material	£M	0.81	2
	Main and support labour	£M	1.25	2.33
Summary	Capital cost	£/MW h	8.46	18.60
	Fixed O&M	£/MW h	0.60	2.09
	Var O&M	£/MW h	0.73	1.59
	Fuel costs	£/MW h	26.45	13.43
	Total generation costs	**£/MW h**	**36.24**	**35.70**

Chapter 2

Market technical challenges

2.1 Introduction

This chapter discusses the key technical issues that are incident upon the transformation taking place in the industry and can constrain the way the industry develops in future. These include

- The agreed requirement to reduce emissions is having a profound effect on the generation plant mix. Because the renewable generation has priority dispatch, it is radically altering the operating regimes and financial viability of conventional generation;
- The system demand prediction process has become more difficult because of the impact of an increasing proportion of embedded generation. Because of the intermittency of renewable sources and the lack of control over other local generation, the normal system demand patterns are no longer valid;
- As the proportion of renewable generation increases, there will be times when it could displace more conventional generation than is acceptable. This is because of generation that is inflexible, like nuclear and Combined Heat and Power (CHP) schemes, and the need to maintain dynamic reserves to manage the changes in renewable output.
- The renewable generation output displaces the use of conventional generation reducing its utilisation to levels below what is financially viable. At the same time, its availability is required at times of high demand when the renewable generation output is low;
- Conventional generation is equipped with governors that support frequency control, and their inherent inertia helps to contain the rate of change of frequency (ROCOF) when imbalance occurs. A lot of renewable sources like wind and solar generation do not inherently provide this support and compound the frequency control problem;
- To meet the targets for emission reduction, it is likely that a lot of transport will be based on electric vehicles. The requirement to keep them charged will add a considerable demand to the network with little diversity and will necessitate distribution network reinforcement;
- The patterns of network utilisation will be influenced by the shift of a tranche of generation from the transmission network to the distribution network. The

direction of flows and loading levels will be impacted by embedded generation output that will in turn compound the problem of system voltage control;

- The development of more cross-border interconnection to meet the European Union (EU) import capacity target of 10% of national system demand will influence the development of generation capacity and network operation. High loop flows resulting from high levels of wind generation in Germany have already impacted on security management in adjacent countries.

This chapter discusses these issues and how they may be quantified to indicate where constraints may arise that need to be accounted for in the wider debate about future development strategy.

2.2   Emission reduction and plant mix

To realise the EU target reduction in overall UK emissions of 15% by 2020, targets were distributed amongst the main sectors with electricity generation given the highest reduction of 34% with transport and heating just 10%. It is planned to meet these target reductions through the introduction of renewable sources like wind and solar and nuclear energy as well as through improved efficiency in use. Because the renewable sources are not yet competitive with conventional generation, governments have been forced to introduce subsidy mechanisms that distort market operation and increases costs. These subsidies may take the form of direct payments through a feed-in tariff mechanism or indirect by introducing charges for emissions of greenhouse gases like CO_2 from burning fuel in conventional sources of generation. There are also subsidies to other forms of generation that are deemed more efficient like combined heat and power schemes or those that result in less overall emissions like biomass, geothermal and hydro. At the domestic level, the application of solar panels and heat pumps is encouraged. Electric vehicles are also seen as an opportunity to reduce emissions from transport, and their introduction will be incentivised.

Governments are also addressing the problem of emissions by encouraging more efficient use of energy by improving insulation and monitoring use through smart meters. This concept has led to proposals for smart cities with active management of local generation sources and demand together with storage. These initiatives have provided mechanisms for more consumer side participation in the energy market and are encouraging a more active role. This, in turn, has spawned the development of consumer groups and aggregators that negotiate tariffs collectively to get better rates. These developments are deemed good for competition. There are also a number of businesses promoting the development of local energy schemes. Some borough councils have been active in introducing CHP schemes to supply public buildings and housing. A key advantage of local schemes is the ability to directly connect to the end users bypassing the local distribution network and avoiding their charges. These network charges can be as much as the energy charges in the distribution tariff and their avoidance can boost the competitiveness of alternative sources. The local schemes can also reduce overall system network losses at the transmission and distribution level. In some schemes, diesel generation is introduced that has an adverse impact on local emissions. Also, there is currently no easy way for players to discover prices within their local environment on a comparable basis, and a number of comparison websites have sprung up to support a switching process. However, most consumers are still on relatively high basic tariffs prompting some governments to propose the introduction of a cap on domestic tariffs. **There is also no competitive market operating for generation at the distribution level** but a general belief that there are opportunities to provide cheaper alternatives to conventional supplier tariffs.

Traditionally, the mix of generation has been matched to its expected operating regime to align with the expected demand pattern. For the generation expected to operate base load, the unit cost was more important than the original capital costs, and this part of the load would frequently be met by nuclear generation. In meeting peak demands of short duration, capital costs are more important than operating costs, and open gas turbines would often be deployed. **The impact of large tranches of renewable generation has meant that the net demand profile is much more variable and difficult to predict with in extreme cases distributors exporting energy back to the grid. The inability to predict an operating regime introduces added investment risk making it more difficult to predict revenues or even establish the data to make an analysis of the optimal plant mix other than much more flexibility will be required.**

The introduction of charges for CO_2 emissions that are variable is shifting the price comparison between different generation sources and also adding to the complexity of establishing the optimal plant mix. Given the current worldwide dependence on coal, an immediate benefit in reducing emissions can be realised by switching from coal to gas-fired generation. The CO_2 content of coal/GJ is 75%, higher than that of gas. In addition, the efficiency of Combined Cycle Gas Turbine (CCGT) generation, at a realised 52%, is higher than coal at around 38% resulting in the emissions/MW h from CCGT generation being around 40% of those from bituminous coal. Table 2.1 shows the cost implications to generation based on a CO_2 price of £10/t.

The impact is less if the coal plant design realises higher efficiencies. The impact on costs based on £10/t is shown in Figure 2.1. At an efficiency of 48%, the emission cost is £6.5/MW h compared to gas at £3.43/MW h. However, coal costs at £2/GJ are usually lower than gas at £3/GJ. Based on the data in Table 2.1, the

Table 2.1   Emission comparisons

Fuel	gC/GJ	tC/mtoe	tCO_2/GJ	Price/GJ	Efficiency	Price/MW h
Lignite/brown coal	25,200	1.169	0.0924	0.924	33	10.08
Bituminous coal	23,700	1.100	0.0869	0.869	38	8.23
Oil	19,900	0.923	0.0730	0.730	40	6.57
Natural gas	13,500	0.626	0.0495	0.495	52	3.43

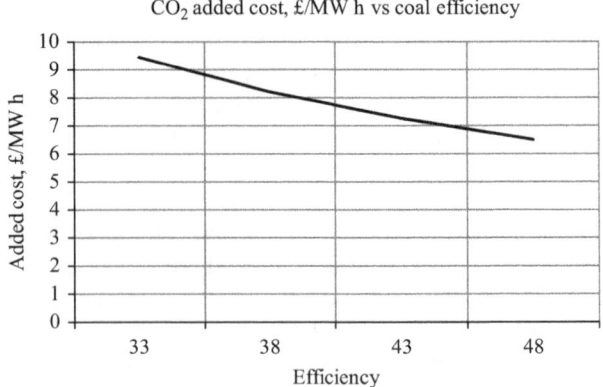

Figure 2.1   CO$_2$ coal cost vs efficiency

efficiency at which the coal variable cost equals that from gas can be calculated ignoring operating support costs.

$$\text{Gas generation variable cost} = 3.6 * (3 + 0.495)/0.52 = £24.19/\text{MW h}$$
$$\text{Coal variable cost} = 3.6 * (2 + 0.869)/\text{eff}$$
$$\text{Comparing} = 3.6 * 2.869 = 24.119 * \text{eff}$$
$$\text{Efficiency} = 3.6 * 2.869/24.19 = 0.428 \text{ or } 42.8\%$$

At a CO$_2$ price of £10/t, coal generation is still competitive on price/MW h if an efficiency of 42.8% is realised, as is possible with modern designs. There is also the option of fitting carbon capture and storage retrospectively adding further complication.

2.3   Demand prediction and reserve

The prediction of total system demand has traditionally been based on tracking the profile of similar days. These would be selected based on the season and day of the working week. These would be adjusted for expected variation from normal weather taking account of temperature and the onset of darkness. Longer term forecasts would take account of the expected development of the country's Gross Domestic Product (GDP) and changes to interconnections. In the short-term algorithms based on an autoregressive integrated moving average package would often be used to track demand in real time. In some circumstances, several algorithms working on different principals have been employed with a weighting given to each depending on the accuracy of recent predictions. These processes were supported by data banks of recorded demand going back several years and included actual demand and associated data on the prevailing weather. It was usual to normalise the demand to average weather conditions for consistency to provide a base for new predictions that could be adjusted to the current weather forecasts.

The intermittency of many renewable sources undermines the traditional demand prediction process and invalidates historical records. Whilst metering is available at the National Control Centre from transmission connected renewable sources, there is also a requirement for the distributors to establish metering to monitor the output from embedded generation that should be made available to National Control. Predicting the output from renewable intermittent sources will need to be based on forecast weather conditions and related to the location of the generation. The requirement for maintain reserves to back up sudden drops in wind output will depend on the installed capacity and the forecast accuracy. The key lead time is often a reflection of the time taken to bring new plant into service and is typically a few hours for conventional gas and coal generation. Open cycle gas turbines can start more quickly but are expensive to operate because of their lower efficiency. They are often used on a temporary basis whilst more economic plant is brought into service. **The operation of conventional thermal generation in reserve mode incurs additional costs in extra fuel burn and wear and tear**.

2.4   Intermittency and curtailment

To maintain secure system operation, there is a need to maintain a proportion of conventional generation operating in reserve mode. This usually means operating part loaded so that it can increase output to maintain system frequency in the event of the loss of output from other generation sources. The proportion of generation required to operate in this reserve mode depends on the largest credible in-feed loss arising from a network or generation fault. The initial rate of fall of frequency is determined by the system generation and demand side inertia. Subsequently, generation operating in reserve mode has to respond immediately to arrest the fall of frequency and avoid more generation being tripped by ROCOF protection relays. Renewable generation like wind turbines and solar does not inherently provide much inertia, and when a high proportion is in service, the frequency becomes more volatile. There is also a need to maintain reserve against a rapid fall in output from renewable generation. The problem becomes more acute at times of low system demand combined with high levels of renewable generation output. There is also usually a proportion of generation that is less flexible or expensive to regulate like nuclear and CHP schemes serving other requirements. Also, small-scale embedded generation may choose to run because of local commitments. This leaves less demand available to take the output of renewable generation sources. In the absence of alternative demand like storage, it may be necessary to curtail the renewable generation output undermining its economic viability. The higher the proportion of renewable generation, the greater the chance there will be a requirement to curtail its output. **The curtailment proportion increases exponentially** as shown in Figure 2.2 for a high and low future demand scenario and including more inflexible generation. It can be seen that the curtailment can be limited, in this example, if the non-synchronous renewable capacity is limited to around 30 GW or 50% for this system with a peak demand of 60 GW. A reserve holding of 30% of wind output is assumed in this example based on a forecast accuracy of three standard deviations of a 10% root mean square (rms) forecast error.

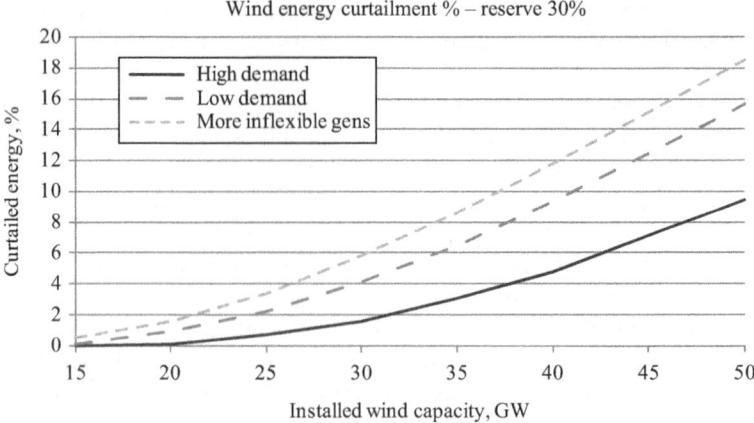

Figure 2.2   Curtailed renewable energy

2.5   Renewable capacity contribution

In planning system development, it is usual to establish a margin of generation capacity above the expected peak demand to cater for plant forced outages or maintenance as well as higher than expected demand. Typically, the plant margin or spare capacity is designed to be around 15% to 20%. A problem with renewable generation is its intermittent availability resulting in a limited probability of it being available at times of peak demand. The expected contribution to capacity is a function of the probability of renewable generation output at the time of peak demand. This will generally be much smaller than the installed capacity. Typical wind generation average load factors are around 25%, and the probable **capacity available at peak times is typically 10%–15% of the installed wind generation capacity**. In the case of solar generation, it may make no contribution if the peak demand occurs during darkness hours. The result is that a high proportion of conventional generation has to be kept available to cater for periods with limited renewable generation output. Unfortunately, this marginal reserve generation has reduced utilisation due to its use being displaced by renewable sources, and the revenue from energy sales may not be sufficient to make it financially viable. **It may be necessary to introduce capacity payments to retain the availability of the conventional generation**. With wind and solar generation, the potential contribution to capacity can be increased if the energy can be stored, but this adds to the cost and incurs additional losses. The process to estimate the probable wind contribution involves the following steps:

- establish conventional capacity meeting security criteria of typically 8 h of shortfall/year;
- reduce conventional capacity but include wind generation and establish that capacity meeting the same security criteria;

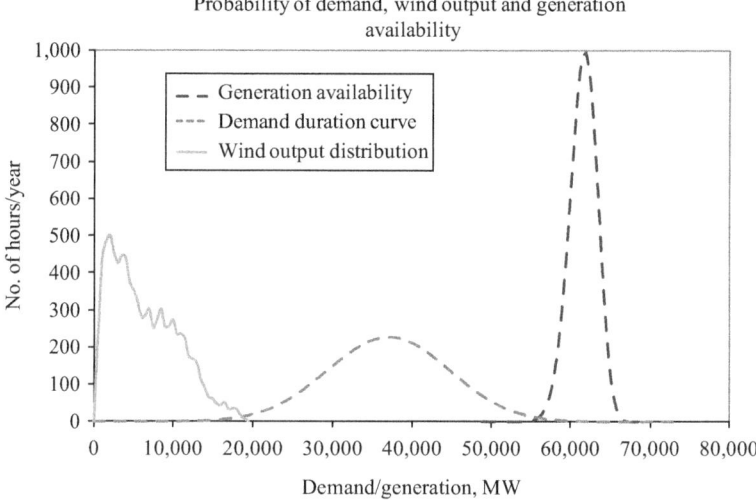

Figure 2.3　Demand and wind output distribution and generation availability

- for each probable wind output level, record the number of hours of shortfall and probability;
- add up the number of hours of probable shortfall for comparison with criteria;
- hence, determine the effective wind capacity contribution as that level resulting in the equivalent shortfall hours to the scenario with just conventional capacity.

A typical output probability function for 25 GW of wind generation installed across the United Kingdom is shown in Figure 2.3 together with a demand level probability function and the probable availability of a tranche of conventional generation. The wind output is based on recorded data and approximates to an equivalent Weibull function as frequently used to represent wind output. It has been found that the demand distribution can be adequately represented by a normal distribution function. The probability of generation outages is calculated from the expected average availability of individual units. With this formulation, the probable wind generation capacity contribution is estimated at between 10% and 15% depending on the load factor of the wind generation.

2.6　Impact on conventional generation

The intermittent nature of renewable generation has a significant impact on the conventional generation that operates in parallel with the renewable sources providing reserve to replace the renewable output when it falls. The effects are that

- The utilisation of the generation will be significantly reduced affecting its economic viability;

- The units will frequently be operating part loaded in reserve mode at reduced efficiency;
- The units will be engaged in frequent ramping to track changes in renewable output using more fuel;
- The units are likely to incur more starts and stops in tracking larger changes in renewable output using extra fuel.

Some generators like Centrica are converting CCGTs to operate open cycle to provide more flexibility to meet system needs. **Where the loading regime affects the fuel burn, it will incur additional costs in wear and tear and fuel and result in more emissions offsetting the saving resulting from the use of the renewable sources.** Figure 2.4 shows the results of analysis of the effect on emissions as the wind capacity on the system is increased. It shows the increases in emissions due to

- more frequent generation start-ups;
- the extra fuel used in ramping up and down to track wind output;
- the extra fuel used in operated at reduced efficiency part loaded providing reserve;
- the reduction in the emission savings due to the need to curtail wind output at times of no load;
- the emission reduction if the above aspects were ignored.

The figure takes account of the reduction in emissions due to the need to curtail wind generation output at times of low load to maintain sufficient conventional generation in service to maintain system inertia and stability. The degradation in emission savings rises to around 10% with a wind capacity of around 50% of the peak demand of 60 GW.

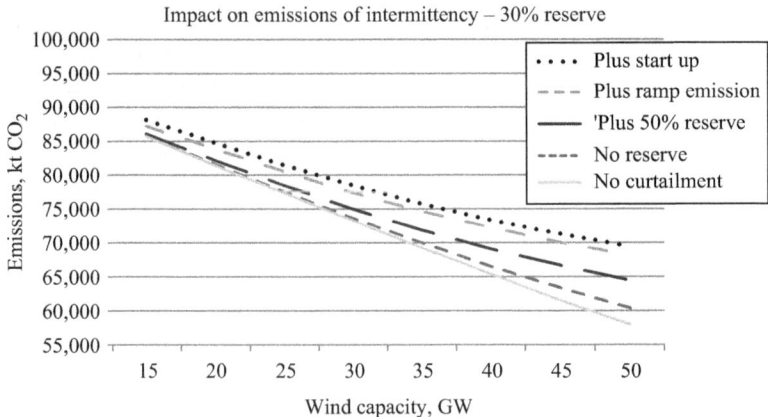

Figure 2.4   Impact of conventional backup generation on emissions

2.7   System inertia and frequency control

The stable operation of the transmission system following a loss of generation relies on frequency regulating generation responding in seconds to release stored energy and arrest the frequency fall. It is common for generation to be equipped with ROCOF protection that will trip generation given a sudden collapse in frequency. If the frequency fall is not arrested, this can result in cascade tripping widening the area of generation shortfall. When the system generation includes a high proportion of renewable generation, then the overall inertia is depleted increasing the chance of ROCOF relay operation. There is a natural reluctance amongst conventional generators to increase settings without a thorough assessment of the potential impact on their plant. Their concerns relate to the effects of torsion on the turbine generator shaft and the reaction of control systems.

As the power system, non-synchronous generation (NSG) penetration increases, the system inertia falls and the system becomes more susceptible to reductions in wind generation output following faults. The sudden demand change will cause rapid changes of frequency. This could cause generation to trip on ROCOF controls leading to cascade tripping and wider system disruption. The current standards applied in the UK Grid Code require generation to accommodate a rate of change of 0.5 Hz as measured over a rolling 500-ms period. This value could be exceeded at higher NSGs, and consideration has been given to increase the standard to 1.0 Hz. Generators are concerned that this increase will subject their plant to higher alternator/turbine shaft torsions and propose that the long-term effects need to be considered. Whereas faults causing voltage dips are system wide in their impact sudden demand, changes may impact on specific machines and be more onerous. The view of system operators is that the NSG penetration should be limited to 50% to avoid generation trips escalating incidents that lead to islanding. The key ratio is the non-synchronous generation in relation to the demand:

$$\text{NSG/demand} = (\text{Wind} + \text{PV} + \text{HVDC import})/(\text{net demand} + \text{DC export})$$

The NSG provides little inertia and post-fault voltage support, and the impact is particularly acute at times of low load with a high proportion of NSG operating. Conventional generation releases stored kinetic energy as the frequency falls that can maintain output for 2–10 s according to its inertia constant 'H' given by

$$H = E_{\text{kin}}/S_b \qquad \text{where } E_{\text{kin}} = 1/2\, J (2\pi f_m)^2$$

S_b is the rated power and J is the moment of inertia.

Renewable generation sources also make a much smaller contribution to fault level that can impact on protection discrimination and voltage flicker. The system reactive to power ratio (Q/P) falls as a result of more embedded generation and more efficient lighting with grid supply points exporting reactive half the time. This coupled with the reduction in reactive support from synchronous generators creates high voltage control problems.

Typically, generation is currently designed to cope with a ROCOF of 0.5 Hz/s as measured over a rolling 500-ms period. Studies of Northern Ireland suggest that following isolation from the rest of Ireland, changes of up to 2.0 Hz/s may occur prior to establishing more interconnections. A ROCOF of 1.0 Hz/s is proposed in the interim. In comparison, the UK Engineering Recommendation G59 refers to a value of 0.125 Hz to avoid nuisance tripping. The Irish have determined that the system non-synchronous generation penetration (SNSP) should be limited to 50% pending the outcome of more studies and new measures. However, it is also recognised that wind farm output could take time to recover leading to high voltage dips with SNSP above 40% with the possibility of cascade tripping of embedded generation.

Options to ameliorate the situation include designing the non-synchronous generation to provide fast frequency response and including power oscillation damping to facilitate fault ride through. Generators usually respond to falling frequency following a system loss incident within 2 s with an aggregated response on the UK system of typically 250 MW/s sustained for 6 s equivalent to 1,500 MW. To contain the ROCOF 0.125 Hz for a 1,800-MW loss, the inertia would need to be 400 MW/s with the frequency fall limited to 49.2 Hz.

2.8 Impact of electric vehicles on system

There is growing concern about the impact of diesel and petrol cars on the air quality in our ever larger more densely populated cities. The Dutch passed a motion banning the sale of diesel and petrol cars from 2025. German buyers receive a €4,000 rebate if they buy electric vehicles. Norway is giving free parking and Value Added Tax (VAT) exemption to electric vehicles and will permit only the sale of hybrid and all electric cars from 2025. The United Kingdom is banning the sale of petrol and diesel cars from 2040. Pollution in cities in India is the worst in the world prompting them to set a target of all-electric car sales by 2030. The French announce that vehicles using internal combustion engines will be outlawed by 2040. Given a typical life expectancy for a petrol/diesel car of 14 years, if sales are banned from 2025, then by 2040 all vehicles will be electric. **This means that from 2025 through to 2040 all transport will migrate to being electric.** BMW expects that 15% to 25% of their sales will be electric by 2025. These timescales are short given the time it will take for the power system to be expanded to meet the extra demand.

The car industry is becoming aware of the direction of these developments and is responding. The Tesla 3 car at a competitive price of £30k has a predicted range of 215 mi. They claim to be producing batteries for less than $190/kW h and expect to produce cars with a range of 200 mi by 2020 for £20k. Ford and BMW expect battery prices to drop to $100/kW h by 2025. The Nissan Leaf offers a viable alternative with a price of £20k, and a range of 155 mi with a 30 kW h battery. The EU is expected to introduce legislation requiring vehicles to emit no more than 95 g of CO_2/kg of vehicle weight from 2020. Cities like Delhi, London and Beijing will have to introduce legislation to curb emissions to protect the health of their citizens.

The car manufacturing sector is competitive, innovative and well financed, and the effect of these incentives is that a rapid increase in the use of Electric Vehicles (EVs) can be expected. There are a number of implications to these developments including

- The impact on emissions;
- The impact on power system demand;
- The impact on local networks.

The emissions from diesel cars have been shown to be worse when the catalytic convertor is cold. After 1 min from starting, it removes 32% less nitrogen oxides than when at full temperature and 13% less after 5 min. With petrol cars, the emissions were 422% higher after 1 min than when at full temperature but less than 1/10 of those from diesel. These figures point to a particular problem in cities where most of the journeys are short. The practice of switching off the engine for short periods, while the vehicle is stationary will delay the convertor reaching full temperature and may increase emissions. Speed bumps often located by schools cause cars to brake and then accelerate resulting in more emissions. **The net impact on emissions of converting to EVs will depend on the emissions from generation used to charge the EV battery** compared to the avoided emissions from using a comparable petrol or diesel car. The emissions from the generation/kW h charging the batteries can be expressed as the level/mile taking account of the charge/discharge cycle efficiency. This figure can be compared to the published technical data on emissions/mile from conventional petrol cars. Calculations show that emissions from electricity generation are less from combined cycle gas turbines operating at the margin to charge the EV but largely neutral if the electrical energy is derived from coal-fired generation.

The impact of meeting just a 10% reduction in UK transport using petrol/diesel with EVs is analysed in Chapter 13. It shows that it would increase generation requirements by 3.5 GW at a 43% load factor producing 13.1 TW h. To realise 100% conversion to EVs by 2040 would require a 40% increase in the power system capability without exercising demand control. The distribution system is usually designed to meet after-diversity demand and relies on there being diversity in the use of the system by consumers. Figures of around 1.5 kW/consumer are often assumed for domestic users. This is reasonable when related to the use of kettles, washing machines and spin driers with natural diversity in their time of use. In the case of electric vehicle recharging, it can be expected that there will be some correlation in use. This is likely to occur during lunch times and just after working hours before normal commercial and industrial demand has subsided. A project 'My Electric Avenue' suggested that a 40% EV take-up would lead to the need for measures to protect against network problems.

Charging from a UK domestic socket at 240 V and 13 amp delivers 3 kW h/h. With a charging time of 3–4 h, 10 kW h could be delivered equivalent to range of around 27 mi that may be sufficient for a local commute. To fully charge an 80-kW-h battery would take around 24 h from a domestic charger installation. A faster alternative is available with a Level 2 charger designed to operate at up to 80 amp at 240 V delivering 20 kW h/h and able to fully charge a vehicle in 4 h.

The Japanese have developed a Level 3 charger capable of delivering 62.5 kW at DC voltages of 50 up to 500 V. Induction charging is another option with the primary coil buried and the secondary one on the underside of the vehicle. This enables charging by parking over the primary coil without direct connections. It is apparent that sustained loads of 3–20 kW h could result from charging electric vehicles with little diversity during lunch periods and early evening. **This could seriously stress a distribution network designed for an after-diversity demand of 1.5 kW per consumer**. Volkswagen, Mercedes, BMW and Ford established a joint project in November 2016 to roll out thousands of fast charging stations across Europe to encourage electric vehicle take-up. These developments are going to have a significant impact on both generation and network requirements. These changes will occur relatively quickly in a relatively slow-moving heavy engineering sector. Table 2.2 indicates the battery capacities of a range of cars.

Indicative of the rate of introduction of EVs, the Norwegian Government plans to ban the sales of petrol and diesel cars by 2025. The UK government are offering grants of up to £4,500 towards new fully electric vehicles, and the cars are also exempt from vehicle excise tax, whereas petrol and diesel cars will pay according to their emissions in the first year. Extrapolating forward, if we assume 3-m electric vehicles in the United Kingdom with an average charge of 40 kW h being charged 100 times, each year then the energy requirement/year is given by

$$100 * 40 \,\mathrm{kW\,h} * 3\,\mathrm{m} = 12\,\mathrm{TW\,h/year}$$

This compares to the 2015 annual energy demand of 330 TW h but could stretch current generation resources and networks without careful management of the phasing of charging/discharging.

It may be possible to apply dynamic pricing via smart metering to encourage car battery charging at off-peak times. There is also the possibility of drawing energy from the batteries at peak times when prices are high. These options could be attractive financially but the distribution transformer and cable capacity may be a limiting factor when all the connected devices are chasing the same price. Also charging stations would have to be bidirectional incorporating invertors capable of producing accurate voltage and frequency control for the feedback to the grid. Charging from home would be the norm but there is also the question of charging while parking for those residents without private driveways. Parking charges in towns normally far exceed the cost of electricity to supply the car, and more remote charging stations may be required furthering inconvenience. Charging on longer

Table 2.2   Electric vehicles

Model	Renault zoe	Mercedes B250e	Tesla model S
RRP	£17,795	£32,670	£61,500
Power, kW	65–68	179	245–444
Battery lithium-ion, kW h	22–40	28	60–100
Kerb weight, kg	1,268	1,725	2,108
Acc. 0–62 mph (in secs)	13.5	7.9	2.7

journeys at motorway service stations could also present a problem with charging times of 30 min, with a limited number of the expensive chargers long queues could rapidly develop. The infrastructure for EV transport is not in place but the main difference is the time taken to charge in relation to finding a petrol station and filling up in 2 min. Some consideration is being given to exchanging batteries at motorway service stations to avoid long delays.

2.9 Distribution challenges

The distribution networks are changing from being essentially passive to active principally due to small-scale embedded generation at local and domestic levels. As network charges can represent half of a consumer's costs, their avoidance or minimisation can be an attractive option. This can be realised by direct wiring from generation to local loads or by minimising the peak demands against which suppliers get charged by the grid. Local generation is often provided as a backup to maintain essential supplies to hospitals, etc. and can provide the framework for a local microgrid. **Some commentators suggest it might promote defection from the grid.** In the main, these suggestions are based on minimising or avoiding the network charges. There is also the engagement of end users through subsidies for renewable generation like solar panels resulting in a shift of **some 10% of installed generation capacity shifting from transmission to distribution networks.**

These developments raise a number of issues:

- The distribution networks were designed as radial, essentially passive, networks and adding generation results in a requirement for active management;
- The simple tariff charges for use of system paid by embedded generators do not reflect the benefits realised by enabling connection to multiple loads and frequency and voltage control services;
- Some distribution networks are realising limits on the connection of generation and managing flows and security throughout the loading cycles;
- Protection systems were designed appropriate to the network construction and is complicated by the introduction of generation with ROCOF protection;
- There is an injustice with some consumers having to pay for subsidies to other users who happen to have the option to fit solar panels;
- The distribution networks have generally been designed based on normal after-diversity maximum demands that will not apply with new loads influenced by wholesale prices.

The full cost of providing and supporting the network will inevitably rise as a result of increasing volumes of embedded generation and needs to be fairly distributed amongst all users. The distortion in network costs that benefit small embedded generators that reduce the supplier triad demand needs to be rebalanced. There is a need for the networks to develop to be able to support the decarbonisation of heating and transport using EVs and heat pumps. This needs to be based on a longer term view of the full impact of these developments on loading, security and optimum voltage levels.

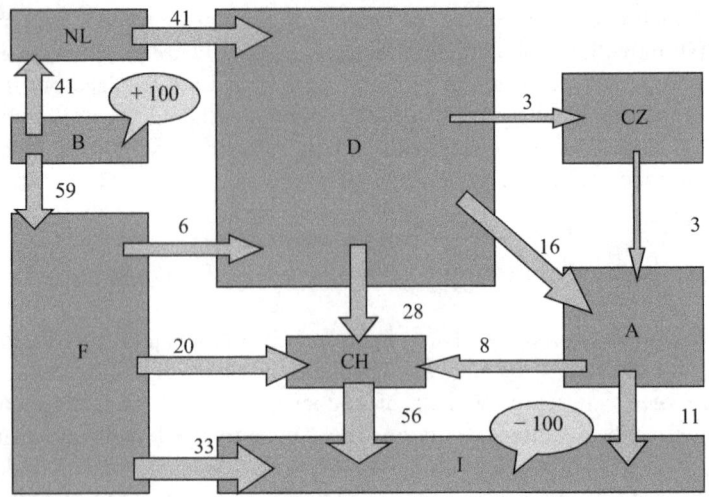

Figure 2.5   European network flows

2.10   Interconnection and loop flows

There is planned growth in interconnection between countries to support trading leading to a requirement to manage capacity allocation. This may be realised through auctions when there is a high demand for access. A problem encountered is that the network flows do not follow a direct point-to-point route but flow according to the impedance of the wider network. This results in loop flows that can affect the loading and security of adjacent networks. An example of the flow of a 100-MW export from Belgium to Italy is shown in Figure 2.5. The cost of the transaction is calculated based on the results of capacity auctions for each flow as shown in Table 2.3 and amounts to €3.8/MW h.

In this example, the loop flows created can be quantified and accounted for. **The developing problem is the impact of variations in the large tranche of wind generation that is not balanced close to source.** This leads to loop flows around the network that are not anticipated and can put security at risk and has resulted in complaints between governments at ministerial level. This is particularly the case in Germany with over 44 GW of wind capacity operational in 2015. Across Europe, there is over 120 GW operational that has the potential to disrupt planned network operation. The problem is exacerbated by the planned increases in interconnector capacity with an EU target of 15% of installed capacity by 2030 that may be difficult to control unless based on DC transmission. Several approaches can be considered to managing the intermittency of renewable generation sources each with different cost implications:

- The use of conventional generation regulating in reserve mode is widely used for system balancing. But, retaining high levels of reserve incurs additional fuel and wear and tear costs;

Table 2.3   Costs of interconnection transfers

Route	Flow, MW	Price, E/MW year	Total, cost/year
BE → NL	41	30,666	1,257,306
NL → DE	41	200	8,200
BE → FR	59	2,190	129,210
CH → AT	8	15,000	98,400
CZ → DE	3	5,090	15,270
CZ → AT	3	15,000	45,000
DE → AT	16	12,000	192,000
IT → AT	11	3,000	33,000
DE → CH	28	44,500	1,246,000
FR → DE	6	26,542	159,252
FR → IT	33	1,512	49,896
Fr → CZ	20	787	15,740
CH → IT	56	1,100	61,600
	Cost/MW h	**3.8**	3,310,874

- Regulating with hydro generation is efficient, provided its use for other purposes is not restricted, and this may be a viable option in countries with untapped hydro potential like China and Africa;
- The association of the renewable source with energy storage capacity in the form of pumped water, compressed air or in heat stores. Generally, the process involves an efficiency loss. In the case of pumped storage, the energy loss is typically 30% associated with the pumping and generation cycles.
- With the advent of smart metering systems, there is increased possibility of using demand side control to effect balancing. Consumers will require an incentive to participate in the form of annual availability payments and reduced tariffs;
- There may be scope to use interconnection to access remote balancing services. This could be remote hydro installations that are prepared to provide balancing services for a fee based on a MW-h charge. There may be scope to take excess energy into the remote system as well as provide make-up when the renewable output falls. As well as the cost of reserving interconnector transmission capacity, there may also be significant network losses associated with remote energy exchanges.

The problem is ameliorated by a good forecast of renewable generation output that enables plant to be brought into service or shut down. The lead time or changing plant status may be around 4 to 5 h or less with open cycle gas turbines or hydro. Initial average rms forecast error was about 17% but has improved to around 10%. This could still result in a worst case error of three standard deviations or 30%. To cater for this level of potential error would require reserve to be available up to 30% of the prevailing renewable output. This could be supplied by part loaded generation or fast start generation or demand control that will incur cost in reserving and calling off services. As well as the inherent energy losses associated with the balancing process, the lost opportunity cost also has to be considered. Storage systems can operate in the primary energy market to buy and sell energy taking advantage of the daily price movement. This process is continuous as opposed to the relatively low availability of surplus wind generation.

2.11 The implication to networks

The windiest sites are often in more remote areas without transmission infra-structure. Connecting the remote generation is expensive both because of the dis-tances and the low level of utilisation to underwrite the investment. There are also significant losses incurred in transporting the energy to demand centres. The use of multiterminal voltage controlled DC is being considered where the distances are large. In China, the majority of wind generation is in remote areas, thousands of kilometres from demand centres. Up to 84% of wind capacity is located in the north and the west of China (known as Sanbei area), while the demand centres are gen-erally in the south and the east due to their industrial advancement. Current network capacity of transferring the wind energy is insufficient, and consideration is being given to ultra-high-voltage electricity transmission to mitigate network congestion.

The stable operation of the transmission system following a loss of generation relies on frequency regulating generation responding in seconds to release stored energy and arrest the frequency fall. It is common for generation to be equipped with ROCOF protection that will trip generation given a sudden collapse in fre-quency. If the frequency fall is not arrested, this can result in cascade tripping widening the area of generation shortfall. **When the system generation includes a high proportion of renewable generation, then the overall inertia is depleted increasing the chance of ROCOF relay operation.** There is naturally reluctance amongst conventional generators to increase settings without a thorough assess-ment of the potential plant impact.

The connection of wind generation installations into the distribution network can adversely affect voltage control. When the wind generation is connected to a point of common coupling with other consumers, their supply can be adversely affected. It may be necessary to establish a more remote point of coupling at a higher voltage but this will add to the connection costs. Consideration should be given to using a higher distribution voltage of 11 or 20 kV to reduce the impact. This network could also support higher capacity EV charging points. There may also be interaction between embedded generation fitted with ROCOF protection and wind farms. The wide variation in the energy flows and their direction makes it difficult to manage the network capacity, and consideration is being given to thermal modelling to enable less conservative capacity assessments based on the recent loading history. Smart networks incorporating the management of wind generation output, storage and demand side regulation will potentially address some of the issues but not the expanding level of renewable capacity. **There will be a need for distribution system operators with facilities similar to those used to manage the transmission networks.**

2.12 Conclusions

There is no competitive market operating for generation at the distribution level to support competition between local generation sources, suppliers and grid-connected generation.

The impact of large tranches of renewable generation has meant that the net demand profile is much more variable and difficult to predict with in extreme cases distributors exporting energy back to the grid. The traditional approach of using historic records is invalidated, and the problem may need to be devolved to DSOs.

It is more difficult for potential investors in generation to predict revenues or even establish the data to make an analysis of the optimal plant mix other than much more flexibility will be required.

The renewable generation output will have to be curtailed at times of low system demand to maintain sufficient reserve and regulating capability to secure the system. The need for wind output curtailment increases exponentially when the non-synchronous capacity reaches around 50%.

The potential contribution of wind generation to the secure capacity available at peak times is estimated on a probabilistic basic to be around 10%–15% of the total installed. This means that conventional capacity has to be retained to maintain plant margins. Because of the low utilisation of this generation, it has been necessary to introduce capacity payments to retain the availability.

The intermittent loading of generation providing backup to renewable generation results in a loading regime that increases the normal fuel burn and incurs additional costs in wear and tear. The extra fuel burn results in more emissions offsetting the saving resulting from the use of the renewable sources.

As the power system non-synchronous generation penetration (SNSP) increases, the system inertia falls, and the system becomes more volatile increasing the chance of ROCOF relay operation. This could lead to reductions in wind generation output following faults and cascade tripping.

The net impact on emissions of converting to EVs will depend on the emissions from generation used to charge the EV battery being below the target emission levels. The loading on the distribution networks due to EVs may not have a lot of diversity. The rapid growth of domestic EVs could seriously stress a distribution network designed for an after-diversity demand of 1.5 kW per consumer leading to the need for reinforcement as the penetration reaches around 30%.

Network costs have increased to accommodate renewable generation that is often remotely sited. Some commentators have suggested that increasing network charges could promote mass defection from the grid where there are cheaper alternatives.

Indicative of the rate of introduction of EVs, the Norwegian government plans to ban the sales of petrol and diesel cars by 2025. This is consistent with policies advanced by other countries and would lead to most vehicles being based on electricity or hybrids by 2040.

There is a developing problem due to the impact of variations in the output of large tranches of wind generation that is not balanced close to source, on interconnector flows.

These technical issues constitute constraints on the future development of the industry that will impinge on economic assessments. They are likely to result in the need to reinforce distribution networks and their control capability to accommodate embedded generation and new EV and heating loads. System operation and demand prediction will become more complicated requiring a greater need for TSO/DSO interaction and cooperation.

Question 2.1 Given data in the table below calculate the emissions per kW h from bituminous coal and gas-fired generation.

Fuel	gC/GJ	tC/mtoe	tCO_2/GJ	Price/GJ	Efficiency	Price/MW h
Lignite/brown coal	25,200	1.169	0.0924	0.924	33	10.08
Bituminous coal	23,700	1.100	0.0869	0.869	38	8.23
Oil	19,900	0.923	0.0730	0.730	40	6.57
Natural gas	13,500	0.626	0.0495	0.495	52	3.43

Using the data from Table 2.2, calculate the average kW h/mi for a Tesla 60 kW h and the emissions/mile when charged from coal and gas-fired generation for comparison with a typical petrol car at 200g CO_2/mile.

Chapter 3
Impact of developments on market players

3.1 Introduction

The changes taking place in the industry are having a profound effect on all the market institutions and participants.

- Large generators are faced with utilisation being lost due to the requirement for renewable generation to have prior system access;
- The distribution networks are becoming much more active than passive as a result of increases in embedded generation and cyclic loading from EVs;
- The transmission system is faced with losing the connection of large-scale generation in favour of local generation;
- Small and domestic consumers are becoming more engaged through the use of smart meters and opportunities to fit solar panels or heat pumps;
- The transition is being promoted through subsidies to promote renewable generation and to install smart meters that is adding costs to consumers;
- There has been a profound decrease in the margin of spare generation capacity with closures due to utilisation being displaced by renewable generation and the perceived risks of building new generation.

These developments have not been planned but are driven by policy makers and developers who see new opportunities. The changes are driven by ambitious targets set by the EU to reduce emissions leading governments to introduce generous subsidies to renewable energy generation principally, for wind and solar. Network charges have increased to accommodate the new energy sources and make up to 50% of end user charges. This has resulted in schemes designed to avoid these charges by using direct wiring to local loads making them financially viable. Local generation schemes can also avoid the losses that occur in supplies via transmission and distribution networks, of around 2% and 6%. Small generators also benefit from reducing the peak charges incurred by their suppliers. This has led to some 10% to 15% of generation capacity being embedded with a proportional reduction in grid connected generation.

In the United Kingdom, there are now some 400 licence holders in the electricity sector including suppliers, generators, network operators and interconnectors. It is difficult to see how a holistic optimal development path can be realised. The Council of European Energy Regulators (CEER) propose there is a need for a market design that allows participants to decide on an appropriate generation mix without subsidies. It does not indicate how an appropriate mix may evolve. It recognises the need to include self-generation (SG) in network planning with SGs accepting the same responsibilities as active power producers. They must have adequate metering, communication and control capability and pay cost reflective tariffs that recognise the value of connection to the distribution network.

This chapter discusses the impact of developments on the market players and the initiatives made by regulators to influence the changes in the interests of consumers.

As the capacity of intermittent wind generation is increased, the utilisation of conventional generation decreases leading to closures and reduced margins and security.

3.2   Impact on generation utilisation

As the capacity of intermittent wind generation is increased, the utilisation of conventional generation decreases. The potential impact has been modelled in a full UK system simulation of operation for year 2020 using actual half hour demand and wind data extrapolated from recorded data. The result is illustrated in Figure 3.1 where it can be seen that increasing the wind capacity up to 25 GW results in the utilisation of marginal generation decreasing by around 10% to 20% with a lot of generation operating at load factors below 40%. This is based on the time connected to the system and a lot of this will be part loaded providing backup for wind data, further reducing the energy supplied.

The impact is higher on the more marginal generation with some of those at the margin not running at all. The reduction in utilisation will have a significant impact

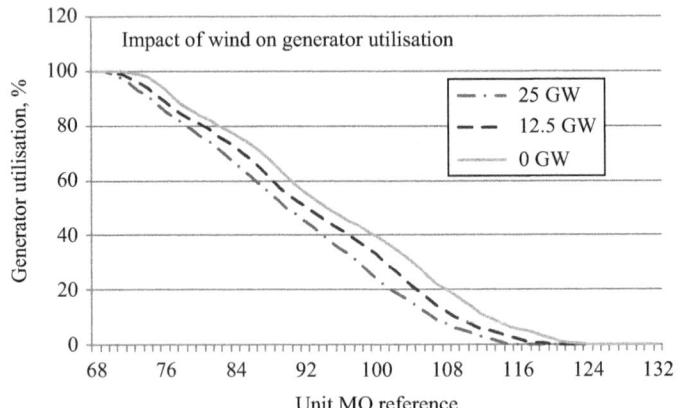

Figure 3.1   Impact on generator utilisation

on generation unit costs. **In parallel, costs are being added to reduce CO$_2$ emissions** and the impact of a price of £20/t has been added for typical gas and coal generation as shown in Figure 3.2.

The implementation of a carbon floor price was designed to disadvantage fossil fuel generation; to reduce risks associated with investment in low-carbon technologies and to advance the UK's climate change mitigation agenda. The impact has been for renewable energy output to reduce the utilisation of conventional generation operating at the margin. Both these developments have a significant impact on the unit cost of generation. Figure 3.2 shows the impact on gas and coal unit prices with varying utilisation with and without adding in a carbon price of £20/t.

It can be seen how coal and gas generation unit prices rapidly increase at the lower levels of utilisation from which they are able to recover fixed costs. Assuming an annual load factor for wind of 25% then the output of 30 GW of wind generation would displace the utilisation of 20 GW of backup fossil generation by the equivalent of 7.5 GW or 37%.

These developments taken together add significant unit costs to a wide range of conventional generation and it is inevitable that some will cease to be economically viable and will close. At the same time, the cost of CO$_2$ emissions has been set in the United Kingdom to escalate further disadvantaging fossil generation and particularly coal that will be driven out of the plant mix. These developments will lead to a fall in the plant margin, whereas the marginal plant is still required to cover periods of high demand when the wind does not blow. It remains to be seen whether demand side response and storage can bridge the shortfalls.

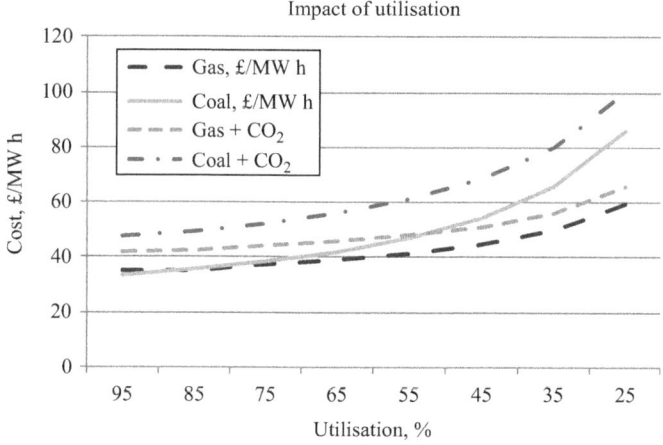

Figure 3.2   Coal and gas unit prices with varying utilisation and CO$_2$

3.3   The impact on generation companies

The rush into subsidised renewable energy has resulted in casualties in the European energy sector that will undermine future investment and lead to capacity shortfalls. Economic theory would argue that business failures are a sign of competition but in this case it's the result of market intervention by government. The problem is that conventional generation is still needed to cater for periods of low renewable output but the reduction in its utilisation makes it financially non-viable.

Rheinisch-Westfalische Elektrizitatswerke (RWE) has a number of problems and the company had to shed a fifth of its UK workforce. The Npower business has suffered issues with its billing system but the big problems are in Germany, where RWE saw its shares tumble from €100 to €10, not least because of its reliance on dirty lignite coal and nuclear energy. In 2014, less than 5% of RWE's power generation came from renewable sources, yet Germany wants 30% of energy to come from renewable generation by 2020.

Earnings' forecasts for RWE tumbled by around 80% with debts climbing to almost €20bn. City analysts forecast that in 2018, RWE's earnings per share will be half the level for 2015 with some analysts forecasting no future dividends. These pressures have less to do with competition than the result of subsidised green energy taking away a proportion of their demand and depressing market prices. The financial failure of coal fired generators is an expected consequence of giving subsidised renewable sources priority access to the system.

British Gas, the owner Centrica, is currently backing away from a misjudged move upstream. Billions were spent on oil and gas fields, but they proved a poor investment when wholesale prices crashed. Now Centrica is refocusing on the millions of customer relationships it has, here and in North America. A big push is expected to sell additional services in a 'Back to the Future' move for the group which once bought the Automobile Association in an attempt to do precisely just that, only to sell out a few years later.

Centrica is expected to focus on its core business reflecting its view of what the future regulatory agenda will hold. It will manage down costs and release capital from the upstream businesses. The dividends were rebased downward in 2015 and the group is targeting a progressive policy, backed by the rising cash flows it expects its restructuring to deliver. On current forecasts, the stock yields around 5.3%. It has also decided to convert some of its CCGTs to operate open cycle to provide the increased flexibility the market needs.

E.ON built a UK energy portfolio, acquiring PowerGen and electricity distribution companies in the Midlands. It faces many of the same challenges as RWE, and its shares have fallen from €50 in 2008 to €8 today. E.ON plans to spin off its conventional generation assets into a unit called Uniper to leave the group more focused on renewable generation. The company will have to carry both the group's financial debt and their nuclear decommissioning liabilities. Dividends have been repeatedly cut, but current forecasts suggest a yield of over 5%, but perhaps only until the next cut.

Iberdrola bought **Scottish Power** that has halved in value since before the financial crisis, but the shares have been recovering. Iberdrola has expanded its renewable portfolio, so is less exposed to the problems facing many other utilities. Even so, only around 20% of its cash earnings come from renewable generation, with 30% still coming from conventional fossil fuel generation. According to consensus forecasts, Iberdrola is set to grow earnings and dividends each year to at least 2019 and the stock currently offers a prospective yield of almost 4.7%.

EDF hit the news when its finance director quit, reportedly in protest at the risks attached to the company's involvement in the proposed new Hinckley Point nuclear station, along with Chinese partners. The £18bn cost, for a business already carrying around €40bn of net debt, could be the final straw. Hinckley Point is to use a new French reactor design, so far sold to only two other projects. Both of these, in Finland and France, are years behind schedule and billions over budget.

The United Kingdom kick-started the project, by guaranteeing the selling price of Hinkley Point's power, at prices that were high then and which now look doubly so. To make cuts with one hand, whilst paying over the market rate to a Franco/Chinese nuclear power syndicate is not easy for politicians to explain. Suggestions were made that the project should be abandoned prompting a review by the prime minister who, after seeking some safeguards, agreed to proceed with a high strike price. Even so some analysts are forecasting dividend cuts and a fall in profits for years to come.

SSE has faced tough trading conditions for some time now, with pressure on earnings in its wholesale and retail divisions. SSE has at least been able to raise the billion-plus pounds it sought by selling down non-core assets, with the latest part disposal of a Scottish wind farm bringing in over £300m. Earnings cover for the dividend is low, and the company has said it may remain so until trading conditions improve, expected in a few years' time. The UK electricity sector has had a largely clean bill of health from regulatory reviews, so SSE has a clearer view of its future from here.

Market forecasts suggest earnings per share for SSE will be lower in 2019 than they were the previous year, but the dividend will still be covered, but only by around 1.2×. A yield of just over 6% suggests the market is giving SSE the benefit of the doubt, but clearly the hoped-for recovery in earnings needs to appear before too long, otherwise SSE will find that the dividend begins to pose a significant restraint on its operations.

All these company assessments indicate an industry driven into crisis by shifts in policy that were not signalled and that are having far reaching consequences. The business impact will be to make investors more cautious of the regulatory risks in the sector that will deter new investment and add to capital costs. A full impact assessment was not undertaken and some policies are now belatedly being re-thought by government and regulators. There is no clear approach to managing the changes being imposed as would take place in any other industry. This book aims to throw light on the optimum pathways ahead.

3.4   Active distribution networks

The development of more embedded generation is resulting in local distribution networks moving from becoming essentially passive to active requiring more control and capacity. This in turn is fostering an interest in cities engaging in becoming smarter. Some examples of local energy developments include:

- Combined heat and power schemes that use waste heat from local generating stations to provide district heating. London uses heat pumps and water from the Thames and Copenhagen pumps heat through 1,300 km of pipes.
- Smart street lighting that operates in response to need triggered by passing traffic or pedestrians to avoid continuous use. Glasgow and San Diego are trialling systems.
- The use of more autonomous local networks to maintain security in the event of faults or capacity shortfalls by restructuring. Lancaster in Texas can completely isolate from the grid and revert to local energy sources, including storage, in the event of grid problems.
- Solar panels have reduced in price by 70% in 5 years and are being widely installed on public buildings. The Netherlands is looking at solar roads using tempered glass with a covering of crystalline solar cells. As a result of a competitive tender, a 100-MW solar installation is being developed in Chisago, Minnesota with the approval of the Public Utilities Commission. Floating solar wind farms are being trialled in China.
- Lithium battery prices have more than halved in recent years and as well as their use in electric cars other applications are being considered. There are suggestions that they could be deployed in small business and even domestic premises to increase the opportunity for demand side response to prices. A pilot installation in 40 council homes in a London borough will use solar to charge the storage batteries drawing down the energy at peak times to avoid high network charges. The supplier, a company called North Star Solar believes that with long-term contracts, the systems will be self-funding and not require subsidies.

These developments will require increases in the capacity of the networks to accommodate the increased flows that will change direction at times of low load and generation output. There will be a requirement to enhance the SCADA systems providing metering and control functionality including the output of embedded generation to manage network security. There will also be a requirement to review protection systems like rate of change of frequency to avoid cascade loss of generation. Automatic switching is likely to be required to manage system incidents and reconfigure networks. Demand side control will need to be monitored and coordinated in conjunction with storage. The shift in network use has not been planned but is being driven by users. The developments are not currently reflected in distribution use of system tariffs that will need to be increased to accommodate the changing use. The control infrastructure will need to be developed on a par with that currently used to manage the super-grid network. **The costs of the control**

infrastructure is expected to add around 10% to distribution tariffs and the capacity of the networks will need reinforcement, particularly to accommodate EV charging and heat pump demand. These costs may shift the balance back in favour of HV connected generation where the control infrastructure and capacity is already in place.

3.5   Transmission utilisation

The capacity of generation connected to the UK transmission network has fallen from 79 GW in 2011 to 68 GW in 2015, a fall of 14%. During the same period, the capacity of generation connected to the distribution system has doubled from 12 to 24 GW. **In effect, some 12 GW of capacity has been transferred from being connected at transmission voltages to distribution**. These developments in part reflect the use of system charging arrangements that favour distribution connected generation that reduces the peak demands that set the bulk supply tariff to suppliers. The consequence is that more of the transmission use of system revenue has to come from the demand side and is distorting cost comparisons between transmission and distribution connected generation. The reduced loading on the transmission network can result in lines operating below their natural load level when they effectively generate Mega Volt Amp (reactive) (MVAr). This can result in high voltages at times of low load necessitating taking lines out of service to reduce the system gain.

At the same time, the transmission network is being extended to connect remote wind farms that are increasingly offshore. To encourage competition, the connection of these sites is put out to tender with ownership external to the transmission company. **This led to there being 18 transmission operators in the UK sector by 2017.**

3.6   Low carbon generation funding

The introduction of a contract for differences (CfD) for low carbon generation was designed to encourage new entry to replace older plant that was expected to close. The generators could sell their energy into the wholesale market as usual but would receive extra payment via the CfD based on the difference between the market reference price and the contract strike price considered necessary to support financing investment in new generation. This financing arrangement removed a lot of the risk resulting from market price volatility and in turn reduced the cost of capital. The government would act as the counter-party and recover the costs of the scheme by a levy on all licenced suppliers and, in turn, consumers. The scheme was arranged so that investors could receive confirmation of a CfD on proof of planning approval and a connection agreement. The scheme was administered on a first come first served basis against a limited budget set within a levy control framework. The CfD payments were set to be inflation linked and were two-way in that if market prices rose above the strike price then the generator would pay into the scheme. The strike prices were set based on construction and operating costs recognising the risks faced by investors. The longer term intention was to move to technology independent auctions for new capacity during the 2020s. The reference price for the contract would be based on the GB day ahead hourly zone price. The contract term would usually be 15 years.

This new scheme replaced the original renewable obligation certificate scheme where suppliers were targeted to ensure that a specified percentage of the energy they sold was derived from renewable sources as evidenced by providing certificates. The value of the certificates was adjusted according to the total capacity delivered to the system. If the capacity target had not been reached then the value of certificates would be increased by redistribution of the penalty payments to complying generators and vice-versa. An advantage of this scheme was that payments automatically adjusted to lead to delivery of the target capacity encouraging suppliers to compete on price as opposed to the arbitrary first come first served arrangement with the CfD scheme. **These schemes have committed consumers to very high energy premiums for years to come and the premium for nuclear energy will add significantly to the burden.**

3.7   Impact on consumers of subsidies

In order to meet climate change targets, governments have introduced subsidies to promote the development of renewable energy sources like wind and solar. The concept was that the subsidies would kick-start the industrial development and that costs would quickly fall enabling subsidies to be phased out. In practice, there has been limited success in reducing wind power costs although solar panel costs have fallen. According to the digest of UK statistics for 2016, subsidies for wind and solar were £4.7bn/year for 1.7% of energy from wind and solar and 6.3% from other renewable sources amounting to 8% in total or 29.2 TW h. The offshore capacity was 2,500 MW and assuming a load factor of 30% the annual production would be:

$$2,500 * 0.3 * 8,760 = 6.6 \text{ TW h}$$

The subsidy cost at £155/MW h less the average wholesale price of £40/MW h is given by:

$$6.6 \text{ TW h} * (155 - 40)\text{£}/\text{MW h}/1,000 = \text{£}0.76\text{bn/year}$$

The onshore wind capacity was 12.5 GW at an average load factor of 0.29 (higher than usual in 2015) generates:

$$12,500 \text{ MW} * 0.29 * 8,760 = 32.7 \text{ TW h}$$

The subsidy cost at £95/MW h is given by:

$$32.7 * (95 - 40)\text{£}/\text{MW h}/1,000 = \text{£}1.8\text{bn/year}$$

The 9 GW of solar panels generate:

$$6 * 0.15 * 8{,}760/1{,}000 = 7.8 \text{ TW h}$$

The cost based on a subsidy £66/MW h if given by:

$$7.8 * (66 - 40) = £0.2\text{bn}$$

The total energy from wind and solar is 47.2 TW h or 12% of the total electricity production at a total subsidy premium of £2.76bn/year. Based on 28 million consumers this amounts to **£100/consumer/year in 2016** and the rises are raising questions about capping prices. Analysis by the Monopolies and Mergers Commission has shown that prices are not generally excessive but can in part be attributed to government policy.

3.8   Total subsidy costs 2030

It is generally reported that future plans to convert to green energy will cost consumers more with reported figures ranging from several thousand pounds a year to a few hundred. It is often not clear what is being referred to with some quotes based on total costs, per person, some per household and some on the additional costs. There are lot of assumptions that have to be made to establish any costs including the future plant mix, fuel and CO_2 costs and capital costs. Table 3.1 shows costs based on a plant mix to meet a low emission green strategy. The nuclear stations are included, based on their Feed in Tariff (FIT) costs, along with conventional generation. The fossil generation gas and coal fuel costs were based on coal at £4.4/GJ and gas at £7.8/GJ with CO_2 at £70/t giving total costs of £10.48/GJ and £11.26/GJ and are summated for all plant types. **The FIT costs include payments for curtailed wind output not included in the TW h provided.** For similar amounts of energy, the renewable costs are several billion higher. Conventional generation costs/year are £18.14bn for 166.7 TW h, whereas renewable costs are £22.24bn for 160.4 TW h.

Table 3.2 compares the costs of meeting all the demand from conventional generation with that from renewable sources by extrapolating the costs as shown in the last column. It can be seen that meeting all energy from fossil sources would cost 35,609 £m/year as opposed to the mixed generation cost of 40,384 £m/year i.e. an additional cost of 40,384−35,609 £m/year = 4,775 £m/year. This equates to £160/year/household and compares with reported government estimates of £200/year when additional network costs are included. If all energy was derived from an extrapolated renewable set of generation then the additional cost would be £9,375m/year. Apportioning these over 30m households **equates to a premium of £312/household/year to support meeting climate change targets in 2030.** Additional costs are associated with network reinforcement to accommodate the connection of remote renewable generation. Although prices are falling, a lot of generation is already contracted at high prices.

Table 3.1 Subsidy premium 2030

2030 technology	Capacity, MW	Capital, £m/year	Fuel/var cost, £m/year	Energy, TW h	Renewable energy, TW h	FIT, £/MW h	FIT costs, £m/year
AGR	1,200	797.3		8.6		92.5	
PWR	11,510	7,466.0		80.7		92.5	
BIT	1,987	240.4		0.0			
CCGT	29,469	1,506.0	5,003.4	29.9			
OCGT	5,579	198.4		0.0			
CHP	4,982	235.9		22.0			
OIL	782	48.7	0.0	0.0			
Gas CCS	2,000	305.2		10.7			
Coal CCS	2,588	621.2	1,719.8	14.7			
Offshore wind	35,956				59.0	155.0	11,948.5
Onshore wind	20,985				34.6	95.0	4,301.0
Solar PV	15,813				18.1	66.0	1,195.8
Biomass	6,736				44.3	92.0	4,075.6
ROR	2,176				4.5	160.0	721.3
Totals	141,763	11,419.0	6,723.2	166.7	160.4		**22,242.2**
		Fossil costs	**18,142.2**	Total energy	327.1	Total cost	40,384.4

Table 3.2   Subsidy premium

	Costs, £m	TW h	Cost all, TW h
Fossil costs	18,142	167	35,609
Renewable costs	22,242	160	45,344
Totals/subsidy	40,384	327	9,735

3.9   Impact on technology type

The UK Emission Performance Standard (EPS) proposed a general maximum target level for emissions from new fossil fired generation of 450 g/kW h. Using the figures shown in Table 3.3, the emissions/MW h can be calculated using the carbon content/GJ figures converted to the equivalent content/MW h and taking account of the efficiency as follows:

$$gas = 0.0495 * 3.6/0.55 * 1,000 = 324\,g/MW\,h$$
$$coal = 0.0869 * 3.6/0.45 * 1,000 = 632\,g/MW\,h$$

It can be seen that conventional coal generation would not comply with the standard and will lead to it being phased out. It was suggested that penalties would apply to non-compliance based on the cost of emissions unless fully equipped with Carbon Capture and Storage (CCS).

3.10   Security of supply

The pace of the transformation taking place across the industry is having an adverse effect on system security. There are several contributing factors to a reduction in the plant margin of spare capacity including:

- Intermittent renewable generation does not provide firm capacity;
- The output from renewable generation reduces the utilisation of marginal conventional generation to the point where it is no longer viable financially;
- The practice of priority dispatch of renewable sources results in an operating regime for conventional generation inconsistent with its design parameters;
- An increasing reliance is being placed on the available support provided by interconnectors that is not embodied in contracts.

There is varying opinion on the capacity contribution that can be expected from renewable sources. The analysis has to be based on the probability of wind generation being available at the time when peaks occur. During cold spells in the Northern hemisphere, when demand peaks, it is often not windy or sunny with little renewable contribution. However, based on a probabilistic assessment using recorded data, a **contribution of around 12% of installed wind capacity has been calculated**. This is of the same order as the figure used in Ireland with a high proportion of wind generation, whereas National Grid Company (NGC) assume a more optimistic figure of 17%.

Table 3.3   Cost of generation

Parameters	Units	Gas	Coal
Capacity	MW	393	398
Capacity net	MW	377	366
Load factor	%	25	25
Annual output	TW h	0.83	0.80
Construction time	Years	2	4
Project life	Years	20	20
Efficiency	%	55	45
Coal cost delivered	£/t		40.279
Coal calorific value	GJ/t		24
Natural gas cost	£/mmBTU	3.842105	
Natural gas cost delivered	Pence/therm	38.42105	
N gas calorific value	kW h/therm	29.32	
N gas grossCV/netCV		1.11	
Cost of dept	%	4.7	4.7
Cost of equity	%	7	7
Inflation	%	2.4	2.4
Debt/equity split		0.8	0.8
WACC ($D/E = 0.8$)	%	6.0	6.0
Site and dev	£m	9.733	27.46
EPC contract	£m	235	500
Electrical conn	£m	0.733	0.733
Gas connection	£m	2.466	
Spares	£m	14.33	31.86
Interest during const.	£m	11.26	23.20
Total investment cost	£m	273.5	583.3
Capital cost	£/kW	696	1,465
Capital cost pa	£/m	23.78	50.71
Consumables	Pence/kW h	0.02	0.113
Labour rate	£/h	16.66	16.66
No. of operators		6	14
Operators labour*1.28	£m	1.12	2.62
Maintenance material	£m	0.81	2
Main and support labour	£m	1.25	2.33
Capital cost	£/MW h	28.78	63.23
Fixed O&M	£/MW h	1.56	4.39
Var O&M	£/MW h	2.49	5.40
Fuel costs	£/MW h	26.45	13.43
Total generation costs	**£/MW h**	**59.28**	**86.45**
CO_2 price £/t	**20**		
Include carbon	tCO_2/GJ	0.0495	0.0869
Added cost/incentive	£/MW h	6.48	13.90
Total cost	£/MW h	**65.76**	**100.35**

The impact on the utilisation of conventional generation was shown in Section 3.2 to be up to 20% depending on the capacity of the wind in relation to system size. The consequence is that plant ceases to be financially viable and is closed reducing the plant margin further, some 4 GW of UK gas-fired generation has been closed as a consequence. At the same time, CO_2 on costs are causing the closure of coal plant further reducing margins. This in turn has reduced the flexibility of the system generation. Wind and solar generation in most countries receive priority dispatch that imposes a lot of regulation on conventional generation balancing the renewable output changes and adding to operating costs. Biomass generation is one of a few renewable sources that can be dispatched.

It is questionable to what extent small-scale embedded generation can contribute to providing firm capacity. It is not directly dispatched to meet peaks but may operate to reduce peak demands and hence use of system tariffs. There is evidence that UK peaks reduced by 6 GW between 2005 and 2015 by running embedded generation. There are currently strong tariffs incentives that are distorting the situation and no certainty that this 'hidden capacity' will be available. There are also questions about the emissions from some of the embedded generation particularly diesel units. Another factor influencing peaks is the short-term demand side regulation. **The consequence of the tight margins is that the system is less robust to any major generation failures**.

The capacity of interconnection is increasing and has the potential to provide support following outages. However, the availability at peaks is questionable during cold spells that may affect the whole of Northern Europe. In order to guarantee capacity, it should be contracted as firm capacity with costs related to its provision.

In analysing security, assumptions have to be made about the availability of generation based on recent performance. Typical figures used are 94% for CCGTs and 82% for nuclear. The equivalent firm capacity attributed to wind is influenced by the potential diversity in output across a country and becomes more important as the capacity increases. The SO will contract for services to manage potential shortfalls including system balancing reserve (SBR), demand side balancing reserve (DSBR), emergency generation and interconnection support. The UK SO services are as shown in Table 3.4 and amount to over 5 GW at a cost over £1m/year:

Table 3.4   Contracted reserve support

SBR/DSBR (GW)	Voltage reduction (GW)	Max. generation (GW)	Emergency intercom (GW)
2.5	0.83	0.39	1.3

The net impact of these developments is that security assessment, and planning developments to manage it, has become more difficult. At the same time, developers and investors see the risks and uncertainty in the market and defer decisions leading to unprecedented low margins of spare capacity. **The maintenance of supplies is now more dependent on the actions of a diverse range of players that are principally motivated by opportunities to make money.**

3.11   Conclusions

As the capacity of intermittent wind generation is increased, the utilisation of conventional generation decreases, leading to closures and reduced plant margins that impact on system security. Most of the large players in the market have suffered from falling revenues and share prices. The German utilities in particular have experienced a collapse in share prices, falling by 80% with earnings/share halved. This has been driven by ambitious targets for renewable generation with a target for 30% of supply by 2020. In parallel, costs are being added to conventional generation to reduce CO_2 emissions. These developments taken together add significant unit costs to a wide range of conventional generation and it is inevitable that some will cease to be economically viable and will close. The generating company assessments indicate an industry driven into crisis by shifts in policy that were not signalled and that are having far reaching consequences.

The United Kingdom introduced an EPS set at 450 g/kW h. It was shown that conventional coal generation with emissions of over 600 g/kW h would not comply with the standard and will lead to it being phased out. This will impact on the plant margin at a time when investors are nervous of the perceived risks and are unlikely to pursue replacement projects. Gas-fired CCGT generation at around 324 g/kW h would comply but if the expected utilisation is low the project may not appear profitable.

In contrast, there is a growth in small-scale generation being connected at distribution voltages. This is in part the result of their potential impact on the use of system charges levelled on suppliers. By generating at the time of peaks, the embedded generators can reduce the distribution demand and the supplier capacity charges. The smaller generators also do not pay transmission charges. As a result, some 12 GW of capacity has been transferred from being connected at transmission voltages to the distribution networks. This is stretching the capacity of the distribution networks with some networks reported as being at their limit. Proposals are being considered to revise the network charging arrangements.

As the distribution networks move from being essentially passive to active, their systems for monitoring and control will need to be enhanced. The costs can be expected to add around 10% to distribution tariffs. The capacity of the networks will also need reinforcement particularly to accommodate EV charging and heat pump demand resulting in costs rising in proportion. These costs may shift the balance back in favour of HV connected generation where the control infrastructure is already in place.

The transmission network is experiencing less utilisation because of the increase in distribution connected generation that sometimes exports back to the grid. Extra residual costs have to be charged to the demand side to cover transmission costs. At times, the circuit loadings are below the natural loading level and circuits have to be switched out to contain voltage rises. The connection of offshore wind farms has been put out to tender with the result that by 2017, there were 18 transmission operators in the UK sector.

Subsidy schemes to promote renewable generation have committed consumers to very high energy premiums for years to come and the premium for nuclear energy in the United Kingdom will add significantly to the burden. Subsidies for wind and solar energy in the United Kingdom for 2016 resulted in extra costs to consumers above normal energy rates, of around £100/year. To meet the climate change targets for 2030, the subsidy premium for UK customers would rise to £312/household/year based on the expected plant mix.

The firm capacity contribution of intermittent generation like wind and solar to the plant margin is questionable. A study based on the probability of generation availability and demand was run both with and without wind generation to realise the same loss of load probability. The study analysed the amount of wind generation necessary to restore the plant margin to the same level based on thermal generation; the result showed that a capacity contribution of around 12% of installed wind capacity could be expected from wind generation.

The maintenance of supplies is now more dependent on the actions of an increasingly diverse range of players that are principally motivated by opportunities to make money, as is their job. Whilst the need for a holistic coordinated approach to system development is generally accepted, how it can be realised in the current liberal environment is less clear.

Question 3.1 The fossil generation gas and coal fuel costs were based on coal at £4.4/GJ and gas at £7.8/GJ with CO_2 at £70/t giving total costs of £10.48 and £11.26/GJ and are summated for all plant types in the table. Assuming that fuel prices do not rise as high then, based on the data in Table 3.1 and assuming that the CO_2 price was £35/t with coal at £3/GJ and gas at £6/GJ, calculate the new gas and coal fuel costs and hence the estimated total fossil costs for comparison with the renewable costs. Assuming that 20 GW of the offshore wind is contracted by competitive tender at £100/MW h and the same level of curtailment applies, find the revised renewable costs on a pro rata basis and the new total.

Table 3.1 Subsidy premium 2030

2030 technology	Capacity, MW	Capital, £m/year	Fuel/var cost, £m/year	Energy, TW h	Renewable energy, TW h	FIT, £/MW h	FIT costs, £m/year
AGR	1,200	797.3		8.6		92.5	
PWR	11,510	7,466.0		80.7		92.5	
BIT	1,987	240.4		0.0			
CCGT	29,469	1,506.0	5,003.4	29.9			
OCGT	5,579	198.4		0.0			
CHP	4,982	235.9		22.0			
OIL	782	48.7	0.0	0.0			
Gas CCS	2,000	305.2		10.7			
Coal CCS	2,588	621.2	1,719.8	14.7			
Offshore wind	35,956				59.0	155.0	11,948.5
Onshore wind	20,985				34.6	95.0	4,301.0
Solar PV	15,813				18.1	66.0	1,195.8
Biomass	6,736				44.3	92.0	4,075.6
ROR	2,176				4.5	160.0	721.3
Totals	141,763	11,419.0	6,723.2	166.7	160.4		22,242.2
		Fossil costs	18,142.2	Total energy	327.1	Total cost	40,384.4

Part II

Review of low carbon generation technology options

The second part of the book reviews the progress on global emission reduction and the technology options to meet targets including the deployment of increasing volumes of intermittent renewable generation, generation based on biomass, the option of large- and small-scale nuclear and generation with carbon capture and storage. The impact on the distribution networks of the growth of embedded small-scale generation and cogeneration schemes is discussed. The advantages and disadvantages of the various low carbon generation technologies are discussed and their expected costs are compared. It discusses the issues related to their respective operating role within a mix of different plant types and identifies any operating constraints that influence the optimal mix.

Chapter 4 – Emissions and Renewable Generation – This chapter focuses on intermittent renewable generation options and the scale of emission reduction required across the world and the progress being made. It provides a practical analysis of potential new wind power schemes and the likely need to curtail output as the renewable capacity is increased. It analyses other options to reduce emissions including biomass, waste to energy; solar and replacing coal with gas-fired generation. It highlights the impact of the growth in capacity of renewable sources on the operation and financial position of conventional generation.

Chapter 5 – Embedded Generation Issues – This chapter highlights the growth of embedded generation connected to the distribution system and the network charging arrangements that is contributing to the transition to an environment with more end user engagement. It discusses the impact on the networks and their management requirements as they become more active. The economics of small-scale Combined heat and power (CHP) and larger cogeneration schemes are analysed and emissions from small-scale diesel installations. The basic costs of embedded generation are compared with large-scale transmission connected generation. The reaction and developing policy position of regulators to the transition is discussed.

Chapter 6 – The Nuclear Option – This chapter discusses the nuclear option for the provision of low carbon energy. The costs and advantages of large-scale installations are compared with those of smaller more flexible modular reactors. Reference is included to fusion generation research and progress on development.

The optimal role of nuclear generation within a plant mix is analysed in relation to other types of generation.

Chapter 7 – Carbon Capture and Storage – This chapter focuses on the option to reduce atmospheric emissions by deploying carbon capture and storage (CCS) to coal and gas-fired generation. The design options are reviewed and estimates are made of the range of expected costs. The status of operational and planned schemes is reviewed together with the motivation for their development. The potential operational role is discussed and the benefit of their flexibility offers in balancing the output from intermittent sources. The results of detailed modelling of the operation of a system with a tranche of CCS generation shows potential cost savings of 17% over a plant mix with a high proportion of intermittent generation while reducing emissions by the same amount.

Chapter 4
Emissions and renewable generation

4.1 Progress on global emission reduction

Faced with the implications of global warming many countries are progressing in the implementation of renewable energy sources. These include principally wind, solar, hydro, tidal and biomass. The IEA reported that worldwide generating capacity in 2016 was 6,400 GW including 1,951 GW of coal and 1,985 GW of renewable with the rest from gas and other sources generating a total of 23,816 TW h. The renewable proportion of total energy supplied was 23% made up of 71% hydro; 15% wind; bio-energy 8% and solar just 4%. Some 67% was produced from fossil fuels. Worldwide total emissions have been estimated at 35,000 $MtCO_2$ of which electricity generation accounts for an estimated 10,363 $MtCO_2$ as shown calculated in Table 4.1 from the production by plant type and typical emission factors per TW h. It can be seen that three quarters of the emission comes from coal generation supporting massive coal mining sectors that would not easily be down sized.

China is probably the largest emitter of green-house gases because a lot of its generation is coal fired. Based on generation production, emissions can be calculated as shown in Table 4.2 where it can be seen that **emissions in China were 26 times those of the United Kingdom in 2015** with a per unit level in g CO_2/kW h 70% higher during 2015 at 553 g CO_2/kW h.

Progress is being made with subsidies to renewable sources but the Chinese National Development Reform Commission struggles with the high costs with shortfalls of over 60 billion yuan/year. Typical unit costs were for solar 0.8–0.98 yuan/kW h; wind 0.47–0.6 yuan/kW h; compared to 0.3–0.5 yuan/kW h for

Table 4.1 World emissions from electricity generation (source IEA)

Generation	Coal	Oil	Gas	Nuclear	Hydro	Other	Totals
% Mix	40.8	4.3	21.6	10.6	16.4	6.3	100
TW h	9,717	1,024	5,144	2,524	3,906	1,500	23,816
Emission factor	0.82	0.63	0.34			0	
Emissions $MtCO_2$	7,968	645	1,749	0	0	0	10,362
% Emissions	76.89%	6.23%	16.88%				100%

Table 4.2   Emissions China/UK

Type	China 553 g CO_2/kW h				UK 329 g CO_2/kW h			
	GW	Energy, %	TW h	Emiss, Mt	Cap, GW	Energy, %	TW h	Emissions
Coal	933	63.5	3,379	2,782	24	22.6	76.4	62.9
Hydro	325	20.6	1,096	0	4.3	2.5	8.5	0.0
Wind	120	6.8	362	0	17	12	40.6	0.0
PV	14	0.3	16	0	6	2.2	7.4	0.0
Nuclear	43	2.9	154	0	9.9	20.7	70.0	0.0
Gas	80	2.7	144	49	34	38	128.4	44.0
Oil	10	3.2	170	112	1.3	2	6.8	4.4
Totals	1,525	100	5,322	2,943	96.2	100	338	111.3

thermal. Based on the renewable costs, the subsidy over low-cost thermal can be calculated as

$$Premium/year = (0.8 - 0.3) * 16 \, TW \, h + (0.47 - 0.3) * 362 \, TW \, h$$
$$= 69.5 \, billion \, yuan/year$$

This is about \$10 billion/year. The capital cost of solar has fallen from 5 to 3 yuan/kW with target installed capacities of 150 GW of solar and 210 GW of wind by 2020 but there are limiting transmission constraints. A consequence of increasing proportions of wind generation is that because of lack of diversity it floods the market reducing the revenues from energy sales while at the same time higher costs in capacity payments are essential to keep conventional plant open to provide backup.

In the USA, fracking for gas has reduced the price of gas accelerating the switch away from coal. Because gas generation emits half as much as coal, this has resulted in a significant **reduction in emissions that is claimed to exceed all Europe has managed with subsidised wind and solar.** The United States has actively pursued the exploitation of shale gas and has become more self-sufficient. It exports gas to Mexico from shale fields in Texas, Louisiana, Oklahoma and Colorado. The development of shale gas has progressed rapidly despite excessive regulation that has been reported to cost US companies \$30,000/year with one developer claiming to need approval from 23 Federal agencies for one pipeline from Pennsylvania that crossed one state line. As the emissions from gas are half those from coal, it appears that the overall strategy will be to progress a mixture of renewable generation with the intermittency managed by gas-fired generation.

Around the same time, the EU had reached a level of 27% of production from renewable generation with a similar figure for Germany with nearly 50 GW of wind capacity by 2016. This was the same percentage as reached in China if hydro is included.

The Renewable Energy Directive requires the United Kingdom to source at least 15% of its total energy from renewables by 2020. To meet this target, the Government has estimated that renewable sources will need to contribute:

- At least 32% of the UK's electricity, with one-third of this coming from bio-mass, of which waste forms a part. Currently, renewables account for 7.4%.
- At least 12% of UK heat requirements. At present, this is less than 1%.
- At least 10% of UK road transport fuel requirements. Current renewable fuel production is less than 3%.

The wind load factor in the United Kingdom was particularly high in 2015 resulting in onshore load factors around 29%. The total production from wind and solar was 47 TW h or 14%. Including other sources like biomass, the proportion increased to 84 TW h or 25% against the 2020 target for electricity of 32%. By 2017, the United Kingdom had installed 12 GW of solar and recorded outputs of up to 8.7 GW during a warm summer day limiting the quantity of conventional generation to manage system operation.

4.2 Impact of managing intermittency on generation emissions

There is increasing public debate about whether a high level of deployment of wind power reduces emissions as much as anticipated. There is concern that the need to run conventional plant part loaded in reserve to balance the wind generation with ramping and more stops and starts uses more fuel offsetting some of the savings due to wind output. This is driven by questions over the reported impact of high levels of wind generation in Ireland and Denmark in particular. There is also concern that initiatives to build demand side response is resulting in aggregators buying 'dirty' embedded generation output and selling it into the balancing market, worsening the overall emissions position.

To estimate the effects of the intermittency on conventional generation requires a detailed modelling approach tracking half hourly the changes in wind generation and solar output and the associated balancing changes in conventional generation. In order to establish realistic models, it is necessary to use actual recorded wind and demand data for a representative year to capture the random variations. The modelling will need to take account of

- the need for additional reserve to backup wind requiring part loading;
- the requirement to ramp up and down to track wind variation;
- the need for additional starts and stops to follow major wind changes.

A particular feature of the model is the need for curtailment of wind generation at times of low system load when to accommodate the wind output would necessitate reducing output from inflexible generation like nuclear and some CHP installations and leave the system with insufficient reserve and inertia to cater for disturbances.

There is also a need to consider the possible utilisation of embedded generation and Demand Side Response (DSR) for balancing where the benefit of avoiding balancing charges exceeds any costs from exploiting local plant flexibility. This may affect emissions from embedded generation.

4.3   Modelling operation with intermittency

The market operation can be analysed using a market simulation based on a merit order dispatch on a half hour basis for each year. The dispatch uses the variable fuel and operating costs with efficiency estimates based on generic data for similar plant taking account of type, age and size. The analysis captures the generation utilisation and fuel burn in base load mode and when operating as reserve. The marginal generation is identified for each half hour enabling the annual Short Run Marginal Cost (SRMC) to be calculated and ramping costs to be estimated. The generation fixed costs are estimated based on the historic capital costs and these are apportioned based on the utilisation of each generation for the year to establish the Long Run Marginal Cost (LRMC). The expected market price is estimated based on a generic exponential function of the SRMC and LRMC and the prevailing plant margin. The overall structure is shown in Figure 4.1 with the generation data processing on the left side with the demand side on the right.

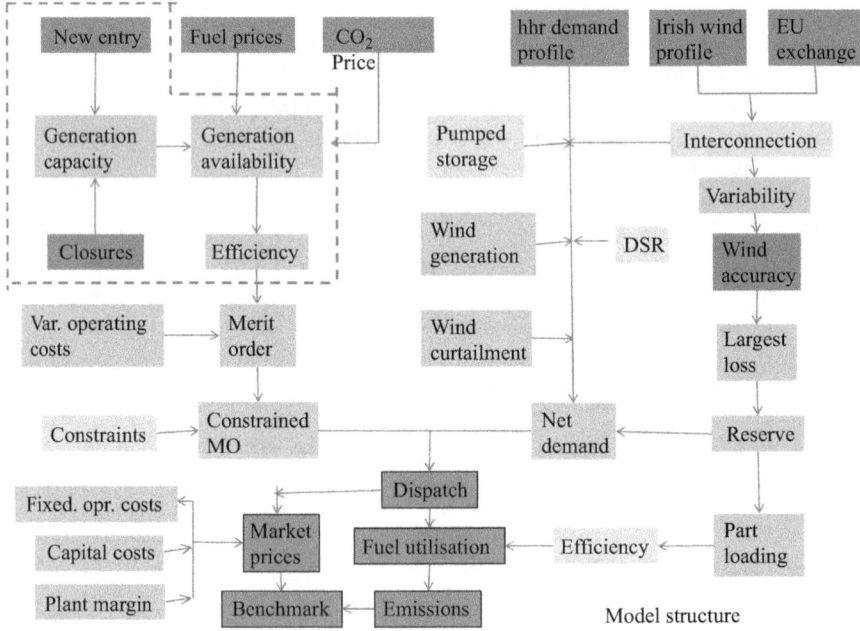

Figure 4.1   Emission model

- The forecast fuel prices are based on a double regression fit to historic short and long trends on a quarterly basis to reflect seasonal variations in particular for gas prices. They are benchmarked against Department of Energy and Climate Change (DECC) annual forecast bands;
- The CO_2 price will influence the dispatch and has been assumed to follow a trajectory to the proposed cap with longer term prices estimated based on the expected least cost of curtailment using carbon capture and storage (CCS), nuclear or renewable sources;
- The generation availability is based on historic outturns and varied quarterly to reflect outage programmes and the varying inter-year plant margin;
- The initial generation data set takes account of NGC published data in the Seven Year Statement (SYS);
- Generation new entry is based on known data in the medium term and beyond on a comparison of expected market prices and new entry costs;
- Generation closures are based on known data in the medium term and life expectancy in the longer term;
- New entry of renewable sources is based on known plans with scenarios;
- Generator variable and fixed operating costs are based on generic data;
- Parameters are included to represent the efficiency reduction associated with part load operation

The demand and wind data is based on actual recorded data for a year. The wind generation output profile is collated for each transmission zone. The future output is then based on extrapolating each zonal value based on known new capacity additions in each zone. The required reserve holding is calculated based on the level of expected output and a variable forecasting error. Curtailment of wind energy output is determined for each half hour by comparing the net demand with the minimum acceptable level and reducing the wind output accordingly to maintain secure system operation.

The dispatch process involves the following:

- Read the half hour recorded demand for the period;
- Net off the wind generation calculated for the period;
- Adjust the wind generation that may have to be curtailed because it exceeds the minimum practical net system demand;
- Set the required reserve level to cater for 4 h ahead wind forecast error set at 51% or 30% based on three std for a 17% or 10% rms variable error level;
- Dispatch available generation to meet net demand plus reserve.

The merit order tables include the columns of the cumulative fuel used at full load and when partially loaded in reserve mode. The latter includes an efficiency degradation factor related to the part loading level that is assumed to be on average at 50%. The fuel consumption numbers are multiplied by the respective fuel CO_2 content in t/GJ to establish the total emissions in kt from fully loaded and reserve generation.

The model was used to quantify the impact of the regulating duty on conventional generation of tracking the variations in the intermittent renewable generation.

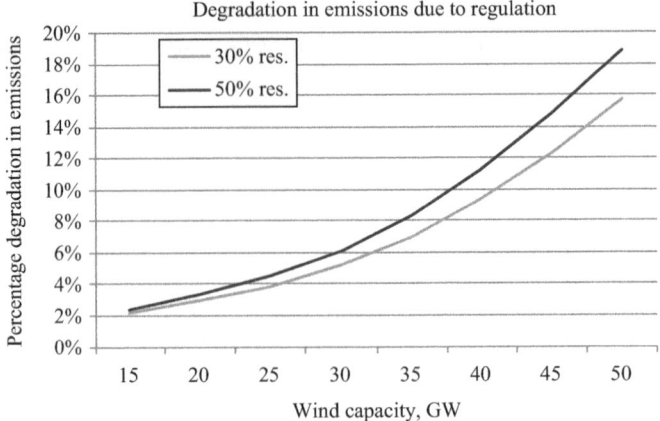

Figure 4.2   Emission degradation – United Kingdom

The results for increasing levels of wind generation are shown in Figure 4.2 with two levels of reserve to cater for forecast inaccuracy. The results will be influenced by the type of plant displaced with emissions from displaced coal being higher than those from gas. For the system modelled, the peak demand was around 60 GW with 5% degradation with 30 GW of wind i.e. 50% of peak. This means that for each 10 GW of extra wind capacity there is around 1% reduction in the expected emission savings due to the regulation required on conventional backup generation. Above 50% wind capacity, the percentage degradation over the theoretical maximum increases more rapidly, in part due to more units starting and stopping and the need to curtail wind generation at times of low load to maintain system security.

4.4   Impact on generation loading

Figure 4.3 shows a generation loading pattern for 2 days having netted off the wind output both with and without the reserve generation. It shows how the number of units changes rapidly to track the wind generation output with the reserve holding varying with the level of wind output. The profiles are unlike the smooth conventional load profiles without wind generation as shown at the end of the graph with no wind output. **It is this continuous flexing of generation that increases fuel utilisation and emissions.**

The requirement for ramping to track demand is based on the changes in wind generation (as opposed to net demand) to quantify the impact of wind. It is based on the change in wind output between successive half hours in MW. A practical ramp capability of 20% is assumed for gas and coal plant. For a 390-MW gas plant, this equates to 78 MW and for a 500 MW coal unit 100 MW. The 20% ramp is assumed to use extra fuel of 584 GJ for coal plant and 219 GJ for CCGT generation. It is assumed that the ramping is predominately undertaken by older gas and coal plant type identified as being at the margin (more modern plant is expected to have lower

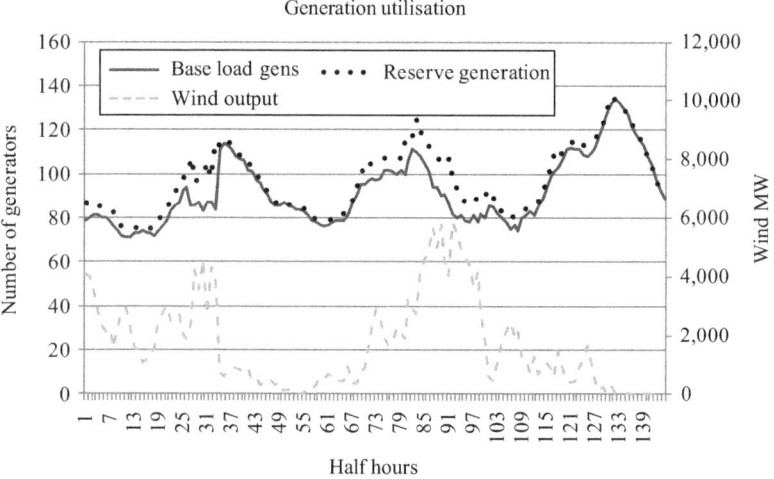

Figure 4.3   Loading profile

values). The additional fuel used and associated CO_2 emission is calculated to meet the wind output change taking account of the fuel type. This figure will vary at different levels of wind output according to the percentage of time different plant is at the margin. With gas at the margin for a higher percentage of time, emissions will be lower. The model also calculates the additional requirement for generation starts and stops due to wind intermittency as illustrated by the changing number of units in Figure 4.3. The additional fuel used in start-up is set as an input parameter for coal at 2,825 GJ and gas at 1,087 GJ.

4.5   Curtailment of renewable output

The need to curtail wind output occurs at times of low load when the wind output exceeds the demand less the conventional generation needed manage stable system operation plus other inflexible generation. In this example, shown in Table 4.3, the minimum system basic reserve generation is assumed to be 2,500 MW and existing inflexible generation like nuclear and CHP is set at 9,000 MW. With a minimum demand of 22,000 MW plus a pumping demand of 1,000 MW, the minimum demand is 23,000 MW. The potential for exports will depend on conditions and wind output of near neighbouring utilities; a figure of 2,500 MW is assumed. The result is that any wind output above 14,000 MW at the time of minimum demand would have to be curtailed. Alternatively, at any system demand below 28 GW, there is the possibility that wind generation would have to be curtailed. The analysis equally applies to solar generation and the combined output of renewable generation during light load periods. The impact is compounded by the need to hold reserve to cater for the sudden loss of renewable output. Assuming a 30% reserve holding for wind implies that only 70% of the demand above the

Table 4.3   Wind curtailment

Wind curtailment at minimum load		MW	MW
Minimum	Reserve generation	2,500	
	Inflexible generation	9,000	11,500
Demand	Minimum	22,000	
	Pumping	1,000	23,000
Exports	Market trading	2,000	
	Inter SO arrangement	500	2,500
	Maximum wind output at minimum demand		14,000
Wind capacity		20,000	
Maximum curtailment MW			6,000

minimum can be met by the wind. The potential need for curtailment increases as the capacity of the renewable generation increases in relation to the system size as illustrated in Chapter 2. **The implication is that there is a practical limit to the proportion of renewable generation before essential curtailment reduces the benefits.** There is also a cost associated with curtailing wind generation in compensating for loss of energy and Renewable Obligation Certificates (ROC) payments or FIT revenue.

4.6   Evaluating wind installations

The wind speed profile can be represented by a Weibull function with parameters k and a where k determines the shape of the curve and is typically 2 and a relates to the wind speed S and is adjusted so that the average coincides with the average recorded for a site. A typical function is shown in Figure 4.4 based on the formulation:

$$\text{Wind speed duration} = 8,760 * ((k/a) * ((S/a)\hat{}(k-1)) * \text{EXP}(-((S/a)\hat{}k)))$$

The wind speed referenced is at a hub height of 100 m in this example.

The wind speed is translated into power output using a turbine characteristic that varies with supplier. In this case, a 5-MW turbine is assumed setting the maximum output as shown in Figure 4.5. The gross energy output is derived as the product of the power at each wind speed and the number of hours it persists through a year.

The net output of a unit located in a wind farm takes account of losses including wake losses of 15%; losses due to outages and maintenance of 11.6%, blade degradation losses of 3% and electrical losses of 0.82%. **The combined effect is to reduce energy output to a net value of 72.3%.** The unavailability takes account of expected failure rates and repair times plus an allowance for planned maintenance. The electrical losses take account of array cable losses and transformer losses.

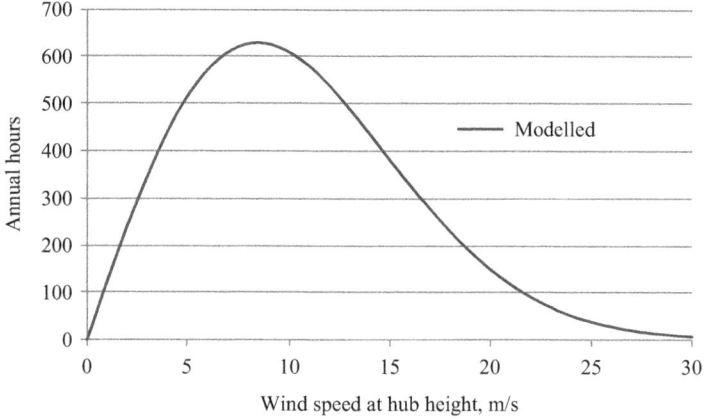

Figure 4.4   Wind speed duration

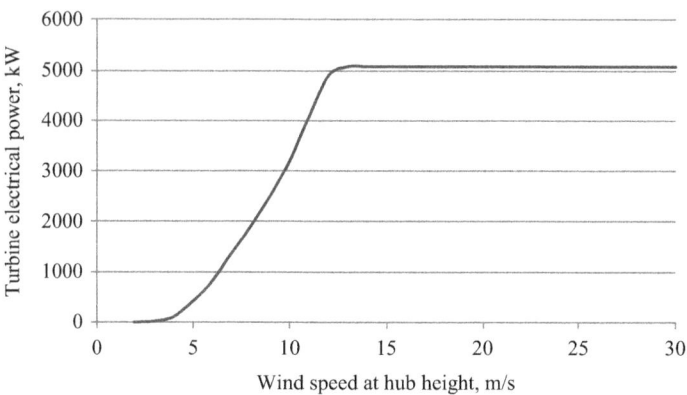

Figure 4.5   Turbine characteristic

There is a degree of uncertainty in the outturn figures due to; mast heights being different to hub height; a range of variation in historic records and future variations that may occur through 20 years. These combine on an rms basis to give a potential error of 0.63 m/s in this example with a hub mean wind speed of 10.6 m/s. The variation around the base case can be represented by a normal distribution and the equivalent energy can be calculated and hence the load factor. Using the load factor figures calculated net of loss factors, the graph in Figure 4.6 can be constructed showing the probability of different outturn load factors. **To manage an investment risk, it is normal to identify that load factor that will probably be exceeded in 90% of cases** that in this case is 37.3 when the cumulative probability equals 0.1. This can be used to estimate the revenues for comparison with costs to assess the value at risk.

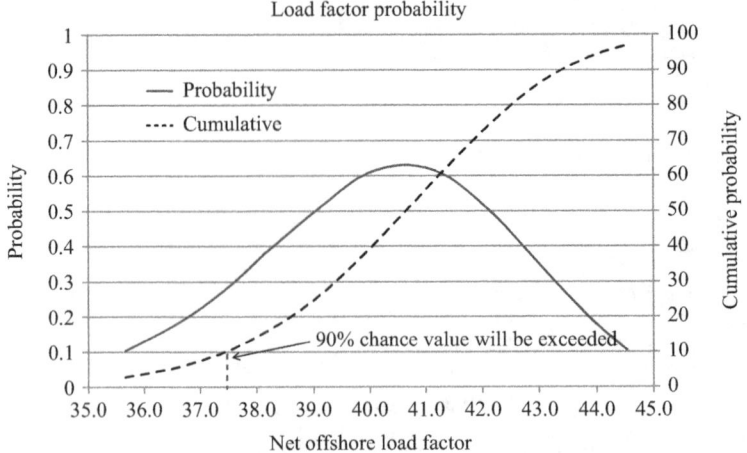

Figure 4.6   Load factor probabilities

Table 4.4 shows a typical cost of onshore and offshore wind installations with total costs similar to the feed in tariffs used in the United Kingdom. In this case, the load factors assumed are consistent with general long-term averages taking account of maintenance and forced outages. If the load factor for onshore wind is 30%, then the unit cost drops to £74.8/MW h and for offshore £116.7/MW h at a load factor of 38%. **This illustrates the importance of the analysis of the load factor in pricing new wind farms in a competitive capacity auction.** There are clear advantages in locating wind farms further from shore where there are stronger winds and they are less visually intrusive but the sea is deeper. The London array wind farm of 630 MW is located in waters around 25 m deep, the Gemini wind farm off the coast of the Netherlands is 600 MW in water 20 m deeper and less visible from the shore but approaching the limit of what's practical. RWE abandoned plans to build a 1.2 GW array in the Bristol Channel in waters 45 m deep with difficult seabed conditions. Gravity-based foundations using concrete slabs have been successfully used but depend on soil conditions as well as the weight to maintain stability. Piling requires a stable rig to drive the piles into the sea bed using a ship that is located in a fixed position using jacks. Suction caissons are an alternative when pushed into the sea bed and the skirt is pumped out to enable suction into the ground.

The latest technology developments include floating wind farms tethered to the sea bed by three mooring lines. The turbine structure is stabilised using three 78-m vertical spar buoys that are filled with 8,000 litres of sea water. Built in Norway, the Hywind floating wind farm supplied in 2017 consists of five turbines of 6 MW each. They are located 15 mi off the coast of Scotland in waters 100 m deep. Although fixed units are only practical up to depths of around 45 m, the floating units could in theory be installed at depths up to 700 m. This opens up wide areas of the oceans that could be windier and more remote mitigating the visual impact of the turbines. It is expected that the higher levels of output will offset the

Table 4.4 Wind installation costs

		Units	Wind on	Wind off
Plant	Capacity	MW	3	3
	Capacity net	MW	3	3
	Load factor	%	25	30
	Annual output	TW h	0.0066	0.0079
	Construction time	Years	1	1
	Project life	Years	15	15
Finance	Cost of dept	%	5.3	5.4
	Cost of equity	%	8.8	8.8
	Inflation	%	2.4	2.4
	Debt/equity split		0.8	0.8
	WACC ($D/E = 0.8$)	%	7.2	7.3
	Total investment cost	£m	4.9	9.6
	Capital cost	£/kW	1,633	3,200
	Capital cost pa	£/m	0.55	1.07
O&M	Consumables	Pence/kW h		
	Labour rate	£/h		
	No. of operators			
	Operators labour*1.28	£M	0.035	0.07
	Maintenance material	£M		
	Main and support labour	£M		
Summary	Capital cost	£/MW h	83.05	135.98
	Fixed O&M	£/MW h	5.33	8.88
	Var O&M	£/MW h	1.2	2.4
	Fuel costs	£/MW h	0	0
	Total generation costs	**£/MW h**	**89.58**	**147.26**

Offshore wind costs

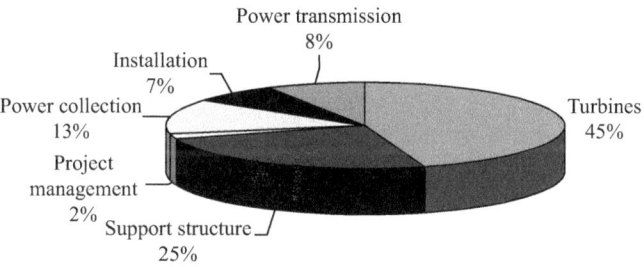

Power transmission
8%

Installation
7%

Power collection
13%

Project
management
2%

Support structure
25%

Turbines
45%

Figure 4.7 Offshore wind costs

additional costs associated with tethering and stabilising the installation. **Each percentage improvement in load factor reduces unit costs by around £3/MW h.**

The makeup of the costs of an offshore wind farm is shown in Figure 4.7 where the support structure is shown to make up 25% of the cost or £2.5 m in this example. **If in deeper water the foundation costs double to £5 m, then the load factor has to increase by 7% to 37% to maintain the same unit cost.**

4.7 Biomass

Biomass generation is classed as renewable generation and attracts premium payments in the form of subsidies either via a contract for differences (CfD) or from the sale of ROC to suppliers. **A biomass station burns woodchips to generate and has limited net emissions in that the trees supplying the wood absorb carbon dioxide through their growth cycle.** The generation output is controllable and supports system security and balancing intermittent renewable generation. In the United Kingdom, a large coal-fired station called Drax has converted several 660 MW units to operate burning woodchips at costs around £700 m/unit.

The wood pellet energy content is around 18 GJ/t or 5 MW h/t ($18*1,000/3,600$). With prices of typically £144/t, the price/GJ $= 144/18 = £8$/GJ. Taking this cost and assuming an efficiency of 36%, the production cost per MW h is given by

$$8 * (3,600/1,000)/0.36 = £80/\text{MW h}$$

Based on a ROC price of £50/MW h plus a Levy Exemption Certificate at £4/MW h, the station would break even on fuel costs at a market power price of £26/MW h ($80-50-4$). To cover other costs and provide a margin an energy price of £40/MW h would provide a profit of £14/MW h. To finance on this basis has risks in that the ROC price could fall if more renewable generation becomes available or energy prices fall. On the positive side, ROC prices could increase if there is a shortfall in renewable generation against targets. The alternative financing arrangement is through a Feed in Tariff with a Contract for Differences (CfD) where the station is paid a fixed price made up to £105/MW h (for the ROC and energy) with added payments above the reference market energy price funded by suppliers. Equally, if energy prices rose substantially, the station would return monies. To qualify for the FIT, the station needs to operate 100% on biomass. Operating with lower proportions of biomass, the payments are reduced to around £85/MW h with production based on 85% biomass.

Given a gross calorific value of 5 MW h/t, taking account of a generation efficiency of 36%, the electricity output would be $5 * 0.36$ MW h/t $= 1.8$ MW h/t of wood chips. Assuming an annual production of 5 TW h, the requirement for pellets $= 5/1.8 = 2.77$ Mt/year.

Estimates of the calorific value of the wood chips vary with some sources quoting 18 MJ/kg, assuming a carbon content of 45%, the emissions can be calculated:

$$\text{kg carbon/GJ} = (1,000 * 0.45)/18 = 25 \text{ kg carbon/GJ} = 25,000 \text{ gC/GJ}$$
$$\text{tCO}_2/\text{GJ} = (25 * 44/12)/1,000 = 0.0916 \text{ tCO}_2/\text{GJ}$$

This is similar to lignite/brown coal at 0.0924 tCO_2/GJ. The emissions/MW h generated, taking account of an efficiency of 36%, is given by

$$0.0916 * 3.6/0.36 = 0.916 \, tCO_2/MW \, h$$

This compares to lignite at 0.99 tCO_2/MW h. A tree is estimated to absorb 22 kg of CO_2/year (NC State University) or 0.44 t over a 20 year life cycle. If a typical tree can be converted to 0.4 t of wood pellets that is used to produce electricity at the rate of 1.8 MW h/t then the CO_2 emission from generation is given by

$$0.4 * 1.8 * 0.916 \, tCO_2/\text{tree} = 0.659 \, tCO_2/\text{tree} \text{ for a production of } 0.72 \, MW \, h.$$

The net cycle emission from using the tree to produce electricity is given by

$$(0.659 - 0.44)/0.72 = 0.30 \, tCO_2/MW \, h$$

The life cycle emissions compare favourably with lignite at 1.01 tCO_2/MW h; coal at 0.82 tCO_2/MW h and gas generation at 0.34 tCO_2/MW h. The absorption by trees varies with the type of tree and the period for which it grows before harvesting but it is reasonable to assume that the emissions during burning relate to the carbon absorbed by the tree plus emissions from fuel used in processing the wood and transporting it to station sites.

4.8   Waste to energy

Rising fuel prices have made it more attractive to convert waste to energy. The energy content is assumed in this example to be 10 GJ/t with conversion efficiency of 27% giving 0.75 MW h/t (27/100 * 10/3.6). The operating costs are principally to cover staff and maintenance with additional costs for firing fuel and handling the raw materials and residual waste. Including capital, the total costs are £41.2/t (£55/MW h) against revenues of £26.3/t for electricity and £31.5/t for the ROC and £6/t from recycled materials i.e. £63.8/t or £85/MW h. The energy content of the waste varies from 6 GJ/t for municipal waste 13 GJ/t for refuse up to 15 GJ/t for general industrial (source UK Dukes). The costs and revenues are itemised in Table 4.5.

Waste to energy provides a vital social service and can be made economic with diligent waste material sorting by residents.

4.9   Solar

There is increasing interest in solar energy in countries with a high number of sunshine hours like Southern Europe and the Middle East. Local authorities are fitting solar panels to their property portfolio in the United Kingdom. Spain is expected to double its capacity. Portugal completed one of the world's largest

Table 4.5 Waste to energy costs

Based on tonnes per annum plant	
Capital cost £ million at April 2006	51
Depreciation period (years)	15
Size k, t/year	240
GJ/t	10
Efficiency	27
MW h/t	0.75
Price/MW h	35
Revenue/t	26.25
ROC/MW h	42.0
ROC/t	31.5

Operating costs (£/t)		Revenues (£/t)	
Capital: £	20.28	Materials	6
Staff: £	6.92	Nutrients	0
Raw materials: £	0.65	Electricity	26.3
Maintenance: £	8.49	Heat	0
Utilities: £	0.54	ROC	31.5
Fuel: £	0.38		
Waste disposal: £	3.93		
Total cost/t input	**41.19**	**Total/t input**	**63.8**
Energy cost £/MW h	**54.9**	Net rev £/t	**22.6**

plants at Amareleja. It includes 2,520 solar panels generating 45 MW at a cost of £250 m (£5,555/kW). Portugal has a target of 31% of all energy from renewable sources by 2020. The industry still relies on subsidies or feed in tariffs to realise 'grid parity' i.e. when it competes with conventional sources. However, it has been predicted that rising fuel and emission costs will enable solar to compete on equal terms within 4 to 6 years depending on technology advances.

Polysilicon is the material used to build most solar cells with the Fluor Corporation expected to be the lead producer. Ascent Solar Technologies are developing the application of copper–indium–gallium–selenide that is claimed to be more efficient than polysilicon. They plan to put their cells into a flexible plastic sheet that boosts connectivity and efficiency by 10% at reduced costs. Other companies like day 4 Energy of Vancouver are focussing on the development of the solar modules. These developments will reduce the costs of solar energy installations and coupled with rising fuel prices will accelerate the time when 'grid parity' is reached.

Figure 4.8 has been constructed based on radiation figures recorded in the UK Midlands for a calendar year. It shows the calculated output profile of 1 GW of capacity for the year. It can be seen to be very seasonal with an overall load factor of about 13% with little output during the winter periods and none when winter peaks occur during the early evening. **It therefore does not make a contribution**

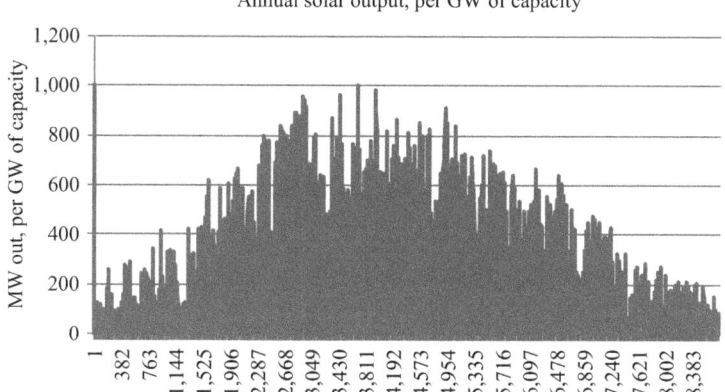

Figure 4.8   Solar output (United Kingdom)

to secure capacity. In summer months, its output can create operational problems in depressing the demand left available to retain in service the flexible plant necessary to sustain stable system operation and inertia and accommodate inflexible generation. There can be expected to be some diversity in output but with around 12 GW installed by 2017 in the United Kingdom, its output could be difficult to accommodate at times of minimum load. There is also unlikely to be much flexible demand side regulation available in northern countries in summer. In hotter climates, desalination and air conditioning loads can result in higher summer peaks with opportunity to exploit the solar output.

Solar is also used extensively to provide hot water as well as electricity and mitigates the problem of managing intermittency by exploiting the hot water storage capacity. In Australia, 11% of households have solar hot water systems. Solar PV also produces about 5 TW h of electricity and wind and hydro 30 TW h contributing towards a total demand of 250 TW h. Some 60% of the energy is generated at coal-fired stations with 105 TW h from black coal and 46 TW h from brown coal with 54 TW h from gas-fired generation.

China has extreme pollution problems estimated to lead to over a million premature deaths a year and is cancelling the construction of new thermal generating plants. **In 2017, the world's largest floating solar power plant was made operational in Huainan city, Eastern China.** The 40-MW plant is situated on a reservoir and is in close proximity to the city. Offshore from Huainan, the plant has been successfully connected with the power grid. Being offshore, it does not take up space and uses less energy than solar farms as seawater acts as a coolant. Sungrow, the world's leading photovoltaic system supplier, has set up the plant. Floating solar power plants can be set up on water bodies such as lakes and seas, particularly near cities where land availability is limited.

Table 4.6   Gas/coal emissions

	Gas	Coal
Carbon, tCO_2/GJ	0.0495	0.0869
Efficiency, %	52	38
Emissions, t/MW h	0.342692	0.823263

4.10   Displacing coal with gas

For countries heavily dependent on coal-fired generation, conversion to gas generation provides a way of effecting some immediate reductions. The emissions from CCGT generation are less because of the lower carbon content of the fuel and the higher efficiency realised by gas-fired generation. Table 4.6 shows that the emissions from gas-fired generation can be as low as 40% of those from coal-fired generation. Significant progress on reducing emissions has been made in the USA as a result of coal generation being displaced by generation burning shale gas. A constraint on reducing coal burn is often the impact on the coal industry and its locally based work force. There may be limited alternative employment locally and unions will naturally resist closures. An advantage of using gas-fired generation is that as well as reducing emissions it provides a flexible source that can be used to track variations in renewable output. It is a readily available option that can be located relatively close to demand centres and subsequently it may be retro fitted with CCS.

Converting all plant using coal to burn gas would reduce world emissions from electricity production by 45% from 10,362 $MtCO_2$ down to 5,698 Mt/year and several times more than the impact of all the renewable output providing a reduction of just 12%. These reductions could be realised by gradually phasing out the use of coal for electricity generation and without subsidies if plants are replaced as they reach their end of life.

4.11   Economic comparisons

The various generation options available to reduce emissions can be compared financially as shown in Table 4.7 based on predicted costs in 2030 rather than subsidised tariffs. The other key factor is the expected load factor. It was shown in Section 4.6 that the combined effect of network losses, maintenance outages, wake losses and blade degradation can reduce the annual delivered energy output of a wind farm to less than 75% of the theoretical level. The load factors used in the table are based on recorded statistics of annual output. In the case of solar, the load factor is also based on recorded solar radiation figures. In the case of CCS, the coal and gas generation efficiency is written down by 18% and 20% to take account of the impact of the process. A charge is added for the transportation and storage of the CO_2/t of £20/t to establish the total costs. Both wind and solar output will vary by country with much higher levels of solar in southern Europe and the Middle East.

Table 4.7 Low carbon generation costs

		Units	Gas/CCS	Coal/CCS	Peaking	Hydro	Nuclear	Waste	Wind on	Wind off	Gas
Plant	Capacity	MW	393	398	120	31	400	25	3	3	393
	Capacity net	MW	377	366	115	30.07	360	23.5	3	3	377
	Load factor	%	85	85	35	33	80	88	25	30	85
	Annual output	TW h	2.81	2.73	0.35	0.09	2.52	0.18	0.0066	0.0079	2.81
	Construction	Years	2	4	2	5	8	2	1	1	2
	Project life	Years	20	20	20	65	15	15	15	15	20
Fuel	Efficiency	%	45.1	35.2	26		38				55
	Coal delivered	£/t		72.8							
	Coal CV	GJ/t		25							
	Gas cost	£/mm BTU	7.8		7.8						7.8
	Gas cost	Pence/therm	78		78						78
	Gas CV	kW h/therm	29.32		29.32						29.32
	GrossCV/netCV		1.11		1.11						1.11
Finance	Cost of dept	%	5	5	5	4.7	5.1	4.7	5.3	5.4	4.7
	Cost of equity	%	8.8	8.8	8.8	8.5	9	8.8	8.8	8	8
	Inflation	%	2.4	2.4	2.4	2.4	2.4	2.4	2.4	2.4	2.4
	Debt/equity split		0.8	0.8	0.8	0.8	0.8	0.8	0.8	0.8	0.8
	WACC ($D/E = 0.8$)	%	7.1	7.1	7.1	6.8	7.2	7.0	7.2	7.3	6.5
Capital	Site and dev	£m	24.3	35.5	10	47.12	860				
	EPC contract	£m	621	919	40		807				235
	Electrical conn	£m	0.733	0.733	0.533		3.5				0.733
	Gas connection	£m	2.466								2.466
	Spares	£m	14.33	31.86	2.86		18.5				14.33
	Interest during const.	£m	11.26	23.20	0.20	11.07	41.00				11.26
	Investment cost	£m	674.1	1,010.3	53.6	58.2	1,730.0	51	4.9	9.6	273.5
	Capital cost	£/kW	1,715	2,538	447	1,877	4,325	2,040	1,633	3,200	696
	Capital cost pa	£/m	64.08	96.03	5.09	4.01	192.89	5.584	0.55	1.07	24.85

(Continues)

Table 4.7 (Continued)

		Units	Gas/CCS	Coal/CCS	Peaking	Hydro	Nuclear	Waste	Wind on	Wind off	Gas
O&M	Consumables	Pence/kW h	0.02	0.113	0.01		0.164				0.02
	Labour rate	£/h	16.66	16.66	16.66						16.66
	No. of operators		6	14	0.5						6
	Operators labour*1.28	£M	1.12	2.62	0.09		2.8		0.035	0.07	1.12
	Maintenance material	£M	0.81	2	0.1						0.81
	Main & support labour	£M	1.25	2.33	0.15						1.25
Summary	Capital cost	£/MW h	22.81	35.22	14.42	46.10	76.46	30.82	83.05	135.98	8.85
	Fixed O&M	£/MW h	0.60	2.09	0.36	3.2	2.75	9.22	5.33	8.88	0.60
	Var O&M	£/MW h	0.73	1.59	0.71	2.06	1.9	18.1	1.2	2.4	0.73
	Fuel costs	£/MW h	65.48	29.78	113.57	0	5	0.5	0	0	53.69
	Total gen. costs	**£/MW h**	**89.62**	**68.68**	**129.07**	**51.36**	**86.11**	**58.64**	**89.58**	**147.26**	**63.87**
	CO$_2$ price £/t	**50**									
	Include carbon	tCO$_2$/GJ	0.0495	0.0869	0.0495	0	0	0	0	0	0.0495
	CCS CO$_2$ trans/store	£/tCO$_2$	20	20		0	0	0	0	0	
	Added cost/incentive	£/MW h	7.90	17.78	34.27	0	0	0	0	0	16.20
	Total cost	**£/MW h**	**97.52**	**86.46**	**163.34**	**51.36**	**86.11**	**58.64**	**89.58**	**147.26**	**80.07**

A key advantage of the CCS generation is in its flexibility to track the variations in the output from intermittent sources. The inclusion of a tranche of CCS also helps to maintain system inertia at times of low load. With some types of design, it is possible to interrupt part of the process to increase the output to meet peak demands. Based on the data assumptions shown in the table, including high fuel prices, it can be seen **that CCS generation is comparable in cost to onshore wind and nuclear and is also significantly cheaper than offshore wind.** The offshore wind would need a load factor of 50% to be comparable in cost to other options.

4.12   Optimal mix of low carbon generation

In considering the optimal mix of low carbon flexible generation, it is important to take account of the potential operator role as well as meeting emission targets. The nature of the load duration curve means that all generation cannot operate fully loaded. Nuclear has a high capital cost but low fuel costs and has minimum unit costs at high load factors. In contrast, peaking plant has a low capital cost but high operating costs with unit costs that do not vary much with load factor. They can be used a few hours to lop peaks. For longer peaking periods, there may be scope to use conventional CCGT stations whilst keeping within emission targets. The emissions from both peaking and conventional gas generation would have to be kept within targets. The coal and gas CCS generation would operate to meet the main variable part of the load duration curve when there is no wind output. The variation in unit costs with utilisation is shown in Figure 4.9. It can be seen that high capital cost nuclear and coal with CCS are cheapest when operating at full load. In contrast, peaking generation can operate at costs lower than any other generation at low load factors below about 35%. Gas with CCS has a lower capital cost and can operate at low unit costs down to below 40% as may be required in

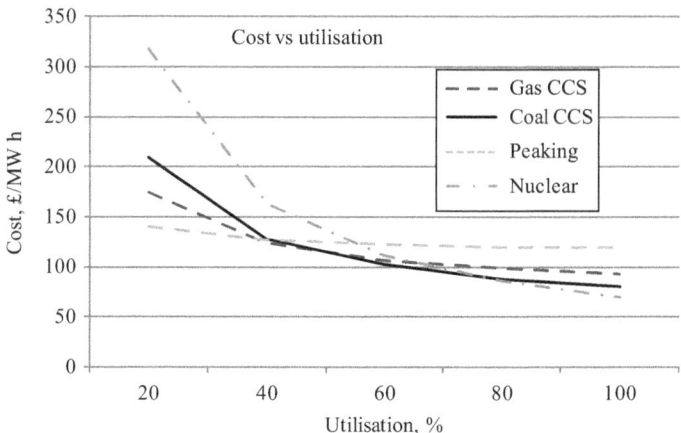

Figure 4.9   Cost of low carbon generation vs utilisation

tracking variations in wind and solar generation. It can be seen from the table that conventional gas-fired generation, without CCS, can operate cheaply at £80/MW h with emission costs of £50 t if emissions can be kept below the target. In practice, some CCGT generation is being converted to operate open cycle to improve its flexibility to track intermittent sources.

4.13   Conclusions

Progress on the reduction of emissions is proceeding slowly across the world by the largest emitters. Emissions in China were 26 times those of the United Kingdom in 2015 but initiatives are being made to expand their nuclear fleet. In the USA, the success of fracking for gas has resulted in increases in gas-fired generation at the expense of coal and it is claimed that their reduction in emissions exceeds that of Europe with all its subsidised wind and solar generation. This is the result of emissions from gas CCGT generation being around 40% of those from coal-fired generation.

There are also questions about the impact on emissions from conventional generation required to complement the variable output from intermittent wind and solar generation. These result from operating in reserve mode part loaded at reduced efficiency; frequently ramping up and down and more unit starts and stops. The impact of these effects has been modelled based on half hour recorded demand and renewable generation for a year. The impact is modest with wind capacity up to around 50% of the peak demand but at higher levels increases significantly with more conventional unit starts and stops required. The other contributing factor is the need to more frequently curtail wind generation at times of low load to maintain operational security. This occurs when the wind generation output exceeds the demand less the conventional generation needed to manage stable operation plus other inflexible generation like CHP and some nuclear. The implication is that there is a practical limit to the proportion of installed wind generation before technical constraints start to undermine the economic case.

The assessment of new wind farm installations needs to fully account for the effect on net output of maintenance outages, system transmission losses, wake losses and blade degradation that can reduce the output to 72% of the theoretical maximum based on wind speed data. It is also usual to take a conservative view of the expected load factor by basing the assessment on that value that will probably be exceeded in 90% of cases. There are clear advantages in locating wind farms offshore where wind speeds are higher. The London array of 630 MW is located in waters 25 m deep. The Gemini wind farm off the coast of the Netherlands is located in water 45 m deep. Limits are reached in deeper waters and RWE abandoned plans for a 1.2-GW farm in the Bristol Channel due to difficult sea bed conditions. However, a floating wind farm (5 * 6 MW) that operates in deeper waters has been located off the coast of Scotland. If the foundation costs double then the load factor has to increase from 30% to 37% to maintain the same unit cost. Wind makes a limited contribution to the provision of firm capacity, estimated on a probability basis to be around 10% of installed capacity.

There are other generation options that reduce emissions whilst providing a degree of flexibility to manage system operation. One near zero carbon emission option is a biomass-fired station burning wood chips. Several 660 MW units have been converted to biomass at the UK Drax ex-coal-fired station. The generation emissions are complemented by the absorption of CO_2 during their growth cycle. Waste to energy schemes are an option that also helps with waste disposal and can provide some flexibility. Solar generation prices are falling but a problem in the Northern Hemisphere is that when the sun is out, demand is generally very low and may not be able to cope with a high injection of solar energy. **Another option to reduce emissions is to convert all existing coal plant to burn gas with just 40% of the emissions from an equivalent coal-fired station. This alone would reduce world emissions from all sources by 25% and more than that from all the worlds wind farms.**

A financial comparison of low carbon generation options shows that thermal generation with CCS can compete with nuclear and wind on price while also offering flexibility. The analysis shows the impact of different levels of utilisation on relative costs and illustrates the importance of establishing a mixture of technologies that match the expected load duration curve at least cost.

Question 4.1 Using the data in the table calculate the unit cost if the onshore wind load factor is 20% and the offshore 25% and also with an onshore load factor of 30% and offshore of 38%.

		Units	Wind on	Wind off
Plant	Capacity	MW	3	3
	Capacity net	MW	3	3
	Load factor	%	25	30
	Annual output	TW h	0.0066	0.0079
	Construction time	Years	1	1
	Project life	Years	15	15
Finance	Cost of dept	%	5.3	5.4
	Cost of equity	%	8.8	8.8
	Inflation	%	2.4	2.4
	Debt/equity split		0.8	0.8
	WACC ($D/E = 0.8$)	%	7.2	7.3
	Total investment cost	£m	4.9	9.6
	Capital cost	£/kW	1,633	3,200
	Capital cost pa	£/m	0.55	1.07
O&M	Consumables	Pence/kW h		
	Labour rate	£/h		
	No. of operators			
	Operators labour*1.28	£M	0.035	0.07
	Maintenance material	£M		
	Main and support labour	£M		
Summary	Capital cost	£/MW h	83.05	135.98
	Fixed O&M	£/MW h	5.33	8.88
	Var O&M	£/MW h	1.2	2.4
	Fuel costs	£/MW h	0	0
	Total generation costs	**£/MW h**	**89.58**	**147.26**

Chapter 5

Embedded generation issues

5.1   Cause of growth

Embedded generation (EG) is the term commonly used to describe generation connected to the distribution network as opposed to the transmission grid. They tend to be smaller in scale and may be linked into the provision of supplies locally. In the past, systems were developed locally often in conjunction with munici-palities but advantage was seen in pooling energy sources through the grid. This opened up the opportunity for larger more efficient units to be built of up to 660 MW coupled to local fuel supplies like coal mines or gas pipelines. There were associated economies of scale in higher efficiencies with lower operating costs and fewer support staff. As a result of these developments, many of the distribution networks became passive, without active generation, and radial in structure taking supplies from grid supply points.

There are a number of changes taking place that is causing a shift from this model with most generation supply from large remote sites to one with a higher proportion of smaller scale generation embedded within the distribution systems. As a result, the distribution systems are becoming more active. In a previous book (published by Wiley in 2009), I predicted that embedded schemes would meet 40% of power energy needs by 2030 (page 255). The reasons for this change include:

* Rising energy prices and the roll out of smart meters have made consumers more aware of energy utilisation and options to contain costs;
* The subsidies for renewable generation has promoted the development of local wind generation and solar panels;
* Subsidies for CHP schemes have made small-scale local schemes financially viable;
* The problem of waste disposal has made waste to energy schemes a necessity;
* Factory based auto-generation schemes combining the supply of heat and electricity offer the opportunity to maximise efficiency and exercise arbitrage between grid and local supplies;
* Emission regulations as applied to larger installations have undermined their competitiveness and forced closures;
* **The network charging regimes offer embedded generators the opportunity to both reduce the host supplier's capacity charge as well as receive capacity payments through a capacity market.**

The development of EG and the priority dispatch afforded to renewable sources undermines the utilisation of conventional large-scale generation utilisation making it financially non-viable leading to premature closure. The subsidies and charging regimes have accelerated this change resulting in capacity shortfalls that equipment suppliers have been ready to take advantage of. The traditional industry has not been quick to anticipate or adjust to the new environment; large generators are suffering financially; uncertainty is curtailing investment; distribution networks are reaching their limits to accept new generation capacity; the transmission networks are having difficulty managing the integration of large remote wind farms; system operation is challenged by the intermittency of the output from renewable sources. Other developments on the horizon that will exacerbate the network problems are the advent of electric vehicles and the electrification of public transport and heat. At a time when the industry needs a coordinated approach to manage the transition, it is becoming more fragmented in the interests of promoting competition and engaging the customer base. It is increasingly difficult for any player to have sufficient knowledge of the future environment to have the confidence to invest. Various attempts to create focus groups and general purpose models do not provide the details necessary to underwrite large-scale investment. This is the case with the development of high cost nuclear stations that have to be underwritten by the government. This chapter discusses the regulatory position and analyses the financial issues of the technology options driving change.

5.2   Network charges

There is an increasing tendency for the development of EG i.e. small-scale units connected to the local distribution system or a factory internal network. This arises for a number of reasons partly related to capacity market payments and network charging arrangements.

- Small generators can compete in the capacity market because of their low capital costs compared to larger more efficient generators. However, where small diesel generators are installed, their emissions are much higher.
- A significant proportion of end user energy cost is to cover the costs of the network. For small domestic consumers, this can amount to 50% of the cost; commercial and small industry 35%; medium size industry 20% and large industry 12%. These charges can be avoided by direct wiring from generation to local demand. Local town councils are able to exploit this option by directly feeding municipal building complexes from their private generation sources. Factory auto-generation schemes similarly feed directly into their industrial processes and may be combined with heat supply.
- The local energy supplier is usually charged for use of the system against a tariff that includes a capacity element that is based on the maximum demand. In the United Kingdom, this charge is based on what is referred to as the TRIAD demand. This is the average maximum demand occurring on three days separated by 10 days during winter months. **EG can be used to reduce these peak demands and share the benefit with the local supplier.**

- The small-scale generators below 100 MW in the United Kingdom do not pay transmission charges.
- System losses at the transmission level are typically 2% and can amount to 6% for transfers though the distribution system. The impact of these losses is minimised by locating the generation close to the demand.

In the United Kingdom, these effects have resulted in around 15% of generation being embedded and displacing large-scale super-grid connected generation. It is claimed that the current charging arrangements are distorting the generation market. These are made up of several elements.

- The transmission use of system charge (TNUoS) includes a location signal, to encourage generation build close to demand areas. The charges may be negative in the range of $-£5$ up to $+£7/kW$ for generation, whereas for demand charges, range from £7 to £20/kW. Embedded generators of less than 100 MW do not pay.
- The location charges are not 'deep' and do not cover all the network costs so there is additionally a TNUoS demand residual charge that was £45.3/kW in 2016 and expected to rise quickly reaching £70/kW by 20/21. As this is based on the net triad peak demand, it may be offset by using EG to reduce the peaks.
- There is a balancing use of system charge that is based on the supplier's net consumption that again may be offset by running EG. The benefit will depend on the load factor of the EG but is typically £2/MW h.

The benefit of reducing the residual charge of £45.3/kW was double the prevailing capacity payment during 2016. The total benefit over the three periods is given by,

$$45.3 * 3 * 1,000 = £45,330/MW$$

This is equivalent to a payment of £30,220/MW h for the three triad half hours. This is 10 times the value attributed to lost load to meet security standards. However, the embedded generator may have to operate for around 20 periods to be sure of catching the peak triads with a benefit equivalent to 2,267/MW h for the 20 periods. Embedded generators may also benefit in relation to the inverse of the demand locational charge. Those generators greater than 100 MW and those connected to the transmission system have to pay TNUoS charges but the residual element is capped at £2.5/MW.

The bias in the charging arrangements in favour of EG is recognised as artificial and distorting. Several suggestions have been proposed to deal with this escalating problem as the number of generators connecting to the super-grid declines and older large generators are closed. This will result in fewer large-scale generators and the residual demand charge will increase. Suggestions to address the imbalance include:

- No new EG should get the benefit from triad demand curtailment;
- No EG receiving capacity payments should be able to also benefit from triad curtailment;
- There is also the need to account for 'behind the meter' generation.

These issues need to be addressed as part of an overall review of active distribution network management and charging that accounts for the benefits of the distribution network in sharing facilities and offering security. There is also the impact on the capacity of the distribution network to accept more generation that it was not designed for. By 2016, the United Kingdom had 27 GW of EG. In some areas, the limit on accepting more generation has already been reached raising the prospect of more network investment and higher charges. Ofgem, the UK regulator, is considering changing the charging arrangements and reducing the residual charge to around £2 from £45/kW. Some developers reported in May 2017 that this could result in up to 2 GW of capacity **not** being built and aggravating the plant capacity margin at a critical time. The sudden changes proposed highlight regulatory risk and will deter future investment.

5.3   Impact on network management

Renewable generation like wind and solar do not contribute to system inertia and makes the power system more susceptible to sudden loss of generation or changes in demand. The frequency changes more quickly and may result in other generation tripping on rate of change of frequency protection aggravating the disturbance. One approach is to run large generators in synchronous compensation mode de-clutched. This issue was discussed in more detail in Chapter 2.

There is also a lack of contribution to post fault voltage support by NSG and there may be a need for more dynamic reactive power support. Synchronous machines are the main contributors to system fault level so that as units are displaced by renewable generation, the fault level falls. In Scotland, with a high proportion of Non Synchronous Generation (NSG) distribution, fault levels are around 10 kVA, some 50% of what was perceived normal at 20 kVA. **This makes the local network more susceptible to voltage fluctuations** as demand and generation output varies. It will also affect the operation of protection systems and their discrimination. There is also the potential for a shift to lower order harmonic distortion impacting on end user equipment.

The operation of EG affects the grid supply point demands with the low reactive demand at times of light load resulting in high voltages. The Q/P ratio has been falling partly due to EG and also more energy efficient lighting. Some grid supply points are reported to be exporting reactive power up to 40% of the time driving up voltages. **This coupled with a reduction in reactive support from a reduced number of synchronous generators has necessitated switching some super-grid lines out of service.**

5.4   Industrial cogeneration complex

Cogeneration schemes are frequently established to meet the needs of industrial complexes for the provision of electricity and heat and have the flexibility to be operated very efficiently. Figure 5.1 shows a typical factory electricity system with

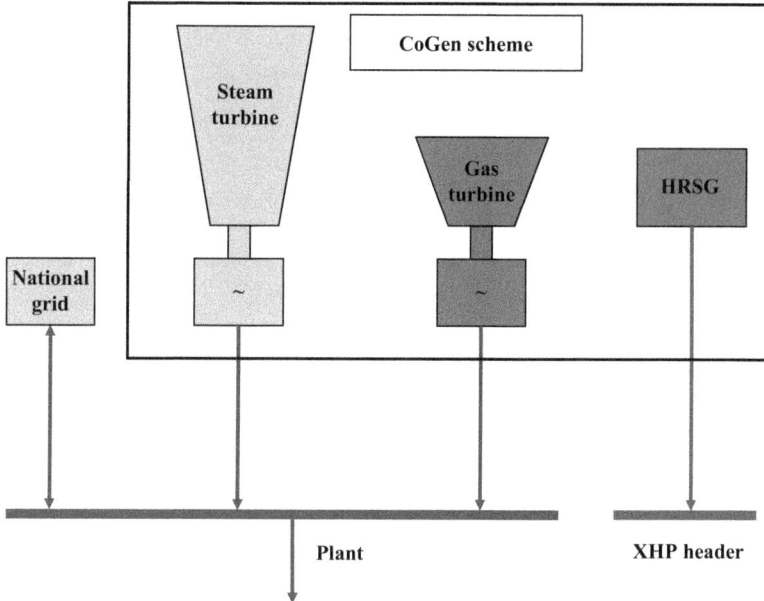

Figure 5.1   Cogeneration scheme

a 70-MW gas turbine generator, a 45-MW steam turbine driven generator coupled with a smaller heat recovery steam generator. The generation system is connected to the national grid enabling the system to import when market prices are low or export surplus energy when prices are high.

The steam system, shown in Figure 5.2 provides 150,000 kW of steam at different pressures to meet the needs of the industrial processes. A group of boilers provides the 230,000 kW of steam to drive the gas turbine as well 80,000 kW to meet the needs of the industrial plant. The exhaust from the gas turbine is used in a heat recovery steam generator with its exhaust feeding into the general HP steam system. This arrangement provides the flexibility to exercise some arbitrage between gas and electricity prices. This arrangement exploits the grid, using it to provide backup supplies and enabling arbitrage with wholesale market prices. The arrangement can be very competitive in that network losses are minimised in transporting energy and capacity charges are limited to the export/import limits. There is also the potential to maximise the use of waste heat to improve overall energy efficiency.

The system may initially be owned and operated by the generation company who will contract for the supply of energy to the plant. Subsequently, there may be an option for the installation to be taken over by the plant operator. The cost of electricity and heat will be linked to the price of gas supplies used to feed the gas turbine and the boilers. As well as the capital cost of the plant, there is also the ongoing cost of operating and maintaining the plant. Although the efficiency of the smaller-scale generating units is lower than that of a CCGT unit, this is offset by

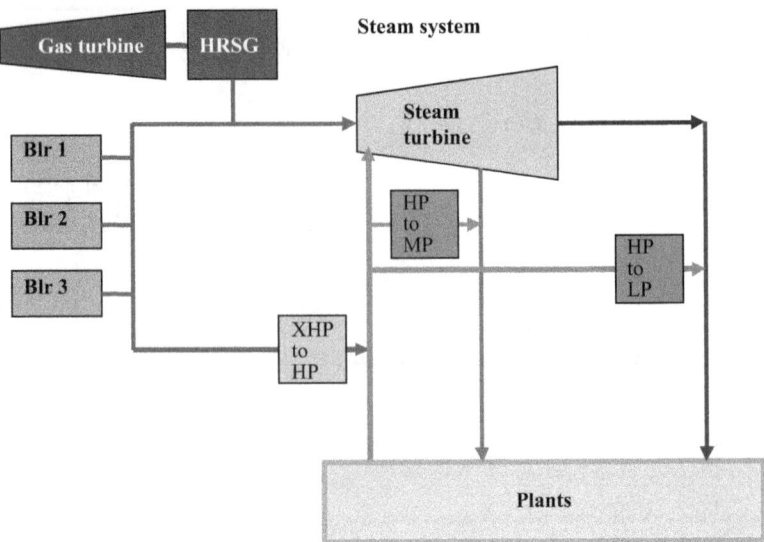

Figure 5.2   Cogeneration steam system

exploiting the waste heat and avoiding grid transmission losses of around 2% and distribution losses of around 5%. A disadvantage is that emissions are local to the industrial complex. The power output net of auxiliary energy is in this example given by

$$70,000 + 45,000 - 1,500 = 113,500\,\text{kW}$$

The electrical efficiency is given by the electrical output over the heat input to the GT

$$113.5/310 = 36.6\%$$

The total heat input is given by

$$230,000 + 80,000 = 310,000\,\text{kW}$$

The thermal efficiency of heat production is given by

$$150,000/310,000 = 48.4\%$$

The overall efficiency of a CHP unit is calculated in Table 5.1 as 85% based on the electricity and heat output over the gas energy input. i.e.

$$(113,500 + 150,000)/310,000 = 85\%$$

The overall efficiency of 85.0% is high in comparison to other options and makes this type of installation sufficiently cost effective to cover the capital and maintenance costs of the installation.

Table 5.1 CHP efficiency

Power of gas turbine 70,000 kW	70,000
Power of back pressure steam turbine 45,000 kW	45,000
Auxiliary power consumption 1,500 kW	1,500
Net power output of plant	113,500
Heat input gas turbine 230,000 kW	230,000
Heat input supplementary firing 80,000 kW	80,000
Process steam output 150,000 kW	150,000
Electrical efficiency	36.6
Thermal efficiency of heat production (only)	48.4
Overall efficiency	85.0

Table 5.2 Micro-CHP

Micro-generation

Cost installed (low)	Euro, 000	115	Running hours	h/year	5,890.0
Cost installed (high)	Euro, 000	140	Electrical production	MW h	589.0
Power production	kW	100	Hot water production	MW h	1,035.4
Electrical efficiency	%	30	Hot air recov. prod.	MW h	117.8
Hot water	kW	167	Avoided fuel in boiler	MW h	1,218.1
Hot air (recovered)	kW	20	MT fuel consumption	MW h	1,963.3
Boiler efficiency	%	85			
			Income with avoided elec.	Euro, 000	51.2
Electricity price	€/MW h	87	Income with hot water	Euro, 000	30.5
Boiler gas price	€/MW h	25	Income with hot air	Euro, 000	2.4
CHP gas price	€/MW h	25	MT gas consumption	Euro, 000	−49.1
Value of hot air	€/MW h	20	MT maintenance	Euro, 000	−8.8
MT maintenance	€/MW h	15	**Total income/year**	Euro, 000	**26.1**
			WACC %		8.0
Operating hours	h/year	6,200	Project life years		15.0
Availability	%	95	Payback range from	5.4	4.4

Electricity price			**Production**	**MW h**	**Costs**	**Euros**
Avoided energy	E/MW h	31.5	Electricity	589	Capital	16,356
CCL tax on energy	E/MW h	3.5	Water	1,035.4	Fuel	49,083
Dx charge	E/MW h	42	Air	117.8	Maintenance	8,835
Tx charge	E/MW h	10	Total	1,742.2	Total	74,274
Total	E/MW h	87	Cost E/MW h			42.6

5.5 Micro-CHP

The data in Table 5.2 relates to a small-scale CHP scheme that is designed to provide hot water and hot air as well as electricity and is fuelled by gas. It has a rated electrical output of 100 kW with 167 kW of hot water and 20 kW of recovered hot air. The unit is assumed available for 6,200 h a year and runs for 5,890 h producing:

- 5,890 * 100/1,000 = 589 MW h of electricity;
- 6,200 * 167/1,000 = 1,035 MW h of hot water;
- 5,890 * 20/1,000 = 117.8 MW h of recovered hot air.

The hot water provided reduces the use of gas in the auxiliary boiler that has an efficiency of 85% equivalent to 1,218 MW h/year (1,035/0.85). The generator has an electrical efficiency of 30% and will use 1,963 MW h/year of fuel (589/0.3). The avoided electricity charge from the local distribution company is priced at €87/MW h including energy, distribution and transmission charges and a climate change levy and equals €51.2 k (589 * 87). The avoided use of fuel in the boiler to produce hot water is €30.5k (1,218 * 25) with gas at €25/MW h. The hot air is valued at €20/MW h worth €2.4k/year (117.8 * 20). The annual fuel cost is €49.1k (1,963 * 25). Based on these figures, the total net income comes to €26.1k/year.

   The costs include capital based on an installed cost of €140,000 and a WACC of 8% over 15 years giving an annual charge of €16,356. The fuel cost comes to €49,083/year (1,963.3 * 25). The maintenance costs at €15/MW h amount to €8.8/year (589 * 15/1,000). The payback period is less than 6 years based on a 2% discount rate. Adding all the energy production, the effective unit cost/MW h is €42.6 based on the total costs (74,274/1,742). This is higher than the avoided energy cost at €31.5, but including the tariff network costs, payback periods of a few years are realised.

   Given the total gas energy used of 1,963 MW h and the total energy production of 1,742 MW h, the effective efficiency is 88% and makes the installation competitive with straight generation options. A counter factor is the higher per unit capital cost of the installation at €1,400/kW. This compares to large-scale CCGT units that benefit from economies of scale and cost around €450/kW with a similar unit production cost. **The key differentiating factor, in the calculation of the payback period, is the inclusion of network costs in the avoided electricity price.** The energy is produced close to the point of use without direct use of the transmission and distribution networks. If the network charges were discounted in the avoided electricity costs, the installation could make a loss. It could be argued that where the network is used to provide full backup during faults or maintenance that some proportion of the network charge should be incurred by the embedded generator. There should also be recognition of the importance of maintaining system frequency and voltages. **As the distribution networks become more active, there will be a developing requirement for system management to maintain security that embedded generators should contribute to.**

5.6   Diesel generation emissions

Diesel generators are the most widely used of small electrical power generating units in factory systems. They are also used in off-grid locations in the developing world due to their low capital costs. A good example is the Sarawak Electricity Supply Corporation installation that produces 68 MW of power from small diesel generating units to supply electricity in rural communities living in remote locations from a national grid. **However, in densely populated areas, diesel engines release many hazardous air contaminants** and greenhouse gases (GHGs) including particulate matter (diesel soot and aerosols), carbon monoxide, carbon dioxide and oxides of nitrogen. Particulate matters are largely elemental and organic carbon soot, coated by gaseous organic substances such as formaldehyde and polycyclic aromatic hydro-carbons which are highly toxic. In 2001, the mortality due to diesel soot exposure was at least 14,400 people out of 82 million people living in Germany. The total amount of GHGs emitted by any system to support human activities directly and indirectly is termed as carbon footprint. It is difficult to get all the required data for every particular GHG emissions due to technical and monitoring problems. There-fore, for simplicity, it is often expressed in terms of the amount of carbon dioxide (CO_2) emitted. The carbon dioxide emissions can be estimated based on the fuel consumption of the generator and its average carbon content. The consumption of 1 litre of diesel emits on average around 2.7 kg of CO_2 usually falling in the range of 2.4–2.8 kg/litre. At optimum efficiency, the generator can produce 3 kW h/litre of fuel. Based on this assumption, the emission factor would be 0.9 kg CO_2/kW h. In practical operation, figures of around 1.27 kg CO_2/kW h are more likely. This compares to the emissions from a CCGT of 0.324 kg/kW h i.e. a quarter as much. **The other consideration is that the emissions from local generation are likely to be close to densely populated areas**.

5.7   Economic comparison of large-scale vs embedded

The costs of EG can be compared with super-grid connected large-scale generation. In this example, a grid connected unit of approaching 400 MW is compared to an embedded one of 400 kW. Typical efficiency figures are 55% electrical for the CCGT unit compared to 42% for the gas engine. This results in the fuel costs/MW h being 31% higher, that in turn results in 31% higher emissions/MW h from the small-scale generator. The CCGT capital costs are around £700/kW for the larger unit compared to around £500/kW for the gas engine. The costs of capital are assumed the same but assuming a shorter life span for the engine the capital costs/ unit are similar. It is expected that there would be economies of scale in providing staff to manage the unit and its operation. Assuming staff monitoring/operating for just 8% of the year, the O&M costs for the smaller unit are still £4.6/MW h higher. Including all these factors, the cost/MW h from the gas engine are 35% higher at £55.08/MW h compared to £40.7/MW h for the CCGT as shown in Table 5.3. Depending on the national scheme for managing emissions, there would be addi-tional costs. Based on a figure of £20 CO_2/t, the added fuel cost to the CCGT for allowances is £6.48/MW h. The additional costs for the smaller less efficient gas engine would be £8.49/MW h **resulting in the small engine unit costs being 35% higher** (£55.08 + £8.49)/(£40.7 + £6.48). It may be that the regulatory system

Table 5.3   Comparison of CCGT and small gas engine

Parameters	Units	Gas CCGT	Gas engine
Capacity	MW	393	0.41
Capacity net	MW	377	0.4
Load factor	%	85	85
Annual output	TW h	2.81	0.00298
Construction time	Years	2	1
Project life	Years	20	12
Efficiency	%	55	42
Natural gas cost	£/mmBTU	4.50	4.50
Natural gas cost delivered	Pence/therm	45.00	45.00
N gas calorific value	kW h/therm	29.32	29.32
N gas grossCV/netCV		1.11	1.11
Cost of dept	%	4.7	4.7
Cost of equity	%	7	7
Inflation	%	2.4	2.4
Debt/equity split		0.8	0.8
WACC ($D/E = 0.8$)	%	6.0	6.0
Total investment cost	£m	273.5	0.2
Capital cost	£/kW	696	488
Capital cost pa	£/m	23.78	0.02381
Consumables	Pence/kW h	0.02	0.02
Labour rate	£/h	16.66	16.66
No. of operators		6	0.08
Operators labour*1.28	£m	1.12	0.01
Maintenance material	£m	0.81	0.0009
Main and support labour	£M	1.25	0.003
Capital cost	£/MW h	8.46	7.99
Fixed O&M	£/MW h	0.60	5.22
Var O&M	£/MW h	0.73	1.31
Fuel costs	£/MW h	30.97	40.56
Total generation costs	**£/MW h**	**40.77**	**55.08**
CO_2 price £/t	**20**		
Include carbon	tCO_2/GJ	0.0495	0.0495
Added cost/incentive	£/MW h	6.48	8.49
Total cost	£/MW h	**47.25**	**63.57**

exempts the smaller unit from CO_2 costs because of its size. Where it is exempt, the small generator unit costs are still 16% higher. In addition, account needs to be taken of the network losses and network tariffs. The losses are not generally specifically allocated but reflected in the tariffs. The super-grid losses are around 2% while distribution losses are up to 6%. Assuming a reduction in the output of the CGGT of 2% from 2.81 to 2.75 TW h makes a difference of only £0.2/MW h. It is not the losses that impact on the costs but rather the complete tariff cost.

The transmission tariffs are usually loaded more on the demand side than generation. In the United Kingdom, the total revenue requirement for the

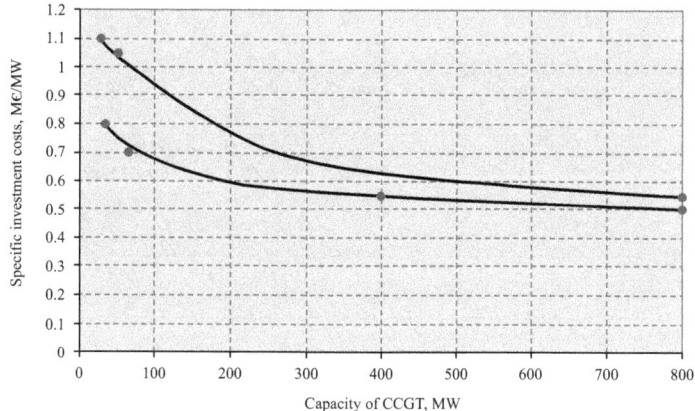

Figure 5.3 CCGT capital costs

transmission network was £2,708 m in 2016/17 with the generators paying £453 m and the demand side £2,255 m or 83%. The resulting generation tariffs are location dependant and vary from −£2/kW in high demand areas up to +£20/kW in remote areas. Assuming an average around +£7.5/kW and transmission connected generation of 60 GW then 60 GW * 7.5/kW = £453 m to cover the generators contributions. Discounts apply to lower load factor intermittent generation.

The demand side charges assuming a maximum demand figure of around 50 GW can be estimated based on a residual charge at £45/kW * 50 GW = £2,250 m. The embedded generators do not incur the transmission charge and avoid the network charge of typically £7.5/kW/year or £7,500/MW/year. Assuming a load factor of 85% then the avoided cost/MW h is given by: £7,500/(8,760*0.85) = £1/MW h increasing proportionally at lower load factors. The other impact is that the embedded generator can reduce the triad peak demand and hence network charges for the supply company by £45/kW of installed capacity. This benefit is normally shared with the generator, so in the case of this 400 kW generator, the additional income would be £18,000/year (45*400). Based on an expected output of 2,980 MW h/year, this equates to £6/MW h that would offset half the extra unit costs of the smaller generator.

The current popularity of small generating units is partly fuelled by the structure of regulation and this will be corrected. The desire for end users to take control of their energy supplies is understandable given the recent rises in costs. But, this needs to be founded on accurate and stable information on the regulatory environment. It is also undesirable to have high levels of emissions in local populated areas and consideration needs to be given to introduce EG charges in concert with banning polluting cars from cities.

Figure 5.3 shows a typical curve of how the capital costs of CCGT generation varies with unit size. It can be seen that a 25-MW unit is 50% more expensive than a 400-MW unit. Figure 5.4 shows how the efficiency improves with larger units.

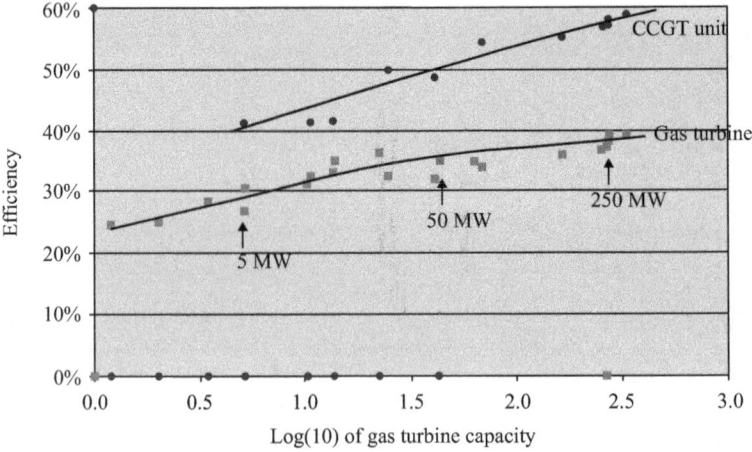

Figure 5.4   Variation in efficiency with size

Again, a 250-MW CCGT unit is 50% more efficient than a 5-MW unit. Both point to significant economies of scale. It is these economies of scale that has led to the development of larger units. Contrary to this, the trend to local smaller units is driven by other consideration. CHP schemes that provide both electricity and heat are an example where high effective efficiency is realised if there is a local heat demand. **Another key factor that can make small generators financially attractive is their opportunity to avoid transmission network charges and mitigate those of the local supplier.**

5.8   EU perspective

The EU has recognised that there are issues with EG that they refer to as self-generation (SG). **The rapid growth is having an impact on distribution networks leading to proposals of a number of measures to more equitably share network costs and manage the impact on the network development including:**

- The SG development must be included in network planning;
- The SGs should accept responsibility as an active power producer;
- Taxes and subsidy costs should be shared by all system users;
- The use of system tariffs should be fully cost reflective;
- The SG should be able to participate directly in the provision of flexibility;
- Adequate metering should enable monitoring of production and consumption;
- Net metering should be discouraged as it does not meet operational requirements.

It is argued that SGs benefit from connection to the wider network that facilitates trading but do not accept responsibility or contribute to the requirement for system services like balancing and frequency and voltage control. The SG is not subject to dispatch and as such, there are risks in assuming it is always available at peak times and it's argued its capacity should be discounted in assessing system peaks and plant margins. This would particularly apply to intermittent sources like wind and solar generation. There is a danger that consumers with SGs are being subsidised by those without it through their bills. This particularly applies to the generous terms provided to the initial solar panel schemes installed in the United Kingdom. These renewable sources are not flexible and sudden changes in their output require extra balancing services that have to be paid for by all consumers.

Some distribution networks in the United Kingdom are already becoming stretched with limited capacity to absorb more generation and manage network security and voltage control problems. At the same time, the super-grid network is experiencing periods of very low load causing high voltage problems with excessive network gain. In the United Kingdom, it has been found necessary to switch out some HV circuits to contain high voltage levels at times of low load. As the distribution networks become more active, they will inevitably need to establish similar control centres as used on the super-grid to manage network operation and security. Rather than the cost being shared by all consumers, it could be argued that the beneficiaries are the SGs and they should bear most of the costs. Demand prediction is a particular problem and SG metering needs to be collected and processed to provide the data essential to support the process. It is also arguable that SGs benefit from the provision of backup security when their installation is not in operation. Emissions and noise from EG can also be a problem when they are sited close to local communities.

The industry needs to rethink the tariffs that should be applied to local generation embedded within the distribution network to better reflect the longer term costs that are incurred in managing their introduction onto the power system. This includes enhanced network management and control costs, the provision of network capacity to export the output to other consumers and security implications during disturbances. Other consumers without SG should not bear these additional costs.

5.9 Distribution security

Distribution design standards are usually defined in terms of the size of load area being supplied. In the United Kingdom, the standard refers to loads in 5 classes ranging from 1 MW or less up to 1,500 MW. For each class, the load supported after the first and second outage into the area is defined as shown in Table 5.4. Compliance with the standard Engineering recommendation P2/6 is a licence condition for distribution companies.

The current standard does not explicitly recognise the impact of demand side regulation (DSR) or EG, whereas their increase is expected to continue. The potential for demand side response to contribute to security needs to be accommodated for the full benefit to be realised. Equally distributed generation can make

Table 5.4   Distribution system security standard

Class	Load MW	First outage	Second outage	Switch Tm	Maintenance	Other
A	≤ 1	Repair time	Nil	N/A	N/A	N/A
B	1–12	3 h	Nil	3 h	2 h	24 h
C	12–60	3 h	Nil	3 h	18 h	15 days
D	60–300	3 h	Restore	3 h	24 h	90 days
E	300–1,500	Immediate	Restore	N/A	24 h	90 days

Table 5.5   Embedded generation security contribution

Type	CCGT	Waste	CHP	Landfill
Contribution % DNC	63	58	40	63

a contribution to security that needs to be recognised. For demand, the DSR MW could be netted off the demand that may result in a change of class or other classes could be established. Judgement needs to be exercised to avoid decreasing security to other customers not involved in DSR. In the case of distributed generators, there is a need to distinguish between those that operate most of the time and those operating intermittently for a limited number of hours. Some judgement is required to estimate the contribution to security that might be expected from different generation types as shown in Table 5.5.

These factors would be used to estimate the potential in-feed to an area. For intermittent generation, a factor can be added based on the probability of the generation being available. **With the expected growth of DSR and EG, it is important to take account of their potential contribution to security of supply to realise the full benefit**. SGs and DSR can also be an effective mechanism to delay transformer or network reinforcement by curbing peak demands. These factors are also leading to an expansion of SGs.

5.10   Conclusions

A rapid growth is occurring in EG or SG as labelled by the EU that the traditional industry has been slow to adjust to. A number of large generators are suffering financially and uncertainty is curtailing investment in new stations. Some distribution networks are reaching saturation in their ability to accommodate more SGs. The distribution networks have become more active rather than passive requiring a more sophisticated monitoring and control infrastructure to manage security and the impact of intermittency. The SGs like wind and solar do not inherently provide inertia making frequency more volatile following system disturbances that can cause local trips. The reduction in super-grid generation reduces the reactive support to manage voltages necessitating switching out some transmission circuits at times of low load. The distribution control framework needs to interact with an increasing number of players and has yet to be established.

One of the reasons for the growth in SGs is related to increases in network charges that in part are the result of the extra funding required to develop the network to accommodate the new renewable generation sources. A contributing factor is the charging arrangements for use of system where the transmission charges to the host supplier are based on recorded peak demands. The embedded generator can reduce these charges by running at the peak times to reduce the import and sharing the benefit with the supplier. The smaller generators also do not usually contribute to transmission costs and may avoid emission charges other than were they added to fuel costs.

Some distributed generation schemes involving combined heat and power can operate at efficiencies of up to 85% where there is a local demand for heat. This applies to both large-scale systems of tens of MWs as well as systems with embedded generators of 100 kW. These systems can be very cost effective and enable arbitrage between local and grid supplies.

A comparison of unit costs shows that the smaller generator costs may be 30% to 35% higher than from large-scale units. An analysis of recorded data confirms economies of scale with larger CCGT units being 50% cheaper/kW and 50% more efficient than smaller embedded units. A key differentiating factor, in the financial analysis of small embedded units, is the inclusion of network costs in the avoided electricity price. At the distribution level, the network costs can equal the energy costs in the tariffs to end users. It is not surprising that energy from local generation appears cheaper. However, the local installation will require backup from the network and will rely on the network for frequency and voltage control. They may also export surplus energy to the grid and it could be argued they should pay for that option. It appears unjust that a neighbour has to contribute to subsidies for the energy from a neighbour's solar panels. Other factors to be considered in relation to SGs are the emissions particularly from local diesel generators that may be located in densely populated areas.

The current popularity of small generating units is partly fuelled by the structure of regulation and this will likely be changed. The desire for end users to take control of their energy supplies is understandable given the recent rises in costs. But, this needs to be founded on accurate and stable information on the developing regulatory environment. The industry needs to rethink the network tariffs that should be applied to local generation embedded within the distribution network to better reflect the longer term costs that will be incurred in managing their introduction onto the power system.

Question 5.1 Given data in the table for a micro-generation, calculate that price of avoided electricity that would result in a payback within 15 years based on the lower capital costs of €115,000.

What would the payback period be if the boiler gas price is doubled?

Micro-generation

Cost installed (low)	Euro, 000	115	Running hours		h/year	5,890.0
Cost installed (high)	Euro, 000	140	Electrical production		MW h	589.0
Power production	kW	100	Hot water production		MW h	1,035.4
Electrical efficiency	%	30	Hot air recov. prod.		MW h	117.8
Hot water	kW	167	Avoided fuel in boiler		MW h	1,218.1
Hot air (recovered)	kW	20	MT fuel consumption		MW h	1,963.3
Boiler efficiency	%	85				
			Income with avoided elec.		Euro, 000	51.2
Electricity price	€/MW h	87	Income with hot water		Euro, 000	30.5
Boiler gas price	€/MW h	25	Income with hot air		Euro, 000	2.4
CHP gas price	€/MW h	25	MT gas consumption		Euro, 000	−49.1
Value of hot air	€/MW h	20	MT maintenance		Euro, 000	−8.8
MT maintenance	€/MW h	15	**Total income/year**		Euro, 000	**26.1**
			WACC %			8.0
Operating hours	h/year	6,200	Project life years			15.0
Availability	%	95	Payback range from		5.4	4.4

Electricity price			**Production**	**MW h**	**Costs**	**Euros**
Avoided energy	E/MW h	31.5	Electricity	589	Capital	16,356
CCL tax on energy	E/MW h	3.5	Water	1,035.4	Fuel	49,083
Dx charge	E/MW h	42	Air	117.8	Maintenance	8,835
Tx charge	E/MW h	10	Total	1,742.2	Total	74,274
Total	E/MW h	87	Cost E/MW h			42.6

Chapter 6

Nuclear option

6.1   The case for nuclear

There are emissions associated with excavating, concentrating and transporting uranium and building nuclear plants. But, according to the International Atomic Energy Agency, the emission rate from nuclear generation is 1%–3% of that from equivalent coal-fired generation. Unlike renewable sources, the generation is available all the time other than during planned maintenance periods. Although capital costs are high, the long range price for energy is comparable with other renewable sources and probably cheaper if account is taken of the cost of managing renewable intermittency. This has necessitated capacity payments to make it financially viable to retain conventional capacity in service to provide backup for when the wind speed is low or there is no sunlight for solar farms.

The market for nuclear technology is expanding with 65 plants under construction in 14 counties with a lot in China, India and Russia. In China, nuclear is seen as the most viable option to reduce emissions. In addition to having 25.5 GW of capacity in service, a further 25 GW is under construction. India is planning to produce 25% of its energy from nuclear by 2050. In 2016, there were some 440 nuclear power reactors operating in 32 countries producing 11% of the world's electricity. In sharp contrast, other countries like Germany and Japan are choosing to reduce their dependence on nuclear. Following the Fukushima disaster in Japan, the energy supply from nuclear was reduced from 28.6% in 2010 to 1.1% in 2015 being replaced by more gas and coal-fired generation as illustrated in Figure 6.1.

There are risks associated with nuclear generation, in addition to the fallout from failure incidents, there is a fuel supply question. Some 60% of the uranium used in the EU to fuel its 130 reactors comes from Russia, the Niger and Kazakhstan but a new uranium mine is under construction in Spain that could eventually supply 10% of Europe's needs.

European energy supplies are still heavily dependent on nuclear generation as shown in Figure 6.2 that projects energy supplied by source through to 2020. France has the highest proportion of nuclear and maintains high utilisation factors through exports to the rest of Europe. Germany has chosen to phase out its nuclear generation but at the expense of becoming more reliant on Russian gas.

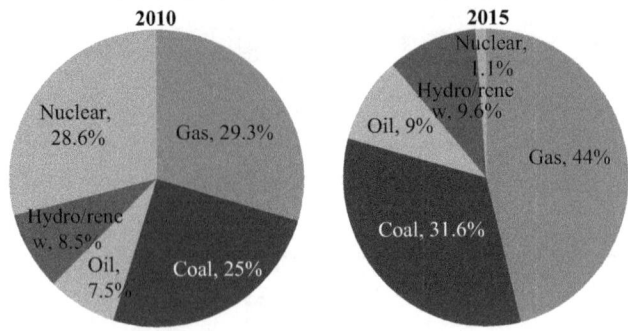

Figure 6.1   Energy mix Japan

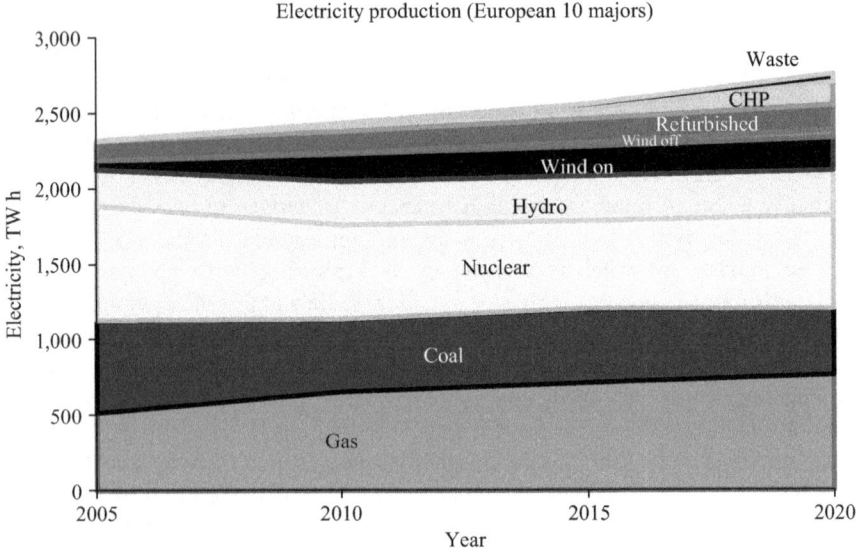

Figure 6.2   European energy sources by type

6.2   Nuclear generation in China

Chinese President Xi Jinping has referred to nuclear energy as important to ensure energy security and tackle climate change. Professor Guo Zhifeng of the China Institute of Nuclear Information and Economics states that whereas the excavating, concentrating, transporting and construction of nuclear plants emit greenhouse gases, the generation process discharges no carbon dioxide. Nuclear generation also has a principal advantage over renewable energy sources in that it is available most of the time unlike wind and solar sources that are intermittent and dependant on weather conditions. The first nuclear plant was built in 1954, and the technology can be regarded as mature even without further development.

Opponents of nuclear power argue that waste product management is too difficult to be dealt with and poses a threat to the environment. The incidents at Three Mile Island, Chernobyl and Fukushima have influenced views on safety. Germany has decided to halt operations at its eight oldest nuclear plants and plans to shut down the remaining nine by 2022. France has also announced plans to reduce its reliance on nuclear reactors. However, research by the Swiss Paul Scherrer Institute concludes that nuclear power plants have caused fewer fatalities per unit of energy generated than other major sources of power. The storage of spent fuel rods has been managed safely, and no plant has been shut down as a result of problems. Past disasters have also led to technology improvements that are making nuclear facilities safer.

China is currently very dependent on coal-fired generation with a limited ability to meet its future energy requirements from wind power and other renewable sources because of transmission limitations. The development of more nuclear generation is considered essential to meet emission reduction targets. **By 2015, China had 27 nuclear generating units in operation with a capacity of 25 GW with another 25 units under construction equivalent to a third of the world total.** By 2020, a total capacity of 58 GW is expected, outpacing all other countries. A third-generation nuclear reactor, the Hualong One, is under development with installations planned in Pakistan and Argentina and possibly Britain. The China General Nuclear Power Group (CGN) has been established to promote the new model and captures a large part of the global market for nuclear reactors. The expansion of nuclear generation in China is likely to do more to reduce emissions than anything Europe does.

The planned 58 GW of nuclear generation operating at 85% load factor will produce energy equivalent to about 250 GW of wind power installations. The emission reduction from reducing coal burn is calculated as follows by

Coal CO_2 content is equal to $0.0869 \, tCO_2/GJ$
The emission/MW h is given by $0.0869 * 3.6 \, tCO_2/MW \, h$
Taking account of efficiency emission $= 0.0869 * 3.6/0.36 = 0.869 \, tCO_2/MW \, h$
Given an annual generation $= 58 \, GW * 0.85 * 8{,}760 = 431 \, TW \, h$
The emission reduction $= 431 * 0.869 = 374 \, mtCO_2$

The saving resulting from the use of nuclear power in China is equivalent to around 20 times the current United Kingdom target annual emission level.

6.3　Large-scale nuclear costs

The capital costs of large-scale nuclear are estimated at £6,913M for a 1,660-MW station with a cost of financing of 710M/year at a WACC of 8.1%. This figure is close to that reported in July 2016 of £7bn for 1,660 MW with a unit cost approaching £80/MW h. By the time the UK govt. confirmed its support for the project, the cost had risen to £9bn for 1,660 MW of capacity as shown in Table 6.1. **This results in the unit cost rising to £88/MW h over a 30-year life and has to be compared against the contract strike price of £92.5/MW h.** The calculations are based on a lifetime utilisation of 80% that reflects the availability figures realised in practice.

Given the dominance of the capital costs, the WACC and assumed lifetime utilisation are paramount in estimating unit costs. Assuming the £9bn capital costs at a WACC of 10%, as opposed to the 8.1% as used in the table, the unit costs rise to £102/MW h. This emphasises the importance of the government underwriting

Table 6.1   Large-scale nuclear costs

Nuclear		Units	Nuclear
Plant	Capacity	MW	1,660
	Capacity net	MW	1,494
	Load factor	%	80
	Annual output	TW h	10.47
	Construction time	Years	8
	Project life	Years	**30**
Fuel	Efficiency	%	38
	Fuel cost	£/MW h	5.8
	Decommissioning	£/MW h	1.5
Finance	Cost of dept.	%	7
	Cost of equity	%	9
	Inflation	%	2.4
	Debt/equity split		0.8
	WACC ($D/E = 0.8$)	%	8.1
Cap costs	Site and dev	£m	3,569
	EPC contract	£m	4,590.0
	Electrical conn	£m	31.0
	Spares	£m	76.0
	Interest during const.	£m	735.0
	Total investment cost	£m	**9,001.0**
	Capital cost	£/kW	5,422
	Capital cost pa	£/m	807.93
O&M	Consumables	pence/kW h	0.164
	Labour rate	£/h	16.66
	No. of operators	4 shifts	48
	Operators labour*1.28	£M/year	8.97
	Maintenance material	£M/year	8.3
	Maint. and support labour	£M/year	4.3
Summary	Capital cost	£/MW h	77.17
	Fixed O&M	£/MW h	2.50
	Var O&M	£/MW h	1.20
	Fuel costs	£/MW h	7.3
	Total generation costs	**£/MW h**	**88.17**

the contract to minimise the risks. With an assumed 25years of full utilisation, as opposed to the 30 years assumed in the table, the cost rises to £92/MW h. Other risks are associated with sourcing fuel supplies and decommissioning. Using an annual output figure of 10.47 TW h, the provision for decommissioning of £1/MW h would generate £10.47M/year or £3,141M over 30 years that is over a third of the original capital costs.

6.4   Consumer costs of nuclear

Large-scale nuclear parks are attractive in facilitating the management of site security as opposed to plant being scattered over several sites as illustrated in Figure 6.3. Hinckley Point C is a large installation estimated to cost £18bn with 33.5% of the funding put up by China CGN and the rest by EdF that is 85% state owned. The station will have a total capacity of 3.2 GW and is expected to start generating in 2025 and supply around 7% of UK electricity needs. It will be guaranteed a strike price of £92.5/MW h by the UK government and inflation linked for 35 years giving a return estimated at 9%–10% to the developers. The National Audit Office estimated in 2013, when the deal was proposed, that compared to forecast wholesale market prices the strike price represented a **subsidy by consumers of £6.1bn. More recent estimates of future wholesale prices indicate the subsidy could cost consumers an extra £27.9bn.**

The design is based on Areva's European Pressurised Reactor (EPR) that has been installed at Flamanville where costs escalated to three times the original budget with several years delay. The capacity of 3.2 GW would, if generating at a load factor of 85%, produce an output of

$$3{,}200 \text{ MW} * 8{,}760 * 0.85 \text{ h} = 23.8 \text{ TW h/year}$$

Figure 6.3   Hinckley nuclear station

If we assume average wholesale prices of £56.5/MW h over the period, the strike price represents a premium of £36/MW h. Over 35 years, this amounts to a total subsidy given by

$$23.8 \text{ TW h} * 35 \text{ years} * £36 = £30\text{bn}$$

This is close to the figure quoted by the National Audit Office but if gas prices continue to be constrained by the development of 'fracking' then comparable gas prices could be as low as £45/MW h during the 2020s. **This would result in a premium of £47.5/MW h and a total subsidy cost of £40bn.** Assuming a national annual demand of 340 TW h/year, the output of Hinckley at 23.8 TW h/year is around 7% of the national energy requirement. Assuming a typical annual domestic consumption of 4.5 MW h, the extra cost of the subsidy for the Hinckley station for each domestic consumer is given by

$$47.5 * 4.5 * 7/100 = £15/\text{year or } £525 \text{ over 35 years}$$

The Chinese General Nuclear Power Corporation (CGN) have worked with EdF on the development of nuclear power in China and have developed their own reactor the HPR1000 (Hualong-1). Together with EdF, CGN is interested in building two more nuclear stations in the United Kingdom at Sizewell C and Bradwell B with the later based on the HPR 1000 design. If two similar stations are built to maintain the nuclear component in the baseload mix, then the total subsidy cost could rise to £90bn (3* £30b) with consumers paying an extra £36/year. There are serious risks in the sector with Toshiba an investor in the Nugeneration consortium reporting a loss £6.7bn for 2016. They planned to build a 3.8-GW station in Cumbria based on a Westinghouse design but this may now be in doubt.

6.5   Operating environment of nuclear

Despite its high cost, nuclear does compete with other low-emission sources of energy because it has a higher availability and can operate baseload. In contrast, intermittent renewable sources require backup. Because of its high capital costs, nuclear is only viable if it can be operated close to baseload. At a 60% load factor (as opposed to 80%), unit costs would rise to £115/MW h, and this limits the potential capacity of economically viable nuclear generation. The priority given to using renewable sources, whenever available, reduces the utilisation of conventional thermal generation making them potentially none commercially viable. This has led to closures and an unprecedented reduction in the plant margin of spare capacity to cater for forced outages and worse than average weather. Shortfalls in capacity can even occur during summer months, and on 11 May 2016, National Grid was forced to issue a notification of inadequate system margin with one plant being paid £1,250/MW h following a plant breakdown coupled with no wind generation output. **It is considered that some 60% of European gas generation may be subject to closure because of reduced utilisation.**

To retain more marginal plant in service, capacity payments have been intro-duced in the United Kingdom. The T-4 capacity auction for delivery in 2018/19 resulted in 49.3 GW of capacity being contracted at a price of £19.4/kW/year. This amounts to a total cost of £956m/year including some 1.2 GW of small diesel fuelled generators with low efficiency and high levels of emissions, three times that of conventional generation. The government had estimated a capacity price of £49/kW/year based on best new entry open cycle gas turbine that would have resulted in a total cost of £2.4bn. The government expectation was that capacity payments would lead to a commensurate reduction in market wholesale energy prices. This is unrealistic given that not all plant receives capacity payments, and if flexible open cycle gas turbines are operating at the margin, energy prices will be twice as high due to their lower efficiency.

The small diesel generators are exempt from environmental regulation and are very polluting to the local atmosphere with emissions of around 1.27 tCO_2/ MW h compared to CCGT generation at 0.28 tCO_2/MW h i.e. about one extra t/ MW h. Given the 1.2 GW of contracted small diesel generators, the extra emissions when running would be 1,200 t/h that priced at £30/t costs an extra £36,000/h. Over a year assuming 1,000 h operation, the additional costs would be £36m with extra emissions of 1.2 Mt offsetting 11% of the saving by the wind generation.

Based on an analysis of recorded data, the current load factor of wind gen-eration is about 24.7%. On this basis, some 15 GW of wind would produce 32 TW h. CCGT generation emits around 0.34 mt/TW h of CO_2. If we assume the wind generation displaces gas-fired generation, then the emission reduction due to wind generation would be 11 Mt/year. Based on Emission Trading Scheme (ETS) prices of £4/t, the cost would be £44m/year. Against this, the FIT for on shore wind is £95/ MW h which compared to 2016 wholesale prices of around £40/MW h represents a subsidy of £55/MW h at an additional cost to consumers of £1,760m/year (£55/MW h * 32 TW h). This is additional to the capacity market costs of £956m and extra emissions resulting from the use of small diesel engines. There are also costs associated from having to curtail wind due to transmission constraints and extra operating costs incurred by conventional generation in tracking wind output with more frequent start-ups and shutdowns and ramping output up and down. Both these costs increase significantly as the proportion of intermittent generation capacity increases above 20%. The additional costs are principally as a result of wind/solar intermittency and the need to retain backup. Taking these factors into account, a tranche of baseload nuclear generation looks an attractive proposition.

6.6 Small modular nuclear

Several companies are competing to provide small-scale nuclear units that can be built in a factory and transported by lorry to site. Rolls Royce and Bechtel are in competition to supply small modular reactors (SMRs). Rolls have submitted detailed designs to the government for SMRs capable of generating 220 MW, that could be doubled up to 440 MW on plants covering ten football fields. **The design**

is based on Rolls experience with powering nuclear submarines and is expected to be a tenth of the size of a traditional nuclear power station and at least a fifth cheaper.

Bechtel and BWX Technologies, Inc. formed Generation mPower, LLC (GmP), an alliance to build SMRs, that are safe, scalable, reliable and a cost-effective alternative to new-generation nuclear facilities as illustrated in Figure 6.4. Bechtel is currently leading design engineering, procurement and licencing services to develop a plant design based on the BWXT mPower^{TM} technology. The reactor is a nominal 195-MWe light water, steam generating, SMR. The plant design includes a fully underground containment structure and safety systems that are located in and protected by underground structures. The UK government pledged £250m to fund a research and development programme the first phase being a competition to find the best value designs.

NuScale has designed an extraordinarily safe integral pressurized water reactor based on light water reactor technology proven in operation for over 50 years. The design incorporates several features that reduce complexity, improve safety, enhance operability and reduce risks.

Critically, it features a unique triple lock passive safety system enabling it to safely shut down and self-cool indefinitely with no operator action, no power and no additional water.

The small, modular design of the technology offers the prospect of deploying nuclear power in a range of locations. It can be entirely factory fabricated and transported to its destination by rail or road, reducing on-site construction.

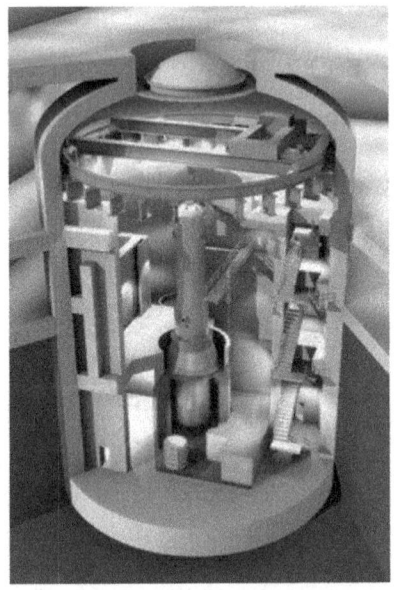

Figure 6.4   SMR (mPower)

The design also means that additional capacity can be added to a NuScale power plant incrementally over time – up to a maximum of 12 modules, with each module producing 50 MWe of low-carbon energy.

The smaller units may also be more flexible than their larger counterparts to help accommodate the rapid changes in frequency that are expected during low demand periods with high concentrations of wind generation. The supplies to auxiliaries also need to be secured against frequency and voltage disturbances. Standards and codes need to be developed to ensure safety and environmental protection.

6.7   Nuclear fusion

Research on the release of energy from nuclear fusion has been taking place on a large scale since 1983 at the Culham Centre for Fusion Energy. The research facility required huge pulses of power of several hundred MWs for seconds from the power system to research the process. The author was engaged in demonstrating the ability of the UK power system to provide the best location for the Joint European Torus facility to meet the supply requirements. The facility required a strong system connection point to ensure that damage did not occur to local generation, and supplies to other consumers were not affected by the pulsing process. The plant located at Culham in Oxfordshire is the world's largest and most powerful tokamak (see Figure 6.5). It is designed to study fusion in conditions similar to those of a practical power plant using a deuterium–tritium fuel mix that is expected to be the basis of commercial plants that could be operating by 2040.

The fusion process mimics that occurring on the sun, it involves heating the fuel mix to temperatures of 100 million degrees Celsius to create the conditions where the nuclei collide and fuse releasing energy in the form of high-energy neutrons. To realise the right conditions, the hot gas plasma is confined magnetically in a ring-shaped chamber called a 'tokamak' in Russian. This device draws pulses of energy from the local power system to drive the experiments. The neutrons pass through to a blanket of denser material surrounding the tokamak raising the temperature to produce steam that can be used to generate electricity in a conventional turbine.

Nuclear fusion has a number of potential advantages:

1.   No carbon emissions just helium
2.   Fuel availability – Deuterium from sea water and tritium from mined lithium
3.   Efficient – 1 kg = 10-mkg fossil fuel
4.   No long-lived radioactive waste

Cost projections claim that fusion generation should provide energy at prices competitive with existing sources particularly when the full impact of CO_2 emissions is factored into prices. It is envisaged as a fully sustainable long-term source of baseload energy.

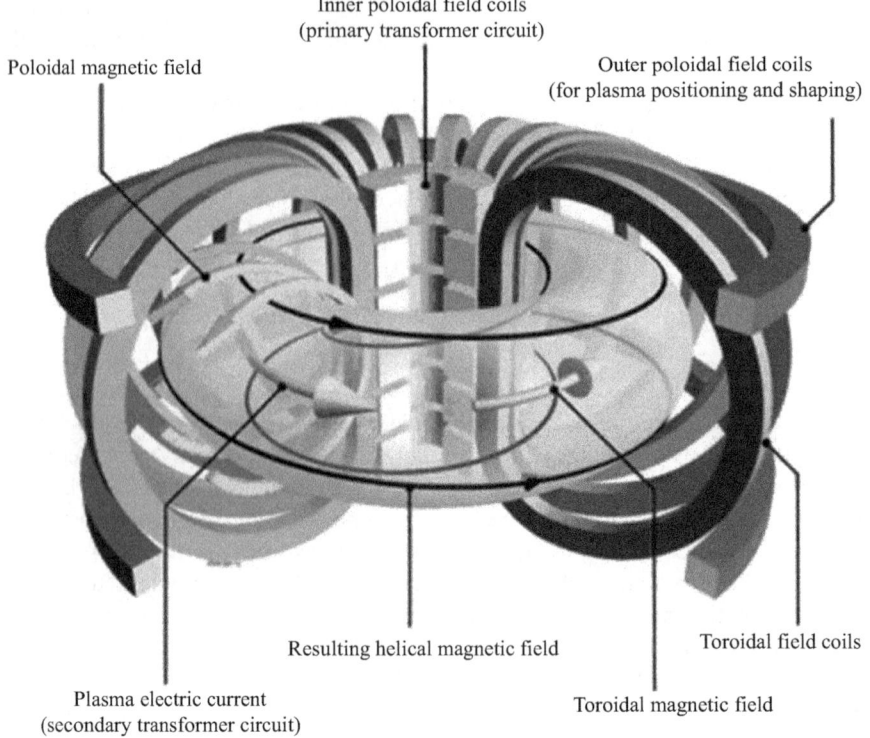

Inner poloidal field coils
(primary transformer circuit)

Poloidal magnetic field

Outer poloidal field coils
(for plasma positioning and shaping)

Resulting helical magnetic field

Toroidal field coils

Plasma electric current
(secondary transformer circuit)

Toroidal magnetic field

Figure 6.5   Tokamak (source EURO fusion)

6.8   Prospects for nuclear

It's time governments realised that energy prices in a capital-intensive industry like electricity generation are principally set by long-term investment decisions rather than short-term market dynamics. Leading economists argued, at the time of privatisation, that the market would encourage the right level of capacity and reduce prices. But, having sold off a key part of the UK infrastructure, state-owned EdF is now able to pressure the United Kingdom to commit to pay twice the current market price for energy for 40 years before they will invest in new nuclear stations. It is not clear who will be obliged to take the nuclear energy but the United Kingdom will look pretty silly and be commercially disadvantaged if it ends up paying twice as much as French consumers. Meanwhile, the EU is actively promoting an integrated European energy market. Member states have to consider their options including building more interconnection to exploit the wider energy market and managing a smaller plant capacity margin making use of demand management and imports. It is no use investing in high levels of security if people cannot afford energy. Before the UK government commit future generations to more stations with these high costs, they have to fully consider the risks and alternatives including shale gas, Liquid

Natural Gas (LNG) imports and coal with carbon capture and storage that could all benefit future energy prices. It is not clear on whose behalf the government is contracting but contractual options could include

- To link the strike price to the prevailing EU power price
- To involve suppliers in an auction of the nuclear capacity with prices and volumes, they are prepared to pay
- To provide the incentive to build through a capacity payment as has been muted
- For the government, to take a contract for differences to make up any perceived shortfall or take any excess profit in the generator revenues derived through the market

It also has to be explained how the contract with EdF will affect expectations in the wider UK market. The United Kingdom is already committed to subsidising wind generation and now is not the time for government to commit consumers to yet more energy price rises without a wider debate and analysis of the issues.

Nuclear generation constitutes a major source of energy worldwide and currently supplies around 30% of Europe's energy. It has high capital costs and needs to be operated at a high load factor to be economic. It has the advantage that it does not emit CO_2 but there are concerns about radiation associated with failures and politically about nuclear technology proliferation. Waste treatment and decommissioning costs are not well established and have recently been revised upwards. Deep burial is economically attractive and may become more acceptable when compared with the potential impact of global warming. The development of the pebble-bed reactor offers potentially cheaper and more modular options.

The economics of nuclear could look attractive if competing fuel prices are driven up by environmental concerns and CO_2 emission charges. The waste management looks less of an impediment if deep site burial is accepted, and it has been suggested that this could be at an undersea location, perhaps off the cost of Cumbria in the United Kingdom. Faced with global warming and the prospect of a countryside blighted by thousands of wind farms, the public are becoming much more receptive to the nuclear option. Press reports have heralded the need for 20,000 wind turbines to begin to address the 2020 target. If nuclear is allowed to decline, there would need to be ten times that number to replace the European nuclear output of some 600 TW h.

6.9   Optimal plant mix

Nuclear generation has a relatively high capital cost but low operating costs. It's most cost effective to operate the plant baseload at the highest possible utilisation. CCGT has moderate capital costs and could be used mid-merit tracking daily demand variations. Open Cycle Gas Turbine (OCGT) have the lowest capital cost but lower efficiency and higher running costs and are often used to meet peak demands taking advantage of their flexibility.

Table 6.2   Plant mix basic costs

Plant type	CCGT	Nuclear	OCGT
Capital cost, £/kW	800	4,165	333
Load factor, %	55	88	15
Output/MW h/year/MW	4,818	7,708.8	1,314
Project life years	20	30	20
Capital cost, £k/kW/year	69.75	405.41	31.43
Fixed O&M, £/MW/year	4,467.6	18,067.5	800
Capital cost/MW h	14.5	52.6	23.9
Fuel price £/MW h	39.8	5	54.4
Fixed O&M £/MW h	0.93	2.3	0.61
Variable O&M £/MW h	0.73	1	0.71
Total unit cost £/MW h	55.9	60.9	79.7
CO_2 emission, t/GJ	0.0495	0.0	0.0495
Efficiency, %	52.0		38
CO_2 costs, £/MW h	6.85	0	9.38
Total cost, £/MW h	62.8	60.9	89.1

Table 6.2 shows a set of costs for CCGT generation, nuclear and open cycle gas turbines. It is assumed that this plant is used to meet a typical annual demand curve with a tranche of baseload nuclear operating at 88% load factor, a tranche of CCGT generation operating mid-merit with an average load factor of 55% and at least 2 GW of open cycle gas turbines operating at an average 15% load factor. The capital costs are based on a WACC of 6% for the CCGTs, 9% for the nuclear and 8% for the OCGTs. The total demand is assumed to be 276 TW h/year, and the total generation capacity is set at 50 GW. It can be seen that the basic generation unit costs are £55.9/MW h for gas, £60.9/MW h for nuclear and £79.7/MW h for OCGTs. Adding in CO_2 charges at £20/t raises the CCGT price to 62.8/MW h and the OCGT to £89.1/MW h. The object of the exercise is to determine that mix of generation that minimises the total costs whilst meeting the constraints.

The optimum solution in this example was found using the Microsoft Solver LP using the data in Table 6.3 for 1 year. It shows 33 GW of gas and 15 GW of nuclear with OCGT set at its lower limit of 2 GW. In this case, the active constraint is the restriction on the total capacity of 50 GW. This may occur in practice when limited suitable generation location sites are available. This optimisation considers just one particular year. It will be necessary to consider costs through the lifetime of the installation with varying fuel and CO_2 prices. In this example, it can be shown that a modest rise in CO_2 prices to £30/t causes the gas prices to rise to £66.2/MW h. Alternatively decreasing the nuclear project life by just 5 years causes nuclear prices to rise above those of gas. This reflects the costs of serving the high capital costs associated with nuclear generation being spread over fewer years. The gas price was set at 60p/therm with a CV of 29 kW h/therm. It can also be shown that an increase in gas prices to 67p/therm causes gas generation prices to rise by £4.7/MW h.

Table 6.3   Optimal plant mix example

Plant type	CCGT	Nuclear	OCGT	
Capital cost, £/kW/year	69.75	405.41	31.43	
Fuel price, £/MW h	39.79	5	54.4	
Fixed O&M, £/MW/year	4,467.6	18,067.5	800	
Variable O&M, £/MW h	0.73	1	0.71	
CO_2 costs, £/MW h	6.85	0	9.38	
Utilisation	55	88	15	
Unit cost, £/MW h	55.92	60.93	79.69	
Unit cost, £/MW h plus CO_2	62.78	60.93	89.1	Totals
Optimal mix, GW	33	15	2	50
Energy, TW h	161	112	2.6	276
Cost £m/year	10,112	6,842	234	17,188

Only a proportion of the generation can operate baseload, and given the limited flexibility of large nuclear, it is usually assumed it will fulfil that role. If large tranches of priority dispatch renewable generation is introduced onto the system, then this would restrict the scope for more baseload nuclear. This example illustrates the sensitivity of the optimal plant mix to data assumptions.

Other constraints on the problem will result from the need to regulate quickly enough to balance variations in renewable generation output that in this case is met by retaining at least 2 GW of fast acting OCGT generation. If 25 GW of wind is included in the mix with a load factor of 30%, then the average output is 7,500 MW less 500 MW available from other provisions equals 7,000 MW on average. With a forecast error of 10% rms, on average 700 MWs of reserve would be required. To cater for the worst case, up to three standard deviations i.e. 2,100 MW of fast acting reserve could be required to cover times of peak output. With a maximum potential output of around 20 GW, there could be a requirement for conventional generation to regulate by this amount to accommodate a change in wind output. **This introduces a new dimension to the plant mix problem with a requirement for a lot of flexible generation**. This generation will also suffer lower levels of utilisation due to its output being displaced by wind. The lower capital costs of gas-fired generation makes its overall economics less sensitive to utilisation than large-scale nuclear or coal. In the simplified example, the whole tranche of CCGT generation has been assumed to operate at a Load Factor (LF) of just 55% with 65.7 TW h of wind output impressed on the system. Without any wind generation, the tranche of CCGT generation would have a healthier average utilisation of 77% with some generation at full load and lower merit generation at around 50%.

6.10   Conclusions

The market for nuclear technology is expanding with 65 plants under construction in 14 counties. In countries like China with growing pollution problems, there is an urgent need to reduce dependence on coal-fired generation that could not be quickly met from renewable sources. By 2015, China had 27 nuclear generating units in operation with a capacity of 25 GW with another 25 units under construction equivalent to a third of the world total. The planned 58 GW of nuclear generation operating at 85% load factor will produce energy equivalent to about 250 GW of wind power installations.

An estimate of costs results in a unit price of £88/MW h and consistent with the strike price for energy from the latest UK station of £92.5/MW h. Based on estimates of future prices from conventional generation sources of £56.5/MW h, this results in a premium of £36/MW h. **Based on the expected output, this would result in an extra cost to consumers over 35 years of £30 billion or around a £1,000/consumer.** If gas from 'fracking' keeps gas prices low, then the subsidy could rise to £40 billion.

The development of large-scale nuclear stations appears to be often fraught with problems and delays. SMRs have been developed and used for many years to power nuclear submarines. Based on this design, commercial nuclear stations are expected to be a tenth of the size of a traditional nuclear power station and at least a fifth cheaper. Advantages are that units could be built in a factory and installed underground. They could also be located and used to meet the needs of more remote areas.

SMRs represent a major opportunity for the nuclear industry to contribute to meeting targets for the world transition to a low-carbon society. A preferred design and licencing requirements need to be established to pave the way for prototype installations. The smaller units could potentially be designed to offer more flexibility than the larger counterparts and increase their potential market share. Against small distributed nuclear stations is the problem of managing security. Large-scale nuclear parks are attractive in facilitating the management of site security as opposed to plant being scattered over several sites.

A longer term option is fusion generation, and experimental installations have been operating to support development for many years, and it is estimated that plants could be operating by 2040. It has been predicted that energy prices would be comparable with existing sources with the added advantages of no long-life radioactive waste and more readily available fuel supplies.

The market for nuclear technology is expanding with 65 plants under construction in 14 counties with a lot in China, India and Russia. In 2016, there were some 440 nuclear power reactors operating in 32 countries producing 11% of the world's electricity. In contrast, a few countries like Germany and Japan are choosing to reduce their dependence on nuclear because of safety concerns. However, given the potential increase in demand for carbon-free electricity to supply the growth in electric vehicles and heat pumps, there is an expanding need for nuclear generation. In particular, it will contribute most economically if operated in base-load mode.

Question 6.1 What would be the impact on the unit cost of extending the life of the nuclear station to 40 years rather than 30 as shown in the table below:

Nuclear		Units	Nuclear
Plant	Capacity	MW	1,660
	Capacity net	MW	1,494
	Load factor	%	80
	Annual output	TW h	10.47
	Construction time	Years	8
	Project life	Years	**30**
Fuel	Efficiency	%	38
	Fuel cost	£/MW h	5.8
	Decommissioning	£/MW h	1.5
Finance	Cost of dept.	%	7
	Cost of equity	%	9
	Inflation	%	2.4
	Debt/equity split		0.8
	WACC ($D/E = 0.8$)	%	8.1
Cap costs	Site and dev	£m	3,569
	EPC contract	£m	4,590.0
	Electrical conn	£m	31.0
	Spares	£m	76.0
	Interest during const.	£m	735.0
	Total investment cost	£m	**9,001.0**
	Capital cost	£/kW	5,422
	Capital cost pa	£/m	807.93
O&M	Consumables	pence/kW h	0.164
	Labour rate	£/h	16.66
	No. of operators	4 shifts	48
	Operators labour*1.28	£M/year	8.97
	Maintenance material	£M/year	8.3
	Maint. and support labour	£M/year	4.3
Summary	Capital cost	£/MW h	77.17
	Fixed O&M	£/MW h	2.50
	Var O&M	£/MW h	1.20
	Fuel costs	£/MW h	7.3
	Total generation costs	**£/MW h**	**88.17**

What would be the impact of the load factor extending to 85% rather than 80%?

Chapter 7

Carbon capture and storage

7.1 The process

The process of carbon capture and storage (CCS) refers to the isolation of carbon dioxide from generation by gas- or coal-fired generation and storing it rather than discharging it to the atmosphere. The CO_2 is extracted and piped to a storage facility. This may be a partially depleted oilfield where it is used to facilitate enhanced oil recovery (EOR). Figure 7.1 shows a typical installation where the natural gas is fed into a reformer and capture plant where it is separated into CO_2 and hydrogen using specially developed membranes. The CO_2 is pumped into the offshore oilfield to enhance oil recovery or an offshore aquifer. The remaining hydrogen fuels a gas-fired power station producing clean electricity without CO_2 emissions.

It would not be very efficient to have dedicated pipelines from individual generating stations to oilfields but rather a network linking coal producing areas with a concentration of stations to a collection of oilfields. There is also the question in the medium and long term of aligning the production of electricity with the need for EOR in local oilfields.

The overall process is less efficient than conventional generation because of the energy consumption of the extraction and storage process. The overall efficiency may be reduced by 10%–20% depending on the design but can be competitive with other renewable generation sources when subsidies are taken into account. The plant output can also be controlled to track demand changes unlike wind and solar generation. **Given the current dependence on coal-fired generation across the world in countries like China and India, CCS offers the opportunity to contain emissions in the short term as it can be retrofitted**.

7.2 Design options

There are various arrangements for extracting the CO_2 resulting from burning coal or gas in power generation including

- Post-combustion flue gas treatment;
- Oxy-fuel combustion where the fuel is burnt in oxygen;
- IGCC (integrated gasification combined cycle) based on coal with CCS.

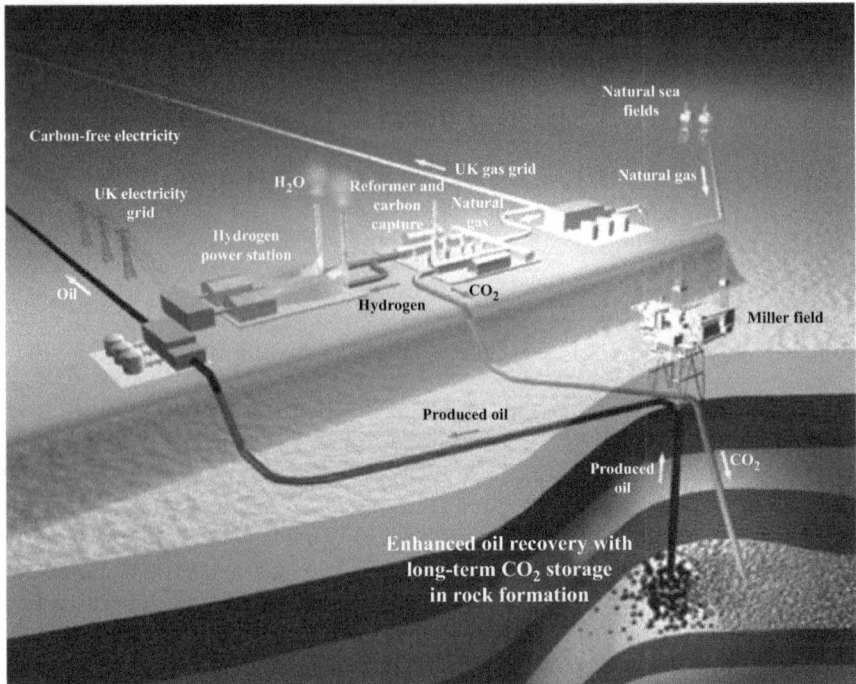

Figure 7.1   CCS installation with enhanced oil recovery (source BP)

The impact of the CCS is to reduce the overall efficiency when compared to conventional generation. The capital costs are higher with the added costs associated with transportation and storage of the CO_2.

Post-combustion Capture (PCC) – There are several alternative arrangements for effecting CCS on gas and coal power plants. In the post-combustion process, the flue gas is fed into an absorber with a counterflow of reactive solvent that absorbs the CO_2. The clean gas may be exhausted to the atmosphere. The CO_2-rich solvent from the bottom of the absorber is fed into the top section of the Stripper where the chemically bonded CO_2 is separated from the solvent. The chemical solvent has to be regenerated by the application of heat in the reboiler reducing the efficiency of the overall process but the costs may be minimised by the use of cheap energy periods. The CO_2 lean solution is passed through a filter back to the absorber as illustrated in Figure 7.2. Various solvents have been investigated with the objective of reducing the energy consumption for solvent regeneration and eliminating the operational difficulties. One option showing promise is a solvent based on amino acids.

Oxy-fuel combustion – In this process, the oxygen required for combustion is separated from air, and the fuel is combusted in oxygen diluted with recycled flue gas rather than air. This oxygen-rich, nitrogen-free atmosphere results in final flue gases consisting mainly of CO_2 and H_2O (water), so producing a more concentrated CO_2 stream for easier purification, transportation and storage as illustrated in Figure 7.3.

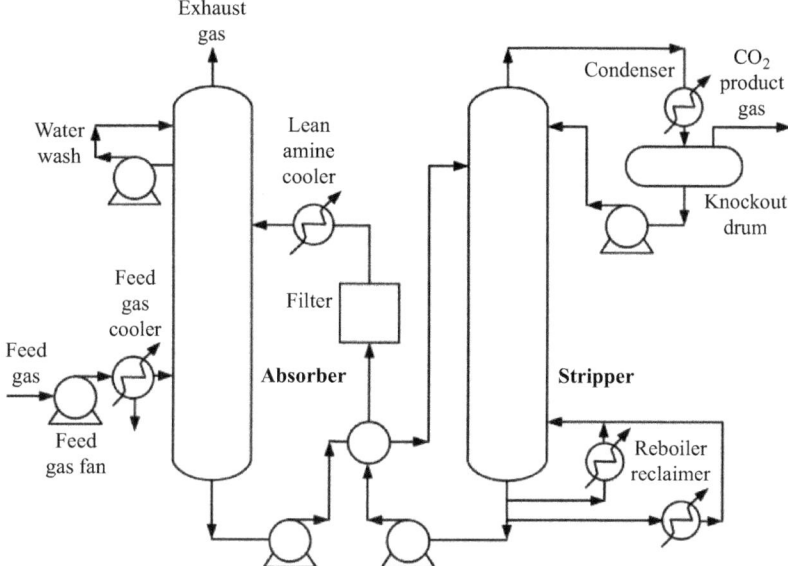

Figure 7.2 CO₂ recovery process (Kohl & Nielsen)

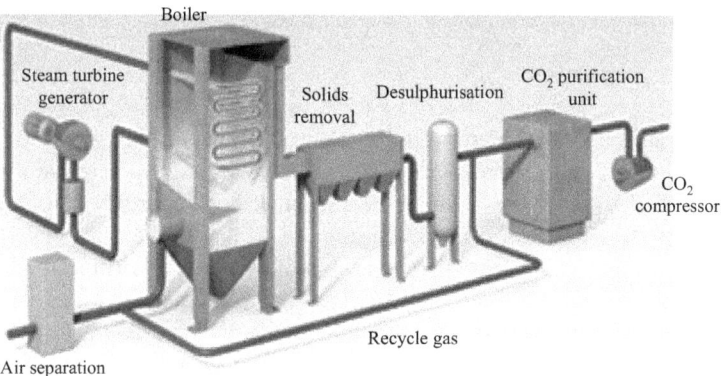

Figure 7.3 Oxy-fuel combustion system (source Costain)

Integrated gasification combined cycle (IGCC) – In this process, coal gasification takes place in the presence of a controlled shortage of air/oxygen, thus maintaining reducing conditions. The process is carried out in an enclosed pressurised reactor, and the product is a mixture of CO and H_2 (called synthesis gas, syngas or fuel gas). The resulting gas is cleaned and then burned with either oxygen or air, generating combustion products at high temperature and pressure. The sulphur present mainly forms H_2S but there is also a little COS. The H_2S can be more readily removed than SO_2. Although no NOx is formed during gasification, some is

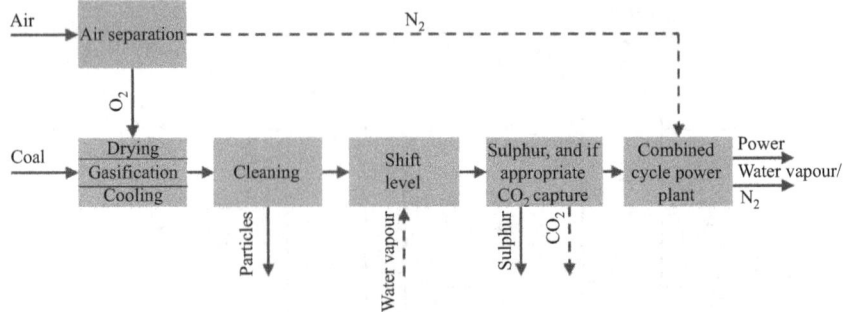

Figure 7.4   Integrated gasification combined cycle (source RWE)

formed when the fuel gas or syngas is subsequently burned. Three gasifier formats are possible, with fixed beds (not normally used for power generation), fluidised beds and entrained flow. Fixed bed units use only lump coal, fluidised bed units use a feed of 3–6 mm in size and entrained flow gasifiers use a pulverised feed, similar to that used in PCC. IGCC plants can be configured to facilitate CO_2 capture. The new gas is quenched and cleaned. The syngas is shifted using steam to convert CO to CO_2, which is then separated for possible long-term sequestration as illustrated in Figure 7.4.

7.3   CCS economics

CCS generation can make a valuable contribution to the generation mix because its output is controllable, and it can be used to help balance the intermittency of a lot of renewable generation. It is also largely carbon neutral and can be used in baseload mode. At times of peaks, if there is a shortage of capacity, its net output can be increased by temporarily interrupting the CCS process. The inclusion of a tranche of CCS generation reduces the system overall capacity needs and the need for increasing amounts of network extension to connect remote wind sites. The CCS can be applied to gas- or coal-fired generation and also biomass when the overall process results in a net absorption of CO_2. The technology used is not new but does need to be scaled up to meet the needs of large-scale generation. The requirement to transport and store the CO_2 will require careful planning to minimise the costs through the use of shared facilities including

- Identifying generating complexes in close proximity to each other that can be expected to remain in merit through the life of the scheme;
- Establishing pipe routes for the CO_2 that can be shared with other generation sources to realise full annual utilisation;
- Identifying storage site options associated with clusters of partially depleted oil/gas fields that can be used through the life of the scheme.

Table 7.1 Typical medium-term CCS costs

Type	Cost, £/kW	Eff. HHV	Eff. LHV
CCGT post-combustion	1,400	45	50
CCGT retro post-combustion	1,000	45	49.8
CCGT pre-combustion	1,600	37	41
CCBT oxy combustion	1,630	35	39
Coal post-combustion	3,000	32	33
Coal retro post-combustion	1,860	32	32.8
Coal oxy combustion	2,350	33	3.5
Coal IGCC	3,400	32	32

These requirements will have a significant influence on the cost of new generation and where it may be best located. Initial costs are likely to be high to establish the basic transmission infrastructure and are likely to need to be established by the regulated transmission sector that has experience of managing this type of infrastructure. A longer term view would need to be taken of the initial cost recovery by the regulator.

Typical plant costs and efficiencies for CCS generation expected in the medium term are shown in Table 7.1.

For initial build stations, a capital cost of £1,715/kW has been assumed for gas-fired stations and £2,538/kW for coal stations in constructing Table 7.2. This shows an estimate of the expected costs/MW h for gas and coal stations equipped with CCS. It includes a cost for CO_2 transmission and storage of £20/t. This can be converted to an added cost/MW h by calculating the emissions/MW h taking account of the efficiency. The added cost is equivalent to a capital cost of £550 m or £52m/year over 25 years for a 400-MW coal-fired station or 42% of this for gas because of the lower level of emissions. Net efficiencies are written down by around 20% from normal values to take account of the energy needs of the process. Based on future fuel price estimates of 78p/therm for gas and £72.8/t for coal, **the costs/MW h including transmission and storage are around £97.5/MW h for gas and £86.5/MW h for coal** and similar to the contracts for new nuclear. Given shale gas developments and reduced demand for coal, it is possible that fuel prices will remain depressed. Based on lower fuel prices of 50p/therm for gas and £50/t for coal unit costs would be between £75 and £80/MW h and very competitive with other low-carbon alternatives. The financial assessment is significantly improved if the station makes use of low-cost local coal or if the CO_2 is used to realise EOR. This requires that the station is located near to oilfields and is expected to run in merit.

The introduction of a tranche CCS generation can potentially have a significant impact on overall costs and end-user bills. Table 7.3 shows the plant mix for three scenarios with increasing levels of CCS gas and coal generation with the main changes highlighted. These scenarios have been simulated using a detailed half hourly dispatch simulation based on recorded demand and wind data extrapolated to the year 2030. The plant parameters and fuel prices are as shown in Table 7.2.

Table 7.2   Medium-term CCS costs

	Item	Units	Gas/CCS	Coal/CCS
Plant	Capacity	MW	393	398
	Capacity net	MW	377	366
	Load factor	%	85	85
	Annual output	TW h	2.81	2.73
	Construction time	Years	2	4
	Project life	Years	20	20
Fuel	Efficiency	%	45.1	35.2
	Coal cost delivered	£/t		72.8
	Coal calorific value	GJ/t		25
	Natural gas cost	£/mmBTU	7.8	
	Natural gas cost delivered	Pence/therm	78	
	N Gas calorific value	kW h/therm	29.32	
	N Gas grossCV/netCV		1.11	
Finance	Cost of dept.	%	5	5
	Cost of equity	%	8.8	8.8
	Inflation	%	2.4	2.4
	Debt/equity split		0.8	0.8
	WACC ($D/E = 0.8$)	%	7.1	7.1
Cap costs	Site and dev	£m	24.3	35.5
	EPC contract	£m	621	919
	Electrical conn	£m	0.733	0.733
	Gas connection	£m	2.466	
	Spares	£m	14.33	31.86
	Interest during const.	£m	11.26	23.20
	Total investment cost	£m	674.1	1,010.3
	Capital cost	£/kW	1,715	2,538
	Capital cost pa	£/m	64.08	96.03
O&M	Consumables	Pence/kW h	0.02	0.113
	Labour rate	£/h	16.66	16.66
	No. of operators		6	14
	Operators labour*1.28	£M	1.12	2.62
	Maintenance material	£M	0.81	2
	Main and support labour	£M	1.25	2.33
Summary	Capital cost	£/MW h	22.81	35.22
	Fixed O&M	£/MW h	0.60	2.09
	Var O&M	£/MW h	0.73	1.59
	Fuel costs	£/MW h	65.48	29.78
	Total generation costs	**£/MW h**	**89.62**	**68.68**
	Include carbon	tCO_2/GJ	0.0495	0.0869
	CCS CO_2 transport/store	£/tCO_2	20	20
	Added cost	£/MW h	7.90	17.78
	Total cost	£/MW h	**97.52**	**86.46**

Table 7.3   Impact of higher proportions of CCS generation

Scenario Technology	2030 Green MW	Fast CCS 1 MW	Fast CCS 2 MW
AGR	1,200	1,200	1,200
PWR	11,510	11,510	11,510
BIT	1,987	1,987	1,987
CCGT	29,469	14,935	11,929.97
OCGT	5,579	12,375	12,375
CHP	4,982	4,982	4,982
OIL	782	782	782
CCS gas	2,000	9,952	12,957.03
CCS coal	2,588	5,518	5,518
Interconnectors	7,600	7,600	7,600
Offshore wind	35,956	12,600	12,600
Onshore wind	20,985	7,400	7,400
Solar PV	15,813	6,106	6,106
Biomass	6,736	4,806	4,806
ROR	2,176	2,176	2,176
PMST	3,356	3,356	3,356
Marine	854	854	854
Total, MW	153,573	108,139	108,139
Maximum demand	59,812.05		
Less wind/PV, MW	72,365	73,579	73,579
Costs, £b/year	41	34	34
Emissions, mtCO$_2$/year	22	29	22

To meet emission reduction targets of 22 mtCO$_2$ based on extending wind generation with a capacity to 40 GW of offshore wind and 21 GW on shore results in an annual cost £41bn based on current subsidy levels. **The same emission target can be met with 12.6 GW of offshore wind and 7.4 GW of on shore with 13 GW of gas-fired stations with CCS and 5.5 GW of coal-fired stations with CCS at a total annual cost of £34bn i.e. a saving of £7bn/year i.e. 17% less on consumer bills.** The middle scenario shows the impact of slower deployment of CCS that allows more emissions and does not meet targets.

7.4   Operational schemes

The Global CCS Institute website reports 22 large-scale CSS installations in operation or construction with the capacity to capture 42 mtCO$_2$/year. In 2017, there were 5 projects reported to be in construction, 7 planned and 11 at the inception stage. These embraced a wide range of industries and captured techniques and storage arrangements. The Illinois industrial CCS scheme is based on a bio-energy scheme with storage in a deep saline formation; the Petra Nova scheme in Texas is based on power station post-combustion capture; a scheme in Abu Dhabi is based on a steelworks using the CO_2 to achieve EOR; the Tomakemai project captures CO_2 from a hydrogen production process; the Kemper county project is based on coal gasification and Gorgon in Australia is based on a Liquid natural Gas (LNG) project. A lot of the early projects incorporated EOR with others based on natural gas processing and hydrogen production. More recent schemes are based on power production.

The W.A. Parish power plant is one of the largest generation facilities in the United States with four coal-fired units providing 2,475 MW (gross) and six gas-fired units with a total capacity of 1,270 MW (gross). The coal-fired units use over 30,000 t/day of coal, sourced from Wyoming's Powder River Basin. The plant is located in rural Fort Bend County within the incorporated area of the town of Thompsons, Texas (southwest of Houston).

The project retrofitted Unit 8 of the W.A. Parish plant with a post-combustion CO_2 capture system constructed within the existing site. Petra Nova is the world's largest post-combustion CO_2 capture system in operation.

The capture facility uses an advanced amine-based absorption technology to capture at least 90% (reported as approximately 1.4 Mtpa) of CO_2 from a 240-MW flue-gas slipstream of the 610 MW (net) pulverised coal-fired generating unit.

Typical coal plant emissions are 0.8 t/MW h
Assuming 90% utilisation emissions/year/MW $= 0.8*8,760 \ *0.9 = 6,307$ t/ year/MW
For 240 MW, with a 90% extraction rate capture/year $= 1,362,355$ t/year

This calculation gives a figure close to the reported figure of 1.4 Mtpa.

A new 75 MW (net) gas-fired cogeneration plant commenced operation in June 2013 to provide peaking power for the electricity grid in Texas. The cogeneration facility provides steam and power to operate the capture plant with any remaining MW h of electric energy sold into the grid.

The purity of the CO_2 put into the pipeline for transport is designed to be greater than 99%. The captured CO_2 is transported via a new 132 km/82 mi long, 12-in. diameter underground pipeline to the West Ranch oilfield, located near the city of Vanderbilt in Jackson County, Texas. The pipeline route includes mostly rural and sparsely developed agricultural lands. Texas Coastal Ventures (TCV) owns and operates the pipeline.

The transported CO_2 is injected into the 98-A, 41-A, Glasscock and Greta sand units of the Frio Formation, approximately 1,640–2,066 m (5,000–6,300 ft) below ground surface and where oil has been produced since 1938. The portions of the West Ranch oilfield in which EOR operations are conducted are currently owned or leased by TCVs. Hilcorp Energy Company, a partner in TCV, is conducting the EOR operations. Nine injection wells and 16 production wells are being used initially for EOR operations. **As many as 130 injection wells and 130 production wells could be used over the 20-year span of the project.**

TCV, in cooperation with the Texas Bureau of Economic Geology, has developed a CO_2 monitoring plan to track the injection and migration of CO_2

within the geologic formations at the EOR site. In addition, to satisfying the monitoring requirements of the Clean Coal Power Initiative (CCPI) (under which the project received federal funding), the CO_2 monitoring programme is designed to satisfy the sampling and testing requirements of the Railroad Commission of Texas certification program for tax exemptions related to use of CO_2 for EOR and use of CO_2 from anthropogenic sources.

In March 2009 NRG Energy's W.A. Parish CCS Project was selected to receive up to US$154 million under the third round of the CCPI, which includes funding from the American Recovery and Reinvestment Act.

In May 2013, the US Department of Energy (DOE) issued its National Environmental Policy Act Record of Decision supporting US$167 million in cost-shared funding for the project under DOEs CCPI program. The project started operation in January 2017.

The Boundary Dam project in Saskatchewan Canada is based on capture from power generation unit 3 with a gross capacity of 139 MW or 115 MW net of the capture process demand. The CO_2 is extracted using regenerative amine technology and pumped through a 41-mi pipeline to be used for EOR from the Weyburn oil facility. **The motivation is to continue to exploit the availability of cheap local lignite coal that is estimated to be economically recoverable for 300 years.** Coal is an important part of the province's energy supply accounting for around 35% of SaskPower's available generating capacity. This unit is reported to be capable of capturing 1.3 million tonnes of CO_2/year. Based on a fuel carbon content of 25,200 g/GJ and an efficiency of 30% with 95% utilisation, the station emissions that are captured can be calculated as follows:

$$tCO_2/GJ = (25,200 * 44/12)/1,000,000 = 0.0924 \ tCO_2/GJ$$
$$tCO_2/MW \ h = (0.0924 * 3,600/1,000)/0.30 = 1.108 \ tCO_2/MW \ h$$
$$tCO_2/year = 1.108 * 139 * 8,760 * 0.95 = 1.28 \ million \ tonnes \ CO_2/year$$

The calculated CO_2 capture rate is close to the reported figure of 1.3 Mt CO_2/year. This is claimed to be equivalent to the emissions from 300,000 cars during a year. Assuming in practical use 387 gCO_2/mi and a utilisation of 11,000 mi/year, the emissions are given by

$$300,000 \ cars * 387 \ g/m * 11,000 \ m/(1,000,000 * 1,000,000)$$
$$= 1.28 \ million \ tCO_2/year$$

It is claimed that the Sleipner CCS Project in Norway, which started in 1996, has prevented almost 17 Mt of CO_2 from entering the atmosphere. The facility was the first in the world to inject CO_2 into a dedicated geological storage setting. Natural gas produced at the Sleipner West field contains about 9.5% CO_2, which has to be removed to get the gas to saleable quality. Instead of venting the separated CO_2 to the atmosphere, where it would add to the greenhouse problem, Statoil, the operators of the field, and their partners decided to inject it down a 3-km-long well and store it in a porous and permeable reservoir rock called the Utsira Sand.

7.5   The case for CCS

The IEA believes, 'CCS is the only technology able to significantly reduce emissions from coal-fired power plants operating or under construction around the world. CCS is also one of the few technologies that can address emissions from industrial processes, including the production of steel, cement and chemicals'. Statistics for 2016 shows that over 50% of the world's energy use is derived from gas and coal with gas at 24% and coal at 28% with renewable accounting for just 3%. **Any reduction in emissions in the short term can only be achieved with CCS.** It would not be practical financially or politically to simply close mines and coal-fired stations. The Intergovernmental Panel on Climate Change (IPCC) have also recognised the importance of the combination of bio-energy and CCS (BECCS) in delivering future 'negative emissions'. The concept is that the bioenergy absorbs CO_2 through the growth cycle that would be released through combustion but instead is captured and stored leading to net negative emissions to the atmosphere. A key advantage of generation with CCS is that can provide flexibility and unlike intermittent renewable sources does not rely on priority dispatch and does not need backup.

However, despite a strong case, there appears to be a loss of momentum in progressing developments with some projects having recently been cancelled. The technology is available as has been demonstrated by the many schemes currently in operation. The director of the IEA believes that CCS will not be optional in implementing the Paris agreement, and faster deployment will be essential.[1] Analysis in Section 16.7 compares different plant mixes and shows that the total annual capital and operating costs of meeting electricity demand could be 17% cheaper with fast deployment of CCS than the option of relying on subsidised intermittent renewable sources with the same level of total emissions. In part, the resistance to progress CCS may be founded on

- a reluctance to retrofit CCS to generation when it will reduce its output and efficiency;
- the absence of the infrastructure to transmit and store the CO_2;
- the high capital cost involved;
- currently, low CO_2 emission costs;
- the regulatory environment.

Although subsidies and priority dispatch are afforded to intermittent renewable sources, governments are cancelling funding for demonstration CCS projects. A further complication is that deployment cuts across engineering sector skill bases

---

[1]20 years of CCS: accelerating Future Deployment.

where EOR is employed. A positive approach is needed to make this a viable option if world emission targets are to be realised.

7.6   Conclusions

Given that, some 50% of the world's emissions result from the use of gas and coal to provide energy, CCS offers a viable alternative to reduce emissions in the short-to-medium term. It will not be easy, financially or politically, to wean countries like China and India away from the use of coal with pit closures, putting miners out of work and closing stations without alternatives being available, particularly when there are estimated to be hundreds of years of cheap coal available that could be exploited. There is the option to retrofit CCS to recently built stations and begin to progress decarbonisation in the short term. It will require positive regulation to ensure that the reductions in station efficiency are more than compensated for by financial benefits from an emission reduction scheme. There will also be a need for support in the establishment of the shared CO_2 transport and storage infrastructure to minimise costs. The overall costs, including provision for CO_2 transmission and storage, are estimated at £97.6/MW h for gas and 86.5/MW h for coal stations based on increases in fuel prices of 50% from values in 2017 that may not materialise with consumption declining. These costs are of the same order as new contracts for large-scale nuclear and cheaper than offshore wind generation. The economics are improved if the CO_2 is pumped into partially depleted oilfields to realise EOR.

The CCS generation can make a valuable contribution to the generation mix because its output is controllable, and it can be used to help balance the output of intermittent renewable generation. The inclusion of a tranche of CCS generation reduces the overall system generation capacity needs and does not require backup like intermittent renewable sources. The CCS generation would take advantage of existing generation sites with transmission connection already established as opposed to the need for increasing amounts of transmission to remote wind sites. Modelling overall capital and production costs to meet demand for a system with a peak of 50 GW shows that annual costs are reduced by 17% with a tranche of CCS generation as opposed to one based on high proportions of subsidised intermittent wind generation.

The Global CCS Institute website reports 22 large-scale CCS installations in operation or construction with the capacity to capture 42 $MtCO_2$/year. The technology components are generally available and one large-scale plant in the USA is capturing 1.4 $MtCO_2$/year. The CO_2 is used for EOR with 130 injection wells and 130 production wells that could be used over the 20-year life of the project.

Question 7.1 Using the data shown in the table, calculate the cost of conventional gas-fired generation assuming an efficiency of 55%. At what CO_2 price would the costs of conventional generation equal that form gas-fired generation with CCS?

	Item	Units	Gas/CCS	Coal/CCS
Plant	Capacity	MW	393	398
	Capacity net	MW	377	366
	Load factor	%	85	85
	Annual output	TW h	2.81	2.73
	Construction time	Years	2	4
	Project life	Years	20	20
Fuel	Efficiency	%	45.1	35.2
	Coal cost delivered	£/t		72.8
	Coal calorific value	GJ/t		25
	Natural gas cost	£/mmBTU	7.8	
	Natural gas cost delivered	Pence/therm	78	
	N Gas calorific value	kW h/therm	29.32	
	N Gas grossCV/netCV		1.11	
Finance	Cost of dept.	%	5	5
	Cost of equity	%	8.8	8.8
	Inflation	%	2.4	2.4
	Debt/equity split		0.8	0.8
	WACC ($D/E = 0.8$)	%	7.1	7.1
Cap costs	Site and dev	£m	24.3	35.5
	EPC contract	£m	621	919
	Electrical conn	£m	0.733	0.733
	Gas connection	£m	2.466	
	Spares	£m	14.33	31.86
	Interest during const.	£m	11.26	23.20
	Total investment cost	£m	674.1	1,010.3
	Capital cost	£/kW	1,715	2,538
	Capital cost pa	£/m	64.08	96.03
O&M	Consumables	Pence/kWh	0.02	0.113
	Labour rate	£/h	16.66	16.66
	No. of operators		6	14
	Operators labour*1.28	£M	1.12	2.62
	Maintenance material	£M	0.81	2
	Main and support labour	£M	1.25	2.33
Summary	Capital cost	£/MW h	22.81	35.22
	Fixed O&M	£/MW h	0.60	2.09
	Var O&M	£/MW h	0.73	1.59
	Fuel costs	£/MW h	65.48	29.78
	Total generation costs	**£/MW h**	**89.62**	**68.68**
	Include carbon	tCO_2/GJ	0.0495	0.0869
	CCS CO_2 transport/store	£/tCO_2	20	20
	Added cost	£/MW h	7.90	17.78
	Total cost	£/MW h	**97.52**	**86.46**

Part III

How are the changes being managed within a market environment?

The third part discusses the market mechanisms in place and needed to manage the emerging developments. It discusses the balancing and capacity markets and the impact of renewable and embedded generation on conventional large-scale generators. It reviews the impact on network utilisation and charging mechanisms. It appraises the potential for cross-border trading, demand side management and exploiting interaction with the demand side and gas, heat and transport systems.

Chapter 8 – Wholesale markets – This chapter discusses how the performance of wholesale markets has been undermined by the introduction of increasing renewable generation capacity that receives priority in dispatch. It analyses the impact on the conventional generators of their reduced utilisation leading to premature closures and tight capacity margins. The market mechanism to match future capacity to a profiled demand is discussed. The single-buyer (SB) model is advocated as the preferred model to manage the changes sweeping through the industry.

Chapter 9 – Balancing markets – This chapter analyses the impact that renewable generation is having on the requirement for reserve in balancing supply and demand. It reviews the various sources of reserve and their costs for primary fast acting reserve based on lithium ion batteries and short-term operating reserve. The additional requirements to manage renewable intermittency are quantified, and the impact of competition on reserve prices in Germany is analysed. The impact of intermittency on interconnection flows is discussed.

Chapter 10 – Capacity markets – This chapter discusses the impact of increasing levels of government intervention in the market undermining investor confidence leading to potential capacity shortfalls and the need to introduce capacity payments. Capacity markets in the United States, Ireland and the United Kingdom are compared and their shortcomings reviewed. The potential contribution of renewable sources and interconnection to the provision of firm capacity is analysed. The regulatory action required to address the shortcomings in establishing the optimum level of future capacity is discussed.

Chapter 11 – Cross-border trading – This chapter discusses the target expansion of interconnection capacity in Europe to facilitate cross-border trading and price convergence. The approach to evaluation of proposed interconnection schemes is analysed in the short and long term. The analysis of flows across interconnected networks is analysed and the mechanisms used to support cross-border trading. The impact of the growth of intermittent generation on unplanned cross-border flows is discussed.

Chapter 12 – Demand side management – This chapter reviews the range of application of demand control schemes that have been applied to larger consumers. It reviews the potential to engage domestic consumers facilitated by smart meters. The cost of storage schemes to make them viable in a market environment is discussed. The potential role of EVs and heat pumps in smoothing domestic demand profiles is analysed. The structure of tariffs to realise the full value of flexibility is discussed.

Chapter 13 – Emissions and interaction with gas, heat and transport – The emissions from all regions of the world are analysed to illustrate their scale. It illustrates those regions based on the use of coal with increasing emissions and those that migrated to gas and renewable sources with decreasing emissions. The relative costs of using heat pumps to reduce the emissions from heating and EVs to reduce the emissions from transport are analysed. The relative costs of using heat pumps to reduce the heating demand and EVs to reduce the transport demand are analysed. The costs of combined heat and power schemes and district heating are analysed. The significant impact of decarbonisation of heat and transport on electricity demand is analysed together with the scope to profile it to minimise future generation and network capacity requirements.

Chapter 14 – Network issues and tariffs – The EU has pressed for ownership of the networks to be unbundled to establish truly independent SOs able to manage operation of the system without any prejudice. This chapter discusses the change in the role of distribution networks to becoming more active with an increasing amount of embedded generation affecting security. The demands are expected to increase with more EVs and heat pumps that will have less diversity than conventional loads. The current tariff structures are analysed to illustrate the market distortion in favour of embedded generation and how they may need to be changed.

Chapter 8

Wholesale markets

8.1 Introduction

The facility to trade is an essential component in realising a competitive market. In practice, this requires

- A sufficient number of trading partners to enable choice;
- A mechanism whereby current and future prices can be discovered;
- A degree of market liquidity that realises price interaction;
- The absence of dominant market power.

In practice, these requirements are less easily met. Attempts at restructuring have usually required the state generating companies to be split into three to five blocks. This was the case in countries like Italy and the United Kingdom, whereas in New Zealand, no company could own more than 5% of capacity. In practice, mergers and acquisitions have depleted the number of competing players with organisations like EdF, E.ON and RWE dominating across Europe. Other countries have chosen to retain national champions like ENEL in Italy, Iberdrola in Spain and Electrabel in Belgium on the grounds of national security. In the United Kingdom, the view of the regulator is that the market is dominated by the 'big six', and various attempts have been made to encourage new entry by placing a requirement on existing generators to contract generation capacity to them.

The mechanism for price discovery relies on exchanges publishing data based on actual transactions through a period. Reporting agencies also canvas data on prices and provide regular bulletins of typical contract prices. Economic forecasting houses will also publish their views on future prices. It is not always clear what information is being reported, and there are often significant discrepancies between reporting sources. This may be due to the period covered with some exchanges not including weekends or reporting volume as opposed to time-weighted data. **In the past few years, prices have become more volatile due to the intermittency of renewable sources not correlated with normal demand patterns.**

Liquidity can take several years to establish and in the United Kingdom took from 1990 to 2002 to realise gas market liquidity. It does take time but the transition can also be impeded by blocking, market power and ineffective regulation. A restricted number of trading hubs clearly benefits good levels of liquidity.

In Europe, only the UK National Balancing Point has meaningful liquidity for gas trading.

New entry will be inhibited if a large indigenous player is able to exercise market power. It can often happen that a single player controls a lot of the plant operating at the margin and setting prices. Referrals to the UK Monopolies and Mergers Commission have not identified collusion but promoted developments to facilitate customer switching to promote competition. A vertically integrated player is less sensitive to wholesale prices as sales are internal from generation to the supply division with wholesale prices as the transfer price but not affecting the company's bottom line.

In order to promote the development of renewable and low carbon emission generation, governments have introduced subsidies and priority in dispatch and levies on fossil fuels that have undermined the operation of wholesale markets. They have also had a dramatic effect on the financial position of some of the generation companies. This chapter reviews some of the developments in energy policy and wholesale markets that are taking place to meet the new requirements resulting from the introduction of renewable sources.

8.2   Market challenges

The government objectives are to meet the trilemma of maintaining affordability and economic competitiveness, energy security and sustainability within a market framework. However, the approaches taken are leading to a number of drivers that undermine the performance of centralised wholesale markets by introducing bias including

- The subsidises paid to renewable generation sources and their priority in dispatch in many jurisdictions;
- The tariff bias in favour of small-scale local renewable generation with contractual terms that reduce supplier network charges;
- The introduction of taxes on fossil fuel burn to discourage its use;
- The encouragement of more efficient use of energy through subsidies;
- Government underwriting contract strike prices for new nuclear;
- Encouraging more end-user engagement facilitating tariff comparisons and switching suppliers.

These measures have had a major impact on the existing generating companies and increasingly on network operators. **As the renewable capacity has increased, it has reduced the utilisation of conventional generation whilst at the same time it is called upon to regulate to balance changes in wind and solar output**. This has reduced the efficiency of unit operation and their financial viability causing early closures. The intermittent nature of wind and solar results in net demand uncorrelated with the inherent profile of the demand. In the case of solar, its output is high during sunny summer days at times of low heating demand in the northern hemisphere. The growth in embedded generation is largely fuelled by gaps in the charging arrangements for use of the network. It operates outside the wholesale market but impacts on the demand particularly at peak times when wholesale prices would be expected to rise. The priority dispatch afforded to renewable sources completely distorts the normal market demand/price relationship and undermines the ability to discover and forecast prices.

The EU identified four serious shortcomings in the operation of the European gas and power markets:

- Excessive market concentration in most countries;
- A lack of liquidity preventing successful new entry;
- Limited integration between markets;
- The absence of data transparency undermining price discovery.

The EU plans to use its full powers, under competition rules, to improve the situation by encouraging the development of interconnection. This process starts with the formation of closely coupled regions that are in turn expected to be coupled to other regions as interconnection is extended. In the liberalised world, funding new interconnection has to be shown to be economic based on expected revenues from leasing capacity on the link. This requires a sustained price differential through the project life. A fundamental difficulty is that this differential may be undermined by another market participant choosing to build generation in the high-cost country offsetting the price differential. This is a feature of uncoordinated development that necessitates adding in a risk element that weakens the economic case and delays new build. The EU enlists support from experts from different countries to review network development proposals and assess their worth in enhancing competition and reducing end-user costs.

8.3   Renewable support schemes

Several mechanisms that have been used to subsidise the development of wind farms and other renewable sources include

- Feed in tariffs (FITs) where the generator receives a fixed payment for energy delivered;
- Contract for differences where a strike price is agreed that payments are linked to;
- The UK Renewable Obligation (RO) scheme where the generator receives certificates that are sold to suppliers who have an obligation to source a percentage of their energy from renewable sources;
- Carbon Tax Emissions schemes where a charge applies to the fuel according to its carbon content;
- A climate change levy (CCL) on energy supplied to non-domestic consumers form non-renewable sources.

FIT schemes are widely used as a simple mechanism to encourage the development of renewable sources. They provide certainty to investors on returns and encourage maximising the output. They are usually linked with priority dispatch. As the capital costs have reduced, the length of some of the new contracts has been reduced in some countries. The FIT scheme may be linked to a contract for differences where a contract strike price is agreed that is linked to general market prices. If the market price is below the strike price, then payments are made up to the strike price, conversely if market prices are above the strike price, then the generator pays the surplus to the counterparty.

To facilitate abatement at least cost, the EU through Directive 2003/87/EC laid the foundation for the European Trading Scheme (ETS) that required each member to translate the requirements into their national legislation. The first phase of the scheme came into force in January 2005 running to the end of 2007 and covered CO_2 only. The second phase from 2008 to 2012 covered all the greenhouse gases. For each period referred to in the article, each member state is required to develop a national plan stating the total quantity of allowances that it intended to allocate for that period and how it proposes to allocate them. The distribution is established by industrial sector and subsequently for particular installations. Some 93.7% of the expected emissions were allocated in the United Kingdom with 6.3% reserved for new entrants. The amount reserved varied from country to country depending on the likely level of new entry. For the 5-year period beginning 1 January 2008, and for each subsequent 5-year period, each member state has to decide upon the total quantity of allowances it will allocate for that period and initiate the process for the allocation of those allowances to the operator of each installation. The EU scheme covers all installations with an output >20 MW thermal.

Until recently, the RO scheme was the main incentive scheme in the United Kingdom for direct subsidisation of renewable generation and requires electricity suppliers to submit a specified number of certificates based on a percentage of required renewable generation (2016: 29%) within their portfolio or pay a buyout price to fulfil their obligation (£44.33/MWh). Suppliers buy certificates from registered renewable generators who receive this payment in addition to their market price. If annual renewable targets are not met, then penalties derived from those that do not meet their commitments are redistributed to increase the effective value of each ROC. The RO scheme therefore provides a form of financial subsidisation for renewable generators to incentivise their market entry.

An advantage of the ROC scheme was that it provided a liquid trading market for renewable generators with financial certainty in terms of receipt of certificates provided once registered for the scheme. The publication of transparent annual Renewable Energy Supply (RES) percentage targets and buyout price indexes also ensures that participants in the market have some certainty in terms of their outturn financial position. A disadvantage of this scheme, however, is that it is over weighted towards guaranteeing renewable generation growth without adequate measures in place to protect and ensure consumers are paying least cost for the delivery of renewable objectives. This issue was highlighted by Ofgem in January 2007 where they concluded that the ROC scheme was a 'very costly way' of supporting renewable electricity generation. As a result, the ROC scheme is currently being phased out and replaced with a new Contract for Differences (CFD) scheme under the energy market reform process that is based on a feed in tariff linked to market wholesale prices.

In addition, the United Kingdom also operates a CCL on energy supplied to non-domestic customers from non-renewable sources. Although the tax is paid by the customers, the tax was until recently offset by suppliers using a levy exemption certificate issued to renewable generators. The advantage of the CCL scheme was that as a demand side environmental tax, it provided a clear financial incentive to consumers to purchase from renewable suppliers. This demand side signal impacts the supply side by incentivising suppliers to source renewable generation portfolios in order to ensure customer retention and growth. A disadvantage of the CCL scheme is that it is an indirect incentive scheme that does not directly provide or deliver any new renewable generation in the United Kingdom per se. It simply incentivises growth in support of renewable companies registered in the United Kingdom who may or may not directly invest in new renewable generation. This issue formed part of the 2015 decision by the UK government to remove the exemption status of renewable generation from the levy.

8.4   Small-scale subsidies

To encourage the development of renewable generation at all levels, subsidies have been established that embrace domestic applications like solar as shown in Table 8.1. The arrangement is established with the local energy supplier. Approved small-scale schemes of less than 10 kW like PV receive a payment of 3.45 p/kW h for renewable electricity generated, a 'generation tariff'. They can also receive payments from their supplier for surplus electricity exported to the grid. The additional costs to the supplier are passed onto all consumers through their tariffs. The terms have been scaled back in the United Kingdom as the uptake has increased and equipment costs have dropped.

It can be seen that the rate for PV schemes drops for larger schemes >1 MW down to 0.33 p/kW h reflecting the fall in basic solar equipment costs. Similarly,

Table 8.1   Subsidies for small-scale renewable generation (source Ofgem)

1 January–31 March 2018	p/kW h	Plant type	p/kW h
Hydro <100 kW	7.58	Wind <50 kW	7.99
Hydro 100–500 kW	6.09	Wind 50–100 kW	5.82
Hydro 500 kW–2 MW	6.09	Wind 100 kW–1.5 MW	3.82
Hydro >2 MW	4.43	Wind >1.5 MW	0.77
Solar PV <10 kW	3.45	Anaerobic digestion <250 kW	6.65
Solar PV 10–50 kW	3.65	Anaerobic digestion 250–500 kW	6.14
Solar PV 50–250 kW	1.6	Anaerobic digestion >500 kW	6.33
Solar PV 250 kW–1 MW	1.44	CHP <2 kW	13.45
Solar PV >1 MW	0.33	Export tariff	4.91

for wind installations, the energy prices drop to 0.77 p/kW h (£7.7/MW h) for schemes >1.5 MW.

8.5   Market performance improvement

Prices may not be considered competitive either because of structural or operational reasons. It may be that there is inadequate liquidity as a result of an insufficient number of players competing at the margin. Alternatively, it may be that the market structure is too complex or that constraints are fragmenting the market. Some measures that may improve performance include

- Invoke anticompetitive legislation requiring the sale of generation to add new players;
- Opt for a SB market model with competition realised through tenders and auctions for generation;
- Simplify the market arrangements and products to enable wider participation and increased liquidity through more interconnection as in Europe;
- Adopt measures to enable more demand side participation enabled by the deployment of smart meters.

Some of the criticisms of the original UK pool were that it lacked liquidity with too much market power in the hands of two major players such that it was sometimes described as a duopoly. There was also ineffective demand side participation to provide a constraint on price rises. The process was also considered to be complex and lacking transparency. This meant that participants could not independently make judgements about outturn to support their bidding analysis. The algorithms employed were designed for operational rather than market use and included a lot of detail to model generation dynamics. The generators for their part resented their plant being dispatch centrally and believed that they knew best how to manage their generation to reduce costs. An open forum was established to debate and agree new arrangements under the auspices of the regulators office.

It was decided that the mandatory pool should be abandoned in favour of a much freer trading arrangement with less central control. When the pool was in place, bilateral contracting was affected through the use of contracts for differences. Part of the new arrangements formalised this arrangement by enabling bilateral contracting directly between generators and suppliers. This enabled more demand side participation and gave the generators the right to dispatch their own generation to meet their contracted demand. The only mandatory requirement was to advise the system operator of the contracted position prior to the time before the event referred to as 'gate closure'. The original arrangement set this at some 5 h ahead but this was progressively reduced to 1 h ahead as experience developed. The notifications were referred to as initial and final physical notifications in that the participants were required to indicate the physical location of injections and associated demand take-offs to enable load flow studies to check system security. This multi-market arrangement is the most popular model applied across Europe.

The overall philosophy was to place risk where it could be best managed. So it meant generators had to manage their own plant dynamic constraints and ensure capacity was available to meet their commitments. Equally, the demand side managed their volume risk in contracting for sufficient capacity to meet their future needs.

The new arrangements also enabled the establishment of exchanges to support price discovery and shorter term trading to enable participants to adjust their positions. Any organisation was able to establish an exchange and initially attracted interest from existing exchanges like the International Petroleum Exchange.

A key element of the new arrangements was the establishment of a balancing market. This was essential to enable the system operator to balance supply and demand in the event and maintain stable system frequency. The balancing market was operated by the system operator who accepted bids and offers to increase or decrease output or demand from their contracted position.

The priority dispatch afforded to renewable intermittent generation sources has undermined the utilisation of marginal generation to the point where it is non-viable financially. **This has resulted in plant margins falling to unacceptably low levels** necessitating new measures to retain the availability of the marginal plant for times when there is no wind/solar. Options include

- The introduction of payments for the provision of capacity/availability. This will encourage generators to minimise the downtime of units and create incentives to retain older plant in service;
- The introduction of loss of load probability (LOLP)-type increments to marginal prices that react to plant shortages in the event to encourage maximum output;
- Enable the SO to contract directly to secure capacity;
- Ensure open access to the system for new entrants.

Several countries have chosen to introduce capacity payments for generation administered through an auction process. Renewable intermittent sources do not provide firm capacity, and their potential contribution estimated on a probability basis may be as low as 10% of the installed capacity. As the capacity of intermittent renewable sources increase, its priority dispatch causes price volatility that undermines the functioning of the wholesale market. Consideration is necessary to identify market arrangements more suitable to the changing environment at both the transmission and distribution level and their interface.

8.6 Single-buyer models

The SB market model has been adopted by a number of countries that see establishing competition in generation as the priority with less benefit derived from introducing supplier side competition. Counties in the Middle East, South Africa and Ireland have opted for this model. The Middle East markets are characterised by the availability of low priced gas and a significant desalination load as in Oman.

The models accommodate generators both with and without a power purchase agreement (PPA). They bid into a mandatory centralised pool a day ahead with half-hourly pricing. The bids are complex and include dynamic parameters like start-up costs, minimum load costs with an incremental cost curve. The pricing is based on an ex post calculation to determine the unconstrained system marginal price (SMP). Constrained on generators get payments but those constrained off do not. Capacity payments are added to the SMP and adjusted to take account of availability margins and benchmark capacity costs. The purchase of fuel is undertaken by the SB and supplied to the stations. The PPA includes for fuel supply with payments based on agreed heat rate models defining how the plant should perform under different operating conditions and loads. The operator of the plant then benefits if better efficiency is realised in practice through improved operation. Payments for availability are based on the proportion of time the plant is available to provide an incentive to maintain capacity.

A power and water procurement company acts as the SB and contracts with generators for capacity and output in accordance with the terms of the power or water contracts. The distribution companies buy power from the SB against a cost-reflective bulk supply tariff (BST) that includes the costs of power procurement activities. The water is sold to the water department at a cost-reflective BST that also includes the costs of water procurement activities. The market design does not explicitly include exports/imports but these are netted off demand by the SB. The provision of ancillary services is covered by the PPAs. True fuel prices are used in the calculation of payments in accordance with the terms of the PPAs with market power mitigation to ensure accurate reflection of marginal costs. Both the generators and distribution companies pay the transmission company for use of the system.

The SB is exposed to risks in setting the BST for the distribution companies a year ahead. If the expected availabilities of gas-fired stations is not realised, then replacement by expensive diesel generation could lead to significant budget shortfalls. The system operator is responsible for scheduling and dispatching generation taking account of the terms of the PPAs. Some of the plant configurations are complex, and station dispatch by operators in real time may not always reflect the price of the actual running arrangements. This can lead to a station being dispatched out of merit leading to disputes about payments to generators. This may be the result of staffing not being adequate to support optimal commercial operation or that the control room facilities are not suitable. The arrangements and procedures may need to be reviewed and recommendations for improvements proposed.

There is scope for development of SB market models including the implementation of improved scheduling and dispatch facilities to move closer to optimum operation, enabling participation in the market by the demand side in line with developments in other countries, the development of a balancing market and the provision of improved system monitoring facilities to mitigate market abuse. **A key advantage with the SB model is its ability to influence the developing plant mix to match the developing demand profile at least cost**. This is important during these times of rapid change resulting from the introduction of emission constraints and subsidised intermittent renewable generation sources.

8.7 A model for 2030

There are a number of key changes taking place in the current energy environment, which necessitate a rethink of the arrangements for realising competition while steering and coordinating developments to meet the trilemma requirements for low emissions, sustainability and low costs to consumers. Participation is being encouraged by an increasing number of small players, down to individual domestic consumers with a very narrow focus. The power system operates as a signal entity in real time, and its development and operation has to be coordinated if the trilemma requirements are to be realised. Flexibility will be a key ingredient to both accommodate intermittent renewable sources and to profile demand to minimise the required generation capacity. For the full benefits of flexibility to be realised, it must influence capacity requirements in generation planning timescales. Rather than the SO defining the generation capacity requirement, this would be determined by a two-sided forward capacity market. **The demand side would need to contract forward for capacity taking account of its capability to smooth overall demand using DSR and embedded generation.** The contracted volumes would be subject to refinement closer to the event when both sides would refine their positions.

In operational timescales, the basic requirement, in the emerging situation, is to schedule and dispatch both generation and demands to balance in the event at least cost. The generation and demand side will each have their own price/volume curve based on the prices generators are prepared to sell at and buyers are prepared to pay. The object is to schedule generation and demand at the point the curves intersect. **This is essentially the basis of the successful Nord Pool model.** What this model does not do is create an environment to support investment analysis in the new dynamic circumstances. In the last decade, the demand and its profile were known, and alternative generation sources could be compared to meet defined operating roles. In the emerging situation, the demand side is also subject to scheduling and dispatch with neither its level nor profile known. The problem is compounded by the output from intermittent renewable sources being impressed on the system and altering the generation side volume cost curve. There is also a disparity in timescales with large-scale generation requiring 5 years or more lead time as with nuclear, whereas the demand side is smaller and more flexible, and participation in demand management can be affected quickly through a control infrastructure. A problem in predicting the demand side volume price/curve will be the rate of deployment of EVs and heat pumps to decarbonise transport and heat. These loads are subject to volume constraints and can be expected to vary through the day.

To support investment timescale analysis and risk management, there is a need for both the generation and demand side to provide indicative forward price volume curves into a medium-term market. The submissions would be in the form of volume/price curves for generation and demand. The SO should also share its

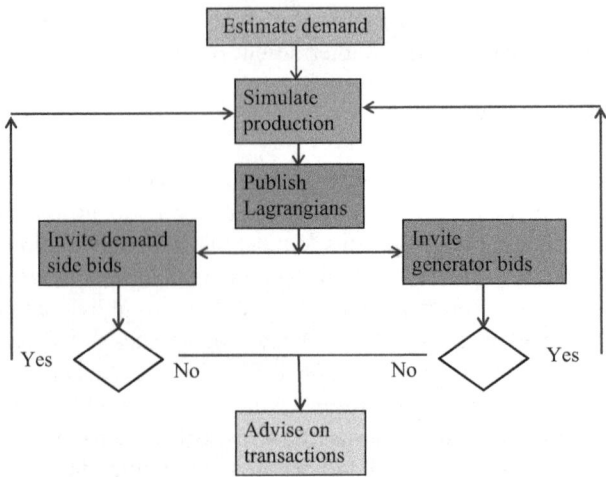

Figure 8.1   Future capacity market process

estimates of demand and renewable generation output to enable both the generation and demand side to plan operations.

It can be shown theoretically how the idealised requirements can be met through the process illustrated in Figure 8.1. The system objective function is to establish that multiplier 'lambda' that results in total generation offers G and demand bids D for energy equating and capacity bids equating to the security level 'beta' required by customers i.e. find:

$f(\lambda_t)$ so that $G_t = D_t$
$f(\beta_t)$ so that $A_t\_D_t \geq f(\mathrm{LOLP})$

Individual generators will seek to maximise their profits, the difference between income and costs which may be assessed independently making use of the Lagrangian multipliers to calculate energy and availability payments.

$$P = \sum \lambda_t * g_{i,t} + \sum \beta_t * A_{i,t} - \sum \left[g_{i,t} * vc_i + \mathrm{STC}_i \right]$$

where g is the individual generator output, A its availability and

$$\beta_t = \mathrm{LOLP}(\mathrm{VLL\_SMP})$$

The demand side will respond to the multipliers so as to minimise their costs.

$$\sum (\lambda_t * d_t + \beta_t * d_t)$$

The overall process will be to adjust the multipliers for each period so as to realise convergence. The sub-gradient method is a technique that may be used where

$$\lambda_t = \lambda_{(n\_1)t} + \alpha(R_t\_A_t)$$

where R = required and A = actual, a and b are constants and

$$a = \frac{1}{a + k * b}.$$

To enable decisions to be made in planning timescales, the figures would need to be published 1 to 5 years ahead. The process would align with capacity markets but augmented by indications of the range of flexibility and the associated costs for generators and the demand side.

The forward market process would be administered by the SB/seller providing the link between the two sectors. The process would involve the system operator to ensure that the developing system can be securely managed and to feed into the process any transmission constraints.

8.8 Availability payments

Several methods have been employed to secure the availability of generation as an alternative to up-front capacity payments. The UK original pool model included an increment to the SMP based on the LOLP and the value of lost load (VLL) to derive the pool purchase price (PPP) i.e.

$$PPP = SMP + LOLP * (VLL - SMP)$$

The LOLP was determined based on the probability of their being insufficient generation to meet the expected demand in any one half-hour. The VLL was set for the year ahead based on the cost to consumers of the loss of supply. In effect, consumers were paying a premium based on the probability of the occurrence of a loss. The advantage of the approach was that the increment adjusted automatically to the prevailing conditions and payments by end users would decrease in the presence of excess capacity. Because the extra payments were only made during periods of shortfall, they would not provide sufficient incentive for new investment.

An alternative approach is to make explicit availability payments to new entry generation for each MW h generated to support financing the investment. The levels of availability payments can be estimated by comparison with the option of the addition of more capacity. If a system has an installed capacity of 4,500 MW with a maximum demand of 4,000 MW and a total annual energy demand of 21,000 GW h, then the cost of improving the plant margin to 15% can be calculated:

- The initial plant margin is $100 * (4,500 - 4,000)/4,000 = 12.5\%$
- To raise it to 15% would require extra generation capacity of x so that

$$100 * (4,500 + x - 4,000)/4,000 = 15$$
$$\text{i.e. } x = 15 * 4,000/100 + 4,000 - 4,500 = 100 \text{ MW}$$

Assuming that the annual capital cost of 100 MW of new generation is £6m, this would be a cost to investors. This can be translated into a new generation hourly availability payment that would equate to the added capital cost of new generation to raise the plant margin by 2.5%. Assuming full availability, the energy payment premium to cover the annual capital cost is given by

$$\text{Availability premium} = £6m/(8{,}760)/100 = £6.8/\text{MW h}$$

Availability payments can be included as part of a power purchase contract arrangement and can be tailored and profiled to match the benefit to consumers of avoided loss of supply. It is left to the sponsors and investors to decide when and if to build new generation capacity. The arrangement is more flexible than a fixed annual capacity payment and gives the supplier and generator options to alter configurations.

When contracting for capacity via PPAs, there is a need to determine how much capacity is worthwhile. **The optimum level occurs when the value of the potential lost load based on the VLL equates to the annual costs of adding new generation.** Figure 8.2 is based on a probabilistic assessment of plant availability and demand level. It shows an example for a system with a peak demand of 60 GW supplied by a number of 500-MW units each with a 90% probability of being available. It shows the benefit in £m/year due to the reduction in the VLL from the addition of one extra 500-MW unit. The addition of extra units reaches a break-even point where the reduction in VLL matches the annual capital costs of a new unit priced at £26m/year, in this example, this occurs in going from 144 to 145 units

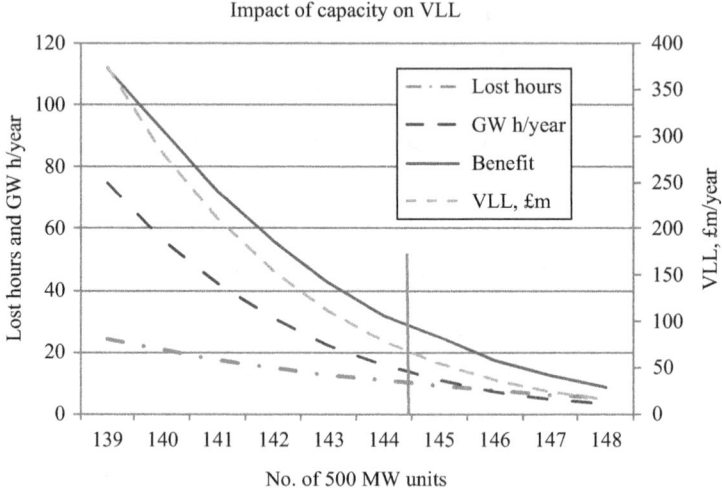

Figure 8.2   Capacity and VLL

when the slope of the VLL equals £26m/unit. The graph also shows the number of hours in which capacity shortfalls could occur and the total VLL.

8.9   Renewable generation impact

The introduction of a tranche of renewable generation impacts on the utilisation of conventional plant and end-user costs. Table 8.2 shows a system with annual energy consumption of 350 TW h supplied from nuclear, coal, oil and gas generation. The emissions are calculated based on standard efficiencies and fuel carbon content and total 166 $MtCO_2$/year. The cost of the emissions based on £20/t is close to £3.32bn that is embedded within the coal generation unit costs of £49.6/MW h made up of basic costs of £35.7/MW h and CO_2 of £13.9/MW h.

If 20 GW of wind is introduced onto the system with priority dispatch, it is expected to displace coal generation. Assuming a 20% load factor, the wind output is given by

$$20 \text{ GW} * 8{,}760 * 0.2 = 52.56 \text{ TW h}$$

The new coal-fired generation is 76.94 TW h (129.5 − 52.56) and represents a 40% reduction in utilisation from a base of 85% and will inevitably lead to marginal plant closures. The coal generation basic cost is based on a capital cost of £1,465/kW with coal at £40/t giving a total cost of £35.7/MW h to which is added a charge for CO_2 emissions of £13.9/MW h based on a CO_2 price of £20/t as shown in Table 8.3. The annual saving in the energy cost from reduced coal production is given by 52.56 TW h * 49.6 £/MW h = £2,607m/year. This saving is offset by the cost of subsidies to the renewable generation at £92.5/MW h for on shore wind and equals £4,865m/year. Based on the unit price difference of £42.9/MW h (92.5−49.6), this adds a premium to all generation of £6.44/MW h (42.9*52.56/350).

The coal generators will see their utilisation fall from 85% to 45% that will result in unit costs rising to £68.4/MW h, and marginal plant will not be able to compete and will be forced to close. This results in tighter plant margins

Table 8.2   Emission reduction

Fuel	gC/GJ	Efficiency	% Prod	tCO_2/ GJ	tCO_2/ TW h	TW h/ year	$MtCO_2$/ year
Nuclear			18	0	0	63	0.0
Coal bit.	23,700	38	37	0.0869	823,263.16	129.5	106.6
Oil	19,900	40	5	0.0730	656,700	17.5	11.5
Natural gas	13,500	52	40	0.0495	342,692.31	140	48.0
			100	Cost £	3,321.635	350	166.1

Table 8.3   Coal cost

Capital cost	£/MW h	18.60
Fixed O&M	£/MW h	2.09
Var O&M	£/MW h	1.59
Fuel costs	£/MW h	13.43
Total generation costs	£/MW h	35.70
CO_2 price	£/t	20
Include carbon	tCO_2/GJ	0.0869
Added cost/incentive	£/MW h	13.90
Total cost	£/MW h	49.61

when the wind isn't blowing and has resulted in some markets being forced to introduce capacity and availability payments to retain marginal plant for when the wind does not blow.

8.10   Impact of renewable generation on market prices

The sudden increase in output from wind or solar generation can have a disproportionate impact on real-time prices. An operational plan will be in place to meet most of the expected demand with conventional generation. Normally, only small volumes are traded close to the event to trim generation to balance demand. Most of the generation will be operating at full load with only a small proportion part loaded to provide reserve. If the prediction of wind output is accurate, then its utilisation will have been planned in conjunction with conventional plant. If the timing of wind generation output increase is in error, then contracted generation would have to reduce output to accommodate it. This will incur costs in part loading at reduced efficiency or if the unit has to be shut down when start-up costs would result when the unit is restarted. These factors can drive down prices in the short term due to contracted generators reluctance to forgo revenue.

A typical generation variable price merit order is shown plotted in Figure 8.3 against the cumulative capacity for the year 2020. It can be seen to exhibit a slow incline through the normal operating range rising by around £10/MW h. The impact of sudden unexpected demand changes due to variations in renewable output will depend on their occurrence in relation to the normal demand trend. It will have most impact when its output is increasing against a normal falling demand curve and will drive system prices down.

The result of an annual dispatch simulation enabled a correlation between wholesale prices and demand net of wind that was 0.736 with 25 GW of wind capacity connected to a 60 GW system. The correlation between wind output and variable price was just 0.1007 reflecting the change brought about by the wind output. The correlation was dominated by the net demand as expected. The simulation did not model the effect of short-term errors in forecasting.

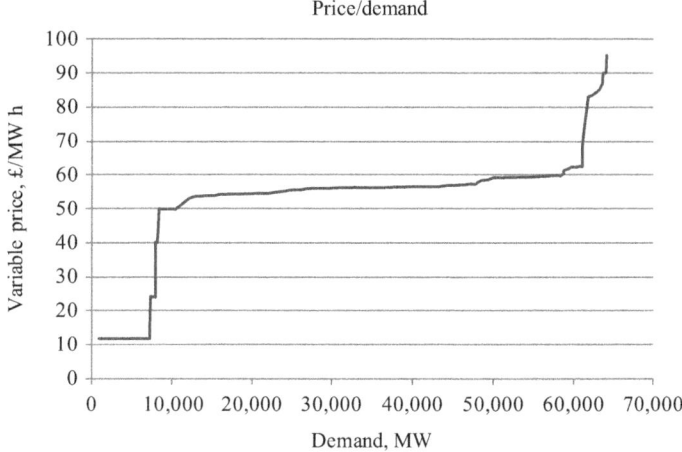

Figure 8.3 Variable price merit order

8.11 Regulatory initiatives

The CEER recognised the impact of renewable generation on wholesale markets and the disadvantages faced by incumbent conventional generators leading to premature closures and low plant margins. It proposed changes to improve the situation in a white paper including

- **The removal of priority dispatch for renewable generation;**
- **The removal of the 90% compensation received by renewable generators when they are curtailed;**
- **The requirement that self-generators pay a fair share of network and system costs rather than avoiding them with net metering.**

Wholesale markets should be based on real-time conditions and prices should be allowed to rise to reflect scarcity. Prices should reflect the consumer VLL without price caps. **The growing requirement for flexibility has been recognised. All participants should face the same incentive to balance their position.** The provision of flexibility services to support balancing should be open to all participants.

Distribution system operators should facilitate the market rather than participate direct. They should not own storage or any infrastructure for EV charging but be transparent on the system needs and service requirements. Their tariffs should be cost reflective and designed to encourage efficient use of the network assets. Their performance should be incentivised through allowed revenues related to efficiency improvements.

End-user market participation is expected to be facilitated by smart meters. Their involvement may be coordinated by aggregators with access to real-time dynamic market prices. Consumer switching should be facilitated with common data formats providing accurate information and fast and free supplier switching supported by comparison tools.

The subsidy to renewable generation coupled with its priority dispatch has virtually destroyed the operation of wholesale markets. It has already led to irreversible premature closures of conventional plant. The network charges that favour embedded generation over supergrid generation has added further distortion. The incentives afforded to existing renewable generation were designed to kick-start implementation. In practice, it has fostered existing wind and solar technology and stifled new innovation. A more holistic approach and longer term view is long overdue. It should recognise the need for flexibility as offered by other low-carbon options like generation with CCS, nuclear and biomass.

8.12 Dispatch process

Generation is usually scheduled and dispatched in merit order. Table 8.4 shows the characteristics of a set of generation available to supply a small system together with their incremental price and minimum and maximum generation. The expected demand is shown for 24 h ahead.

It can be shown that the dispatch of the generation to meet the demand profile in merit order will result in the following:

The baseload price for day (24 h time avg)	23.79
The volume weighted average 24 h price	23.85
The peak price (hours 8 to 19 inclusive)	24.17
The off-peak price for day (hours 20 to 7)	23.42
The super-peak price hour 16	24.50
The generators that operate in baseload mode (run all hours)	1 to 4
The daily % utilisation of the peak gen.	7.25

Table 8.4   Generation data and demand profile

Generation merit order	Incremental price, £/MW h	Min gen, MW	Max gen, MW	Cum, MW	Demand, MW	Hour	Demand, MW	Hour
0				0	1,520	1	1,785	13
1	18	200	500	500	1,470	2	1,800	14
2	20	150	400	900	1,400	3	1,850	15
3	22	150	350	1,250	1,250	4	1,870	16
4	23	150	300	1,550	1,300	5	1,975	17
5	24	150	290	1,840	1,350	6	1,840	18
6	24.5	100	250	2,090	1,450	7	1,780	19
7	26	100	250	2,340	1,600	8	1,700	20
8	26.5	40	90	2,430	1,650	9	1,670	21
9	27	40	90	2,520	1,750	10	1,650	22
10	27.5	40	90	2,610	1,800	11	1,600	23
	Total cap.		2,610		1,825	12	1,580	24

Wind MW	Hour	Wind MW	Hour
25	1	100	13
45	2	130	14
65	3	100	15
90	4	70	16
110	5	50	17
150	6	80	18
180	7	90	19
210	8	65	20
190	9	55	21
210	10	25	22
200	11	20	23
150	12	10	24

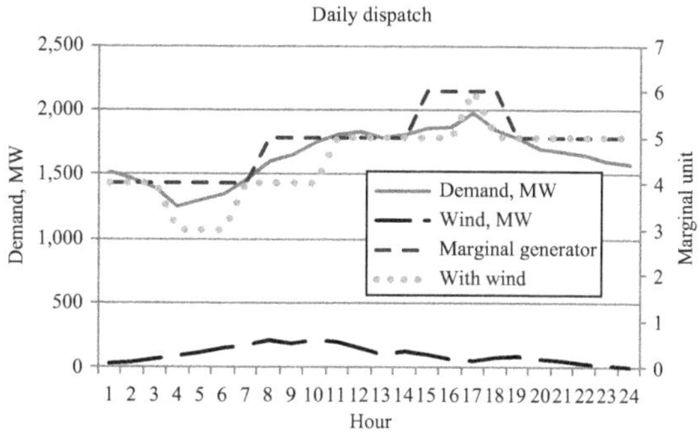

Figure 8.4   Impact of wind on dispatch

If a 250 MW tranche of wind is introduced, with priority dispatch, the conventional generation has to be redispatched to balance and meet the demand net of wind. Figure 8.4 shows the wind output profile with the graph showing the original marginal generator used to meet the demand and that used to meet the demand net of wind output. It can be seen that because the wind output does not correlate with the inherent demand profile that the role of the marginal generators suffers significant change. In the original dispatch generator 4 ran baseload i.e. all hours but in the revised dispatch is not required during an overnight period. Rather than shut down, it may be accommodated by part loading other plant but at the expense of higher operating costs. The reduced utilisation will impact on profits and may lead to some generators having to close. It can be seen from the 'with wind' dispatch profile that the operating regime changes for a number of units. In particular, unit 6 is dispatched for only 1 h and in practice this may not be possible if the unit minimum on-time is greater than 1 h. It could be accommodated by an open cycle gas turbine, and the plant mix may have to include a tranche of more flexible generation to track wind changes. This would be at the expense of efficient operation or met by paying end users to participant in demand management schemes.

8.13   Impact of CO_2 charges

In order to reduce emissions, charges are being levied on fuels in relation to the level of CO_2 emissions resulting from combustion. This adds a premium to electricity prices according to the level of emissions with those from coal being more than twice as high as those from gas as shown in Table 8.5.

The figures can be translated into $tCO_2/MW\,h$ taking account of the efficiency and CO_2 content/GJ:

$(1 * 1,000/3,600)$ is equivalent to 0.2778 MW h

Table 8.5 CO_2 emissions from generation

Fuel	gC/GJ	tCO$_2$/GJ	Effic. %	tCO$_2$/MW h
Nuclear		0		
Bituminous coal	23,700	0.0869	38	0.8233
Oil	19,900	0.072967	40	0.6567
Natural gas	13,500	0.0495	52	0.3427

To calculate the number of GJ to generate 1 MW h, multiply by 3,600/1,000 and divide by the efficiency e.g. for coal:

$3.6/0.38 \text{ GJ/MW h} = 9.47 \text{ GJ/MW h}$

Emitting $0.0869 * 9.47 \text{ tCO}_2/\text{MW h} = 0.8233 \text{ tCO}_2/\text{MW h}$ generated

The added cost is found by multiplying the t/MW h by the charge levied on emissions. Using the data shown in Table 8.6, the impact of varying fuel and CO_2 prices has been analysed as shown in Figure 8.5. The *x*-axis shows the gas price in p/therm; the coal price in £/t and the CO_2 charge in £/t are to the same numerical scale. It can be seen that without the CO_2 that coal is competitive with gas through a range of fuel prices. The addition of CO_2 has more impact on coal prices due to its higher carbon content and becomes uncompetitive. **This will eventually lead to the closure of all coal generation unless it is fitted with CCS.**

Across Europe, the CO_2 price is determined by the ETS. Market arrangements were considered the best mechanism to realise the least cost approach to abatement of greenhouse gas emissions, and the ETS was introduced to facilitate this. The mechanisms provide producers of CO_2 with allowances related to their normal annual emission levels. These allowances are issued free but are set to be less than the producer's historical requirement thereby encouraging abatement or incurring penalty payments. There is also the option to trade and purchase allowances from other producers with a surplus realised through the early application of abatement. This process is designed to encourage the abatement to take place where the cost is least with the allowances being progressively reduced through subsequent periods of the scheme.

The ongoing arrangement proposed was that the market would set the price that participants would have to pay if they exceeded their allocation. The payments would be to buy allowances from those participants that had succeeded in reducing emissions below their target or from the government, in this initial implementation.

Directive 2003/87/EC laid the foundation for the ETS and required each member to translate the requirements into their national legislation. The first phase of the scheme came into force in January 2005 running to the end of 2007 and

Table 8.6   Impact of CO$_2$ charge on gas and coal

	Units	Gas	Coal
Capacity	MW	393	398
Capacity net	MW	377	366
Load factor	%	85	85
Annual output	TW h	2.81	2.73
Construction time	Years	2	4
Project life	Years	18	25
Efficiency	%	54.2	37.6
Coal cost delivered	£/t		50
Coal calorific value	GJ/t		24
Natural gas cost	£/mmBTU	5	
Natural gas cost delivered	Pence/therm	50	
N gas calorific value	kW h/therm	29.32	
N gas grossCV/netCV		1.11	
Cost of dept.	%	4.7	4.7
Cost of equity	%	7	7
Inflation	%	2.4	2.4
Debt/equity split		0.8	0.8
WACC ($D/E = 0.8$)	%	6.0	6.0
Site and dev.	£m	8.733	17.46
EPC contract	£m	111.33	245
Electrical conn	£m	0.533	0.533
Gas connection	£m	2.466	
Spares	£m	14.33	31.86
Interest during const.	£m	10.26	23.20
Total investment cost	£m	147.7	318.1
Capital cost	£/kW	376	799
Capital cost pa	£/m	13.60	24.80
Consumables	Pence/kWh	0.02	0.113
Labour rate	£/h	16.66	16.66
No. of operators		6	14
Operators labour*1.28	£M	1.12	2.62
Maintenance material	£M	0.81	2
Main and support labour	£M	1.25	2.33
Capital cost	£/MW h	4.84	9.10
Fixed O&M	£/MW h	0.60	2.09
Var O&M	£/MW h	0.73	1.59
Fuel costs	£/MW h	34.92	19.95
Total generation costs	**£/MW h**	**41.10**	**32.72**
CO$_2$ price, £/t	**20**		
Include carbon	tCO$_2$/GJ	0.0495	0.0869
Added cost/incentive	£/MW h	6.58	16.64
Total cost	£/MW h	**47.67**	**49.36**

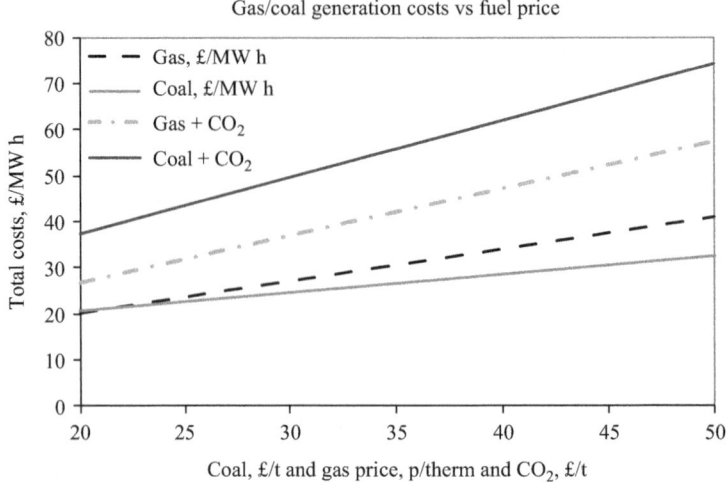

Figure 8.5   Gas/coal comparison + CO₂

covered CO_2 only. The second phase from 2008 to 2012 covers all the greenhouse gases. For each period referred to in the article, each member state is required to develop a national plan stating the total quantity of allowances that it intended to allocate for that period and how it proposes to allocate them. The distribution is established by industrial sector and subsequently for particular installations.

The clean development mechanism (CDM) set up through the 1997 Kyoto protocol enables credits to be gained by sponsoring the development of clean solutions in the fast developing countries like China. The board regulating the CDM strictly defines which types of project are allowed and how they demonstrate compliance. The credits are usually created through 'smokestack' solutions that can prove a reduction in emissions or efficiency improvements. There is also non-regulated activity where companies are voluntarily seeking to improve their image through the pursuit of a green policy. This in turn is supported by carbon offset retailers who trade in the provision of carbon reductions without regulation. Many projects in this sector have not been audited to confirm that reductions are truly being realised.

The EU's LCPD requires that all new plants with a thermal capacity >50 MW and operational before 1987 must meet emission limit values (ELVs). This defines limits on the emission of SO_2, NOx and dust that had to be realised by 2008. For a typical coal plant, the limits amount to 400 mg/m³ of SO_2, 500 mg/m³ NOx and 50 mg/m³ of dust. There is an option to opt out of the scheme and take a derogation of limited life. This limited the plant to 20,000 operating hours starting in 2008 and ending in 2015 when the plant must close. Implementation is either through a system of ELV for individual plants or by establishing a National Emission Reduction Plan (NERP). The NERP defines the total emission limit for the country with allocation effected through a trading system.

8.14   Conclusions

The performance of wholesale markets has been undermined by the introduction of increasing levels of renewable generation capacity that receives priority in dispatch. Prices have become more volatile and less predictable due to renewable output changes that are not linked to demand levels. The renewable generation output may be increasing when the demand is falling making prediction of the demand and marginal prices very difficult. The result is to place a premium on flexibility in generation and demand.

As the renewable capacity is increased, the utilisation of marginal plant is reduced to the point where it becomes non-viable and closes. However, it is still required to meet demand when the renewable output falls, and to retain its availability, it has been necessary to introduce capacity markets. At the same time, fossil-fuelled generation is subject to additional charges related to CO_2 emissions causing coal-fired generation to be prematurely closed. In contrast, the network charging arrangements have promoted the development of small embedded generation connected to the distribution system. These market distortions have had an adverse effect on generation companies that has been recognised by regulatory authorities who have proposed a number of changes including

- The removal of priority dispatch for renewable generation;
- The removal of 90% compensation when their output is curtailed;
- The requirement for self-generators to contribute to system costs.

In the current environment, governments are faced with a spectrum of objectives including securing future supplies, building up renewable energy to meet emission targets whilst maintaining fuel diversity and containing end-user bills (the so-called trilemma). **The view of the author is that in a rapidly changing situation, the SB model offers the best approach to meeting these diverse objectives.** The market would act as an interface between the generation side and an active demand side. The process would be augmented by a two-sided future capacity market including flexibility ranges with costs. It would realise the benefit from competition in generation and provides a mechanism to manage the evolvement of the plant mix and the capacity margin. It would also act as the counterparty for large-scale capital developments like nuclear or hydro. Rather than the SO defining the system generation capacity requirement, it would be established by the demand and generation side contracting directly. **The demand side would need to contract forward for capacity taking account of its capability to smooth overall demand using DSR and embedded generation.** This is essential to realise the full benefits of demand smoothing in reducing capacity requirements.

The SB model is appropriate where there is a need to attract new entry to bolster the plant margin but at the same time ensure fuel diversity and supply security. It also affords the opportunity to retain tariffs on the basis of average rather than marginal costs to meet wider social objectives. Competition in this environment can be managed through auctions for new capacity as well as for additional tranches of energy import at each annual, monthly and the day-ahead stage. New entry can be fostered by the sale of existing generation clusters backed by PPAs, for a limited period, pending wider market liberalisation. A large residual state generator can be retained to manage balancing and also provides make-up and spills options for smaller independent power producers to balance their commitment to end users.

The SB model is not too difficult or costly a mechanism to establish and can be implemented relatively quickly. Subsequently, the market can be developed further by enabling some direct contracting between generators and end users. This could initially include supply to works on the same site or the definition of a percentage of output capacity that should be traded directly to provide price discovery. The security that this direct trading option affords may appeal to large end users operating continuous processes.

Question 8.1 The table below shows the marginal generator at each load level based on the generator data set shown in Table 8.4. Given the wind output shown in the table, identify the new marginal unit from dispatch at each load level and the associated marginal price. Hence, find the average marginal price for the day.

Hour	Hourly demand, MW	Marginal gen. no wind	Marginal gen. price, £/MW h	Wind output, MW
1	1,520	4	23	75
2	1,470	4	23	135
3	1,400	4	23	195
4	1,250	4	23	270
5	1,300	4	23	330
6	1,350	4	23	450
7	1,450	4	23	540
8	1,600	5	24	630
9	1,650	5	24	570
10	1,750	5	24	630
11	1,800	5	24	600
12	1,825	5	24	450
13	1,785	5	24	300
14	1,800	5	24	390
15	1,850	6	24.5	300
16	1,870	6	24.5	210
17	1,975	6	24.5	150
18	1,840	6	24.5	240
19	1,780	5	24	270
20	1,700	5	24	195
21	1,670	5	24	165
22	1,650	5	24	75
23	1,600	5	24	60
24	1,580	5	24	30

Chapter 9
Balancing market

9.1 Market process

This chapter discusses the operation of a balancing market and the provision of ancillary services to manage secure system operation. In particular, it appraises the reserve requirements to manage the impact of intermittent generation.

Up to the point of 'gate closure', market participants are free to trade bilaterally. At the point of gate closure, the market participants have to declare their contracted volumes to the TSO. This is usually an hour before the event when the registered parties qualified to make returns to the TSO have to make their submissions. They are commonly known as balance responsible parties and they return programmes of generation/demand to the TSOs for the period ahead [in the United Kingdom called final physical notifications (FPNs)]. There will inevitably be some errors in the submitted predictions that the TSO will need to address. This is realised through a voluntary balancing market that operates to keep the system in balance up to the event. Bids are invited from market participants to provide rapid increase/decrease of generation or demand. The TSO will buy these increments and decrements of power to affect system balance and manage frequency. The selection will be made on an economic basis while respecting any technical limitations. Those bids called within each period will be used in establishing the system buy or sell price. These prices will be charged to participants found to be out of balance having taken more or less energy than they declared at 'gate closure'. This is established from metering data collected by the central data collection agent (CDCA).

The **system buy price** (SBP) for each period is calculated from the total cost of energy from the accepted offers calculated from the product of the volume (QAPO) and the price (PO) divided by the total energy with some secondary adjustments as shown in the following equation:

$$\text{SBP}j = \{\{\Sigma i \Sigma n\{\text{QAPO}n\ ij * \text{PO}n\ ij * \text{TLM}ij\}\}/\{\Sigma i \Sigma n\{\text{QAPO}n\ ij * \text{TLM}ij\}\}\} + \{\text{BPA}j\} \tag{9.1}$$

where Σi represents the sum over all BM units and Σn represents the sum over those accepted offers QAPO with prices PO giving the weighted energy cost. (The term 'TLM' is an adjustment to take account of losses.) This is divided by the energy (QAPO) to give a price per MW h. The term BPA is the transmission company's buying price adjustment (BPA).

Similarly, the **system sell price** (SSP) is determined from the sum of the product of volume (QAPB) and bid prices (PB) of accepted bids divided by the total energy as shown in (9.2) to give a price in £/MW h:

$$SSP_j = \{\{\Sigma i \Sigma n \{QAPBn\ ij * PBn\ ij * TLMij\}\}/\{\Sigma i \Sigma n \{QAPBn\ ij * TLMij\}$$
$$+ UESVA_j\}\} + \{SPA_j\} \tag{9.2}$$

where Σi represents the sum over all BM units and Σn represents the sum over those accepted bids with the term SPA, the transmission selling price adjustment. The system sell and buy prices are effectively the average volume weighted price of bids and offers as used. This data is used to calculate the imbalance cash flow and the system operator BM cash flow and requires the following additional data to enable settlement:

- From the TSO – FPNs, bid/offer data, acceptance data, balancing service volumes and any adjustments;
- From the CDCA – the BM unit metered volumes, the inter-connector metered volumes and Grid Supply Point (GSP) metered volumes;
- Transmission Loss Multipliers as used to allocate losses.

The data from the TSO consists of dispatch instructions with 'from' and 'to' times and these are recorded as spot times and associated MW values. This applies to FPNs as well as the bid/offer and acceptance data. This spot data is translated into settlement period half hour integrated values based on linear interpolation between the spot values. Figure 9.1 shows the UK imbalance volumes in MW h and the coincident buy/sell prices in £/MW h for a day in February. The interaction between the imbalance and prices can be clearly seen.

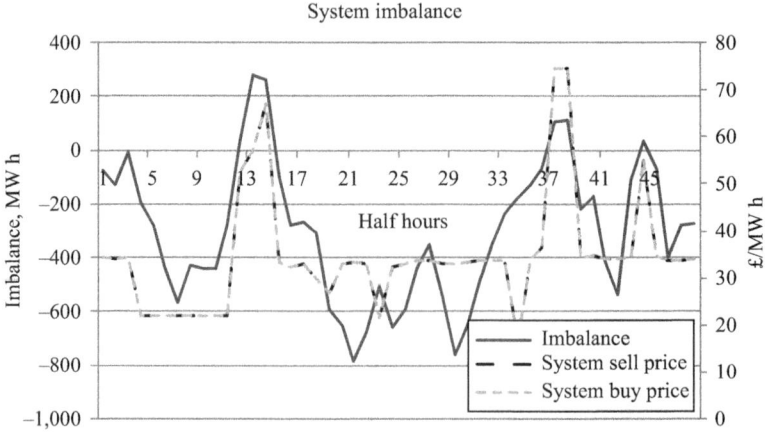

Figure 9.1   Balancing market price and volume

9.2   Ancillary services

The SO also needs to procure other services to ensure secure and stable management of the primary system generally referred to as ancillary services. These may be procured through auctions or bilateral contracts for a month ahead or a year. Three classes of ancillary service can be defined:

- Class 1 – includes reactive power provision and frequency response services that are generally mandatory and essential to manage system incidents like generation loss;
- Class 2 – includes black start and fast start generation usually established by bilateral agreements and used to rebalance the system after a disturbance;
- Commercial – includes Short Term Operating Reserve (STOR), fast reserve, start up, fast deload and maximum generation services that are usually established by annual contracts through a tender process and used to restore economic operation.

The SO defines the requirements for these facilities and calls them off to maintain secure operation or restore services. There may be incentive schemes in place with the SO to encourage limiting the added costs.

9.3   Reserve requirements

This section discusses reserve requirements in the context of the balancing market. **The growth of renewable intermittent generation is having a significant impact on the system requirements for reserve and balancing supply and demand.** The scale of the requirement for additional reserves by the system operator depends on the lead times and the accuracy of forecasting. The requirements are over and above the more traditional requirements for different classes of reserve to cater for system incidents and manage frequency. Three classes of reserves are usually identified including:

- Primary reserve to contain frequency deviations following sudden changes in generation, imports/exports or demand this is usually provided by part loaded generation able to respond within seconds via governor action or demand response triggered by low frequency relays;
- Secondary reserve that can be used in minutes to restore frequency to normal values following an incident or sudden change in generation/demand. This could be provided by pumped storage hydro, part loaded generation or gas turbines within 2 to 10 min operating in response to Automatic Generation Control (AGC) signals;
- Tertiary reserve is maintained to restore reserve levels and reoptimise economic operation and inter-connection flows from around 15 min onwards using standby generation through normal manual or automatic dispatch mechanisms.

The system operator has responsibility for operating the system and will normally take the lead in defining requirements. There is usually a statutory requirement to keep frequency within prescribed limits that in the United Kingdom is ±0.5 Hz with operational limits set to ±0.2 Hz. These limits are set to maintain the quality of supply and provide design standards for appliances and generators. The primary reserve level is usually based on catering for the largest credible loss of generation or system infeed. Where it is provided by part loaded generation, there is a practical limit to the level of response realisable in the short-term via governor action. The governor droop is defined as that frequency fall in per cent that would result in 100% change in output and is typically 4% or 2 Hz. A practical fast output increase of 20% is typical for thermal plant by making use of boiler stored energy backed up by fuel supply control. This would be realised by a frequency fall of 0.4 Hz. This means that to realise an output increase of 1,000 MW for a 0.4 Hz frequency drop, some 5,000 MW of generation would be required operating with free governor action. The amount of regulating reserve on a system is usually referred to as the system gain in MW/Hz and in this case would be 2,500 MW/Hz.

In interconnected systems, shared reserve agreements are usually established that take advantage of the likelihood that incidents will not occur simultaneously. This enables each country to meet only a proportion of the total reserve requirement. An illustration of the assessment of the market volume for **primary reserve** can be found by analysis of the situation in the Netherlands. The Dutch contribution to frequency control was set at 744 MW/Hz which represents 3.6% of the agreed 20,570 MW/Hz system gain that applied to the complete interconnected UCTE grid of which the Netherlands is part. This results in an expected contribution of 37 MW (744*0.050) from the Netherlands in the event of a 1,000 MW generation failure with a 50-MHz frequency drop. (1,000 MW generation failure is expected to cause the frequency to drop by some 50 MHz i.e., 1,000/20,570 * 1 Hz.)

A maximum generation failure of 3,000 MW is adhered to in the synchronous UCTE system, which implies that the Netherlands is expected to deliver at least 109 MW primary reserve (3.6% out of 3,000 MW) for a 0.15-Hz frequency drop. Assuming contracted generation is operating with free governor action with a 4% droop (i.e. delivering rated 100% output for a 4% frequency drop), then for a 0.15 Hz drop (0.3%) in frequency, 7.5% of rated output would be delivered (0.3/4). As the requirement is for 109 MW, this implies 1,453 MW of generation is able to respond via governor action or in response to signals from a central load frequency control system (LFC).

The standard governing the minimum requirements for **secondary/regulating reserve** varies with the season, the day of the week and time of day depending on prevailing system conditions. Generally, more reserve is required at times of rapid demand change to ensure sufficient dynamic capability to be able to follow the system demand. There is equally a requirement for downward regulating capability, particularly during the light load periods of summer. There is then a need to maintain generation that is able to reduce its output if demand falls lower than expected. Reserve is usually set at around 300 MW for control reserve and 600 MW

Table 9.1   UK reserve (source NGC)

NGC reserve	Total cap., MW	De-rated, MW	Cost, £m/year	Cost, £k/MW
Balancing reserve r1	215	65	1.3	6
Supplemental r1	600	600	10.8	18
DSBR 2	300	112	1.3	11.6
SBR 2	1,874	1,784	23.1	13
Total	**2,989**	**2,561**	**36.5**	**12.2**

for other reserve giving some 900 MW in total. Emergency reserve is additional and usually set at 300 MW. In contrast, Germany contributes 600 MW to the UCTE requirement of 3,000 MW with 2,000 MW of secondary and 2,500 MW of tertiary reserve as determined by the SO.

In the United Kingdom, being a more isolated system, it caters for a loss of up to 1,800 MW with a basic secondary reserve of 2,525 MW and 500 MW to cater for variations in interconnector transfers, plus provision for wind backup. The requirement for emergency or contingent reserve is assumed to grow in proportion to the system energy demand. This is consistent with providing backup for loss of regulating reserve following contingencies. Table 9.1 shows typical UK costs for demand side balancing reserve (DSBR) and supplemental balancing reserve (SBR) with an average cost of £12.2k/MW/year.

9.4   New sources of reserve

There are a number of potential sources of reserve that can be categorised according to the reserve type:

- Primary reserve – because of the speed with which response is required this has to be triggered automatically. The requirement is usually met using spinning part loaded generation operating with free governor action or demand tripped by low frequency relays; more recently, battery stored energy that can be released within a second is being trialled.
- Secondary reserve – within this time scale, operating generation may be dispatched to change output via an AGC system. It is also possible to use fast start open cycle gas turbines to increase output and hydro/pumped storage;
- Tertiary reserve – the slower timescales open the opportunity for more demand side participation and small embedded diesel and gas generators.

There are costs associated with using conventional generation to provide reserve; operating at part load reduces the efficiency; the reduced utilisation means unit costs have to be higher to recover the fixed costs; additional fuel costs are incurred in stopping and starting units and in ramping up and down. **The primary reserve for immediate frequency containment following incidents is the most expensive** at around £20/MW/h. **Reserves to enable manual frequency restoration are around £6/MW/h with dispatched replacement reserve at £3.5/MW/h.**

The advent of electric vehicles offers a new alternative when their batteries are connected to the grid. They could be triggered to generate rather than store energy when instructed perhaps through a smart meter system. Typical electric car batteries have a capacity of 50 kW h **with a million vehicles fully charged containing 50 GW h**. Assuming a discharge output of 2 kW, up to 2 GW could be fed back into the networks if all users participated. The 2-GW output for 1 h would only deplete the car fleet charge from 50 to 48 kW h.

NGC submitted an enquiry in 2016 for fast acting battery storage systems to provide enhanced frequency response (EFR). The specification called for response times of less than 1 s with full output sustained for only 15 min. They subsequently contracted for 200 MW with 8 providers for systems based on lithium-ion batteries with capacities in the range from 10 to 50 MW. The total cost was £66m with an average of £328/kW of capacity (£66m/200,000) for 4-year contracts. The effective storage unit cost prices over the 4-year period ranged from £7/MW up to £12/MW of EFR per hour of service with 20 h/day of availability (328,000/20/365/4 = £11/MW for each hour of service). The application was focussed on providing primary frequency response rather than balancing supply and demand and has a limited energy capacity of 50 MW h that is insufficient to cater for the larger changes in wind generation.

Aggregators see the BM area as offering opportunities but currently have to participate through the relevant supplier that manages the demand and metering of the balancing market unit (BMU). The implication is that the aggregated demand controlled needs to be aligned with the supplier customer base and it would be complicated to enable aggregators to participate directly. This could result in the financial arrangements becoming convoluted. A supplier may have contracted to buy 100 MW of generation in the wholesale for supply to its consumer base. The aggregator may contract with the supplier to effect demand side management of up to 20 MW that is sold into the balancing market. The supplier would receive payments from the BM for the provision of the services and would want to retain compensation for lost revenue from the 20 MW of lost supply. In addition, the controlled customer base would want payments for participating in the scheme with the aggregator retaining the residual.

9.5   Short-term operating reserve procurement – STOR

This is a tranche of reserve generation or demand reduction procured from system users to support system operation. **It is based on the provision of reserve with 10 min notice with delivery for up to 2 h and could be classed as tertiary reserve.** In 2013/14, NGC procured 3,097 MW at a cost of £58.3m in availability payments. Of this capacity, 2,534 were committed with 560 MW flexible. The actual average daily availability was 2,376 MW. The payment for availability was £5.83/MW h reduced to £4.94/MW h including cheaper longer term contracts. The

Table 9.2   STOR periods

Period	7.1	7.2	7.3	7.4	7.5	7.6
Season, h	251	1,207	384	362	1,059	599
Season, days	27	103	35	33	97	57

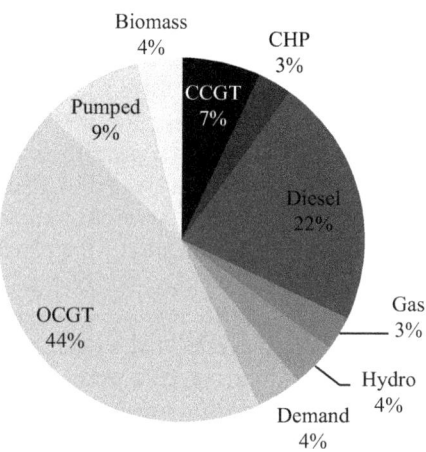

Figure 9.2   Source of STOR UK

average utilisation costs were £191/MW h. The required availability is defined by the number of hours in six periods of the year as shown in Table 9.2.

There are 352 days in total covering 3,862 h or approximately 11 h/day. The availability payments can be estimated as follows:

$$3,097 \text{ MW} * £4.94/\text{MW h} = £58\text{m}$$

The utilisation during the 2013/14 was for 292 GW h of energy at prices ranging from £130 up to £300/MW h. Based on average unit costs of £138/MW h, for those generators used, the total utilisation cost was £40m giving a total STOR cost/year of £98m. The services for STOR are procured by auctions that are usually over-subscribed **with a third of those tendering not receiving contracts.** The requirements are complex and limited transparency discourages participation.

Figure 9.2 shows a breakdown of the contracted sources of STOR during 2013/14. It can be seen that most of the provision is by OCGTs with **a significant proportion from small embedded diesel generators that will have relatively high levels of emissions.** There are two categories of providers; 1,000 MW of larger units like Closed Cycle Gas Turbine (CCGT), OCGTs and pumped storage connected to the super-grid and registered in the balancing market; around 2,000 MW of smaller non-BMUs connected to the distribution network acting as demand side providers. The units are instructed via the balancing market or a specific STOR dispatch system. The potential capacity for energy delivery is 2,376 MW * 3,862 h = 9,176 GW h of which only 292 GW h was used, equivalent to 3.1%. Assuming that the 292 GW h could have resulted in consumer lost supplies, these have been avoided at a cost of £98m that is equivalent to a value of lost load of £335/MW h or £0.3/kW h.

9.6 Impact on reserve requirements due to intermittent renewable generation

Additional reserves are required to cater for rapid changes in the output of intermittent renewable generation. The amount depends on the installed capacity of the generation, and the output forecast error. The load factor of the generation will influence the running periods and hence the time for which reserve has to be provided and hence the costs. The UK system operator assumed **that the RMS forecast error for 4 h ahead would gradually improve from 17% in 2010/11 to 13% in 2015/16 down to 10% in 2020/21**. Assuming a normal distribution and taking three standard deviations as an indication of the worst case error gives reserve requirements of up to 51%, decreasing to 39% and eventually 30% by 2020/21. The reserve provision for the intermittent generation will vary with its output through the year. The expected average output is based on the load factor of 30% less 500 MW reserve provided to manage interconnector flows. In calculating the costs of reserve provision for a year, the forecast percentage forecast error of the average output of the wind generation is used. The reserve energy is priced at a premium to normal market prices of some 60% in this example and a utilisation of 30% is assumed consistent with the load factor. The extra costs incurred by the balancing generation results from part loading generation that causes a reduction in overall efficiency and also the additional fuel used in ramping output. The lower levels of utilisation also results in fewer unit sales from which to recover the fixed operating and capital costs. Table 9.3 shows a calculation of the average reserve requirement for 3 years with increasing wind capacity coupled with improvements in forecasting accuracy. The rise in energy costs with increased wind capacity is partially offset by improvements in forecasting output.

Table 9.3   Reserve for wind generation (UK)

Year		2010/11	2015/16	2020/21
Reserve for wind		Based on 4-h forecast error		
Wind capacity	MW	4,000	12,000	20,000
Expected average output-500	MW	700	3,100	5,500
Wind LF	%	30	30	30
Forecast RMS error (std)	%	17	13	10
Marginal wind effect 3* std	%	51	39	30
Average MW reserve	MW	357	1,209	1,650
Reserve energy used	TW h	0.93	3.13	4.28
Price premium	£/MW h	24	30	36
Total cost	**£m**	**22**	**94**	**154**

9.7 Balancing market Germany

Problems have been reported due to high concentrations of wind energy in Northern Germany where variations in output can directly affect the cross border flows into the Netherlands. It has been claimed that this is impacting on the security of the Dutch network. Whilst new transmission capacity into the South of Germany is planned, it can be expected that there will continue to be additional regulating requirements in the Netherlands to accommodate these disturbances from time to time. In that they are not planned, these disturbances will be managed by Tennet using emergency power. It is assumed that the proportion of the variation exported will be in relation to relative system size of around 5:1 between the Netherlands and Germany. There will not be much diversity between Dutch and German wind power output variations and the effects could be cumulative. The need to accommodate cross border transfer errors is likely to result in increased requirements for emergency reserve of some 20% in Germany as intermittent generation capacity is increased. **The requirement to provide reserve to cater for variations in interconnector flows may also be due to contingencies but more often it will result from variations in renewable capacity output that has not been balanced locally.**

It has been reported (NEON) that whereas wind and solar power in Germany has increased by 200% from 30 to 90 GW, over the period from 2008 to 2015, the total reserve requirements fell by 20% from 5,900 to 4,990 MW (Hirth-Ziegenhagen 2015) including primary, secondary and tertiary. The reserve figures quoted for Germany relate to managing combined wind and solar. A slight negative correlation in output between the two is recognised and we would therefore not expect balancing requirements to increase in proportion to the combined capacity. The calculation of reserve requirement for wind and PV can be considered separately.

Prices in the secondary and tertiary reserve balancing market in Germany fell by more than 50% between 2013 and 2015 but increased by 20% in the primary reserve market. This is despite relatively small changes in the volume of the requirements. The primary reserve requirement between 2008 and 2015 increased from around 600 to 752 MW, whereas the secondary and tertiary fell modestly from 5,300 to 4,200 MW including both secondary and tertiary in similar proportions. This is despite the installed wind capacity increasing from around 20 to 40 GW during the same period. It implies that forecast accuracy by the SO has improved significantly or there is marginally less provision to cater for the more extreme circumstances. There are a number of factors that complicate drawing comparisons:

- The forecasting error is not reported but will clearly affect the requirement. It could be assumed that errors are reduced from a STD of 17% in 2010/11 down to 10%.
- The wind average output is less in Germany at 17.6% than in the United Kingdom with a load factor of 25.4%; hence, they have a lower average output to cover with backup reserve.

Table 9.4   Reserve for wind in Germany

Year		2008	2015	2020
Reserve for wind		Based on 4-h forecast error		
Wind capacity	MW	20,000	40,000	60,000
Expected average output	MW	3,520	7,040	10,560
Wind LF	%	18	18	18
Forecast RMS error (std)	%	17	13	10
Marginal wind effect 3* std	%	51	39	30
Average MW reserve	MW	1,795	2,746	3,168
Reserve energy used	TW h	2.73	4.18	4.82
Price premium	€/MW h	30	38	45
Total cost	**€m**	**82**	**159**	**217**

Based on these assumptions and applying the same logic as used to review the UK reserve provision, the German secondary reserve for wind comes to around 2,746 MW in 2015 as shown in Table 9.4. Allowing 1,800 MW for tertiary reserve, this equates to a total of 4,546 MW (1,800 + 2,746 MW) and close to the reported actual provision of 4,185 MW in 2015. During the same period, the primary requirement increased from 600 MW to around 752 MW.

The overall system reserve costs have fallen in proportion to the reserve requirement during the period as shown in Figure 9.3 based on data from the TSO for payments for available capacity. The breakdown for the 2008 total of 826m€/year was, 135m€/year for primary, 454m€/year for secondary and 237m€/year for tertiary. These figures compare to a total of 337m€/year for 2015 including 135m€/year for primary, 166m€/year for secondary and 44m€/year for tertiary. The large fall in costs occurs in the payments for secondary and tertiary reserves.

The implied unit costs in 2015 are 22€/MW/h for primary, 9€/MW/h for secondary and 2€/MW/h for tertiary with total costs given by

$$752 \text{ MW} * 22 * 8{,}760 + 2{,}000 * 9 * 8{,}760 + 2{,}185 * 2 * 8{,}760$$
$$= 145 + 157 + 35 = 337 \text{m€/year}$$

The secondary reserve costs of €157m is comparable to the figure of €159m derived for 2015 as shown in Table 9.4. The unit costs are of the same order as those in the United Kingdom. **The decrease in proportional costs over the years is principally due to increased competition through weekly and daily auctions.** With a reduction in demand due to wind energy displacing conventional generation coupled with reduced demand due to recession, **more players have spare capacity to participate in the balancing market. The number of suppliers active in the secondary market increased through the period from 5 to 33 while in the tertiary market, the number increased from 23 to 45.** It is noted that, in contrast, there is an increase in the cost of provision of primary reserve where response requirements are more difficult for factories and aggregators to meet. The cost of

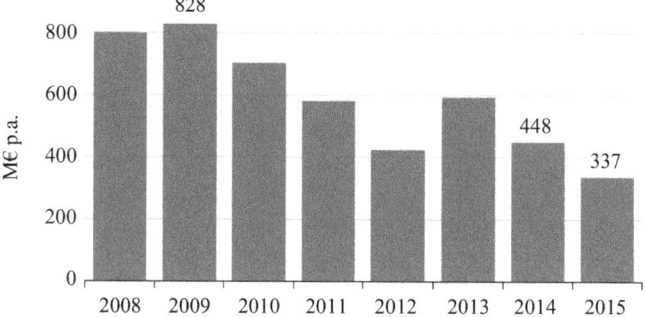

Figure 9.3 Balancing power market size Germany

fuel used in providing reserve also fell during the period and will affect reserve holding costs. There was also less utilisation of reserve reflecting an improvement in forecasting accuracy.

9.8 Balancing USA

As with many other countries, the increased capacity of intermittent wind and solar energy adds to the problem of balancing supply and demand in the USA. **At times, the solar energy output leads to negative energy prices in some southern states.** This has prompted the development of a battery/gas turbine hybrid unit to provide a rapid response to changes in demand whilst maintaining a smooth load on the gas turbine peaking plant. The hybrid unit developed by General Electric for Southern California Edison consists of a 50-MW gas turbine coupled with a battery with an energy capacity of 4 MW h with a maximum output of 10 MW. The unit will earn extra revenue from the provision of spinning reserve with an immediate response from the charged battery backed up by the gas turbine. It enables the gas turbine to be operated more efficiently whilst meeting the stringent requirements of supplying immediate reserve. It is likely to be cheaper than the alternative of running generation part loaded at reduced efficiency to provide spinning reserve.

9.9 Conclusions

The growth of renewable generation is having a significant impact on the requirement for reserve in balancing supply and demand. The SO defines the requirements for these facilities and calls them off to maintain secure operation and restore services. The primary reserve for immediate frequency change containment following loss incidents is the most expensive at around £20/MW/h reflecting the need for very fast response. Reserves available for automatic or fast manual dispatch to arrest frequency fall are around £6/MW/h and manually dispatched reserve to restore normal frequency cost around £3.6/M/h based on less stringent response requirements.

An enquiry by the SO for fast acting primary reserve for frequency deviation containment elicited responses based on lithium-ion batteries leading to 200 MW being contracted with 8 suppliers at costs averaging £328/kW. The specification called for full output for just 15 min and the facilities could not contribute to support wind energy balancing over hours.

The sources of STOR (classed as tertiary) include: OCGTs 44%, diesel 37%, pumped storage 9%, CCGTs 7%, biomass 4%, hydro 4%, CHP 3% with just 4% from the demand side. The annual costs for the United Kingdom 60 GW system were £98m in 2015 with £58m for availability and £40m for utilisation. It is worth noting that the tender round was oversubscribed with a third being unsuccessful. This indicates a competitive market operating with a limited requirement. This has caused prices in Germany to collapse by 50% for secondary and tertiary reserve although the capacity requirement has barely changed. It is envisaged that as the use of EVs grows, they could provide an additional source of reserve when connected to the grid. A million vehicles each with a battery capacity of 50 kW h contains 50 GW h of energy that could provide tertiary reserve with an output of 2 GW that over 1 h would deplete the total charge/vehicle from 50 kW h down marginally to 48 kW h/vehicle.

The requirement to provide reserve to manage intermittent renewable generation output is a function of the forecast accuracy. The RMS forecast error in scheduling timescales is expected to improve from around 17% initially to 13% by mid-decade down to 10% by 2020/21. Based on three standard deviations, the equivalent reserve requirements would be 51%, 39% and 30% of the average wind capacity output. The output depends on the load factor, assuming 30% for 20 GW of wind the average output would be 6,000 MW less 500 MW catered for by other provisions = 5,500 MW requiring 30% reserve i.e. 1,650 MW. The same assessment applied to Germany gives results similar to their provisions based on lower load factors of 17.6%.

There is also an increase in reserve provisions to cater for variations in interconnector flows. This is likely the result of variations in renewable generation output that has not been balanced locally. Germany is a known offender in this capacity where variations in the output from its large wind capacity alter planned interconnection flows and disturb the secure operation of the adjacent networks in the Netherlands and Poland resulting in complaints by government. The high levels of renewable output also impact on market prices. At times, the solar energy output leads to negative energy prices in some southern states of the USA.

Question 9.1 Given data in the table calculate the cost in €/MW h of delivered wind energy in catering for the provision of reserve to manage intermittency in each of the 3 years.

Year		2008	2015	2020
Reserve for wind		Based on 4 h forecast error		
Wind capacity	MW	20,000	40,000	60,000
Expected average output	MW	3,520	7,040	10,560
Wind LF	%	18	18	18
Forecast RMS error (std)	%	17	13	10
Marginal wind effect 3* std	%	51	39	30
Average MW reserve	MW	1,795	2,746	3,168
Reserve energy used	TW h	2.73	4.18	4.82
Price premium	€/MW h	30	38	45
Total cost	**€m**	**82**	**159**	**217**

Chapter 10

Capacity markets

10.1 Regulation

Regulators have a difficult task of creating a framework that secures future electricity supplies at a level that meets the needs of a spectrum of end users. In liberalised markets, this is becoming more difficult due to increasing uncertainty, partly because of the priority given to renewable energy sources, which deters investors in conventional generation. The renewable energy displaces conventional generation output reducing their utilisation to levels that may not be financially viable. This has prompted some government agencies to consider the deployment of capacity markets to encourage new entry by providing a stable income. This involves fixing a level of security on behalf of consumers and determining the level of capacity required to realise this in practice. A complication with renewable generation is fixing the capacity contribution that can be expected from intermittent sources like wind and solar. There is also a need to estimate the potential capacity contribution from increased levels of international interconnection and demand side regulation. This chapter discusses some of the issues facing regulatory and government agencies in creating a commercial environment that encourages the development of levels of security that match consumer expectations and budgets. It discusses

- The normal approach to estimating system security and the appropriate target plant margin of capacity over expected peak demand;
- An approach to estimating the optimal level of security based on the value of lost load to consumers compared to the costs of securing more generation capacity;
- Reviewing an approach to securing future capacity through an auction process;
- Taking account of the intermittency of renewable generation sources based on probability functions;
- Determining the level of capacity support that may be available through interconnectors with adjacent utilities;
- Evaluating the impact of increased demand side market participation through the application of smart meter technology.

10.2   Security analysis

The margin of spare capacity over that required to meet the peak demand is referred to as the plant margin and is usually expressed as a percentage. It is set to contain the number of h/year that a shortfall in capacity would occur due to loss of generation availability, higher than expected demand through investment timescales and worse than average weather. The loss of generation availability may be represented by using typical plant historic records of availabilities. The probability of r generators being unavailable from a population of n generators is given by

$$P_o^r = \frac{n!}{r!(n-r)!} \cdot P_o^{(n-r)} \cdot (1 - P_o)^r \tag{10.1}$$

where P_o is the unit availability; r is the number of generators unavailable; P_o^r is the probability of r units being unavailable. The availability includes the effect of planned outages as it is during the outage period that shortfalls often occur during a cold period.

It has been found that a good approximation to an annual demand profile can be represented by a normal distribution curve of the form:

$$H_{(D)} = \frac{K}{C\sqrt{2\pi}} \exp\left(\frac{\left(\frac{D}{s} - m_c\right)^2}{2C^2}\right) \tag{10.2}$$

where H is the number of hours during the year for which a particular demand level exists, K is a constant, C is a variable set depending on the width of the function, s is the standard deviation and m_c is the mean value. Figure 10.1 illustrates the nature of the functions derived for a system with a mean demand of 37 GW and

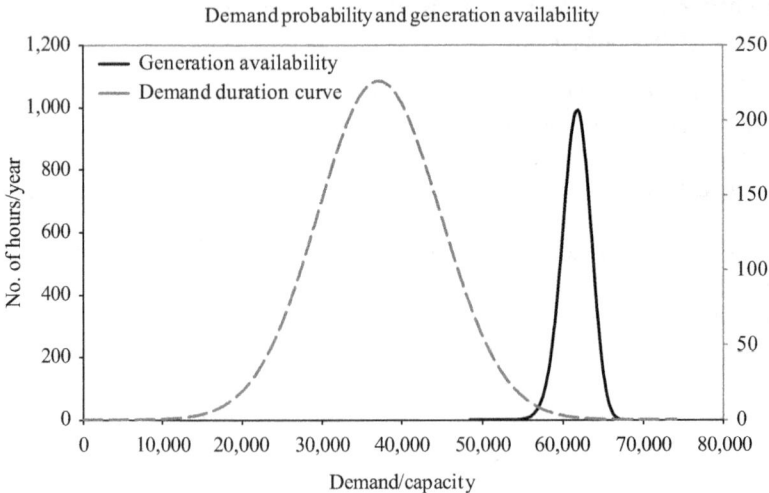

Figure 10.1   Demand and generation availability

peak of 60 GW fed with 137 notional 500 MW units each with an average availability of 0.9. The overlapping area in this example shows where shortfalls may occur.

The loss of load probability (LOLP) is then given by the sum of the number of hours that the demand exceeds the available generation capacity. Each generation availability level can be compared with the demand profile to establish the cumulative number of hours that would result in a shortfall. Given the probability of that availability, the number of hours of potential shortfall can be calculated and summated in a spread sheet to give the total number of hours during a year. The MW shortfall can also be summated and used to evaluate options to manage demand to cover the shortfall. The summated shortfall hours is given by

$$\sum h = \sum_{1,1}^{t,n} \left(H_{n,t}(D > G_n^t) \right) \tag{10.3}$$

10.3   Defining the capacity need

The security analysis theory is illustrated with an example. The target level of security is usually expressed as the number of hours/year of potential shortfall in generation to meet demand. A typical figure is 8 h/year and generally relates to the probability of super-grid connected generation being unable to meet the total super-grid demand. A key factor influencing the security is the prevailing plant margin, this being the percentage of excess generation capacity over that required to meet peak demand. A typical value would be around 20% to take account of loss of generation availability, worse than average weather and demand forecast error. Figure 10.2 has been constructed based on a probabilistic model of a system with a

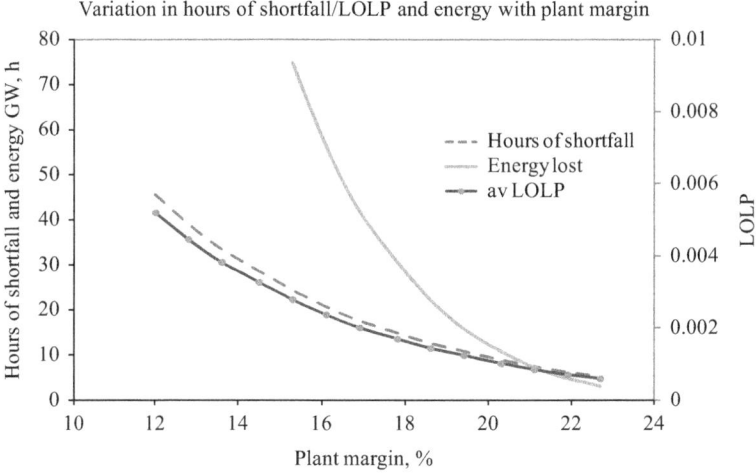

Figure 10.2   LOLP and plant margin

peak demand of 60 GW with an annual energy demand of 325 TW h supplied by a tranche of 146 conventional generators of 500 MW (73 GW) with an assumed overall availability of 0.85. The demand is represented by a function that determines the probability of the demand being at different levels and the generation by the probability of a number of units being unavailable. The figure shows the hours of shortfall reducing to around 8 h/year with a 20% plant margin (72 GW) with the equivalent LOLP shown on the secondary axis. The graph also shows the associated potential value of unserved energy in GW h/year that increases exponentially with lower plant margins. The availability represents that due to planned and forced outages to also take account of the potential for shortfalls to occur during the maintenance programme when cold periods occur before generation is returned to service. The overall generation availability is used to determine the probability of a number of units being out of service for different periods of the year. With the same scenario, if the annual generation availability were to be improved to 0.9 then the same security standard would be realised using 137 units of 500 or 4,500 MW (6.2%) less capacity. This percentage improvement in availability of 5.9% (0.05/0.85) matches the reduction in required capacity of 6.2% (5,400/73,000) expressed as a percentage of the total installed capacity. Some commentators base margin analysis on the derated capacity of generation based on normal availability figures as opposed to nameplate values. On this basis, the 73 GW with a derated availability to 85% provides a margin of just 3.4%. This approach does not lend itself to probability analysis including all the variables.

The loss of load in this analysis relates to that resulting from generation shortfall that could occur without SO intervention. There is also a potential loss due to network disturbances that may be higher at the distribution level but affecting fewer consumers and an overall consistent approach is needed taking account of embedded generation. It is pointless having high generation capacity margins when more demand is likely to be lost due to network failures.

10.4   Wind generation contribution

The analysis of the probable wind contribution can be based on recorded data or a mathematical model. Individual wind turbine output profiles are shown to follow a Weibull function and this could be used as representative of aggregate wind generation output as shown in Figure 10.3. The process steps are

1. Establish shortfall hours based on conventional generation capacity
2. Reduce conventional generation and add in wind generation
3. Include each level of wind output in turn to supplement conventional generation
4. Record the number of hours of capacity shortfall times the probability of each wind output
5. Repeat the process for each wind output level

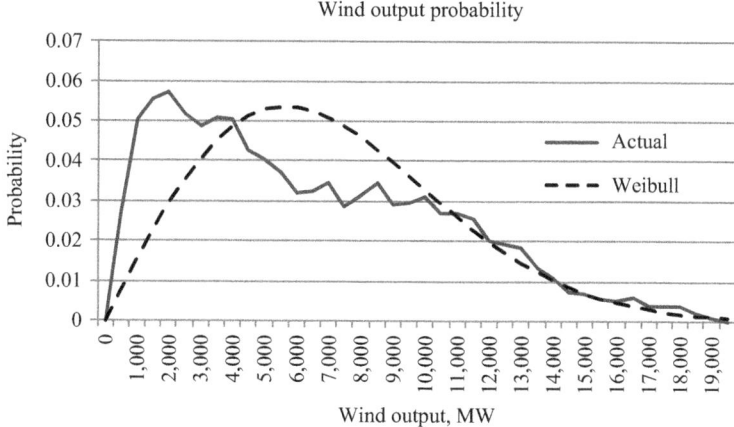

Figure 10.3   Aggregated wind probability

6.   Summate all the hours of shortfall for comparison with the initial shortfall hours with no wind
7.   Adjust total wind capacity until the original security level is matched
8.   The displaced conventional generation gives a measure of the wind capacity contribution

This analysis indicates a wind capacity contribution of around 15% of installed capacity with load factors (LF) of 30% falling to 12% with 25% LF. An alternative approach is to simulate an annual half hourly dispatch of generation both with and without the wind generation output subtracted from demand. This will show the generation that is not used when the wind energy is included compared to that without wind for a specific set of data for annual demand and wind generation. In this example, it was found that 25 GW of wind capacity displaced around 3.4 GW of conventional generation indicating a firm capacity contribution of 13.6% (3.4/25). This is consistent with the results of the probabilistic analysis and is in the same range as the figures used in Ireland who have experience of a large tranche of wind generation. Given the future expected conventional generation and wind capacity and expected load factor the reliability can be predicted.

10.5   Economic analysis

In determining the target generation capacity and margin, consideration needs to be given to the impact on consumers. This is represented by attributing a value to the lost load (VLL). This is usually expressed as a cost/kW h with average values around €10/kW h (£8.5/kW h). This is dependent on the customer and type of load with values ranging from around €5–15/kW h. Given the potential cost to

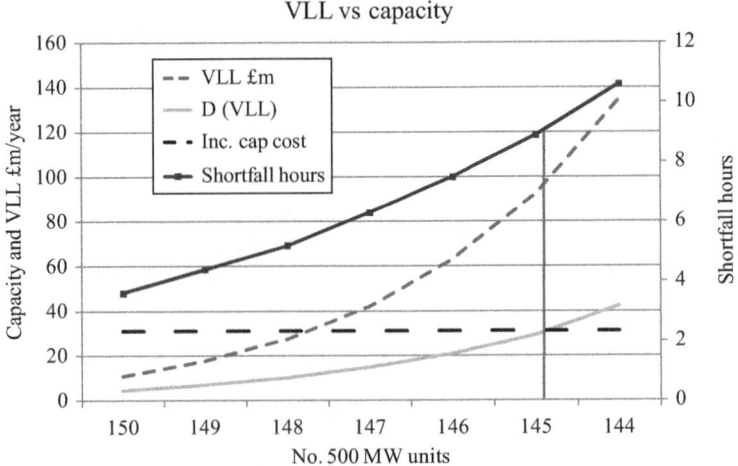

Figure 10.4   Optimal capacity

consumers of unserved energy and the annual cost of providing new capacity, the optimum can be established. This occurs when the costs of an increment of new capacity equates to the reduction in unserved energy priced at VLL.

The results of a typical analysis, based on the same scenario, are shown in Figure 10.4 and include the incremental value of lost consumer demand [D (VLL) shown as a solid line] and the incremental cost of new capacity (as a horizontal dashed line) estimated at £30m/year/unit. In this example, the two lines intersect at the cost of £30m/year with 9 h of shortfall per year as read from the secondary axis. With a 145 units, the VLL/year is £92.4m reducing to £63.1m with 146 units with the change at £29.3m/year matching the annual fixed cost of the new unit at £30m. The limitation of this general approach is that it is difficult to distinguish between the security needs of a small business reliant on supplies for its communication and computing infrastructure and the retired pensioner living in fuel poverty and having to switch off appliances to avoid high costs. Also, the analysis is based on one-off contracted new capacity rather than a general capacity payment to all generators.

It is generally assumed by economists that general payments for capacity will be offset by a commensurate reduction in wholesale energy prices. In practice, the developing requirements to balance wind intermittency are most economically met by open cycle gas turbines. This is because they have low capital costs and are flexible but their running costs maybe 50% higher than combined-cycle gas turbines due to their lower efficiency. In the absence of alternative balancing options, like hydro, open cycle gas turbines (OCGTs) are likely to often set a high marginal price. Bid monitoring to check the bid legitimacy compared to expected actual running costs is used in some market implementations but would not help in this situation.

10.6   Capacity procurement

Mechanisms need to be established to secure capacity provision. Capacity auctions may be used to foster competition in the capacity market and they may be operated on a descending clock format. Pricing is often based around typical OCGT costs, less their expected revenue, with a maximum level set by new entrants with existing generators taking the minimum value. It is necessary to establish what happens to unsuccessful bidders. A lead time of a few years is necessary to take account of the time taken to establish consent and construct a station. It is also desirable to facilitate demand side participation, but different lead times may be appropriate reflecting the nature of their business. A system of penalties is considered necessary to ensure high levels of delivery in the event for both the demand and generation side.

There is the option for the SO to contract for specific capacity additions to meet security requirements rather than introduce an overall system capacity market. This has the advantage of minimising the additional cost and containing the influence on the wholesale market. This is a particularly relevant approach where the requirement is to balance intermittent generation like wind. It is analogous to the SO contracting for support through interconnectors.

10.7   The impact of interconnection

With increasing levels of international interconnection, there is a tendency in plant margin analysis to attribute capacity contributions based on interconnection ratings. There is a risk in this process of double accounting capacity with adjacent utilities both assuming they can get support from their neighbours. Across Europe, countries could be experiencing similar adverse weather patterns with coincident high demand levels with little spare capacity to provide support. Some support may be expected where one of the utilities is enjoying high levels of availability but they may not be prepared to reduce their margins. There is also the requirement to have interconnector transmission capacity available that may have already been reserved for trading. Arrangements can be put in place to provide support with contracts that recognise the cost of maintaining capacity available. **Support can only be guaranteed when firm capacity contracts are in place** and these are frequently used to maintain secure margins as an alternative to building more capacity locally.

With the development of active distribution networks with more embedded generation, it is becoming increasingly difficult to predict what generation capacity will be available from within the distribution networks. Smaller generation is not subject to central dispatch and metering and may not be available to the grid control engineer. If details of the installed capacity by technology type are available then typical availabilities may be attributed to estimate their potential contribution. The load factors shown in Table 10.1 can be used to estimate the probable capacity running and contributing to security. The average volume weighted load factor of 43% is dominated by gas and combined heat and power (CHP) generation.

Table 10.1 *Embedded generation (UK 2015)*

Zone	Capacity MW	Biomass	CHP	Diesel	Gas	Hydro	Thermal	Waste	Wave	Wind	Total
1	Scott/South.	7.1	0	14	0.2	147	14.4	26.9	26.9	11.2	248
2	Southern	85	216	34	489	0	0	0	0	0	824
3	CE electric	215	227	60	365	5	276	77	0	204	1,429
4	North West	18	195	38	141	0	65	11	0	426	894
5	Eon central	189	162	35	185	0	76	134	0	321	1,102
6	UK Pow. SE	106	534	105	105	0	231	222	0	666	1,969
7	Scott. power	12	22	0	0	91	14	76	0	365	566
8	North West	3	29	0	159	111	0	35	20	330	701
9	North Wales	0	390	12	0	79	0	94	0	398	973
	Total	635	1,775	298	1,444	433	676	676	47	2,721	8,705
	Load factor	**0.5**	**0.6**	**0.29**	**0.5**	**0.5**	**0.35**	**0.46**	**0.2**	**0.286**	
	Total GW h	2,782	9,329	757	6,326	1,897	2,074	2,722	82	6,818	32,786
Emissions	ktCO$_2$	0	2,078	568	2,505	0	1,707	0	0	0	6,859

10.8   Demand side regulation

The advent of smart meters offers the option to exercise a level of control over end user demand that could be used to shift demand from high to lower periods and reduce the overall peak capacity requirement. It is expected that the option should be available to enable end user participation in a capacity market. It is more difficult to monitor and confirm delivery, when a load reduction is invoked. It is also difficult to predict the reduction that may be realised when other factors such as adverse weather may cause demand that is not controlled to be increased. There may also be effective substitution of the controlled demand by other appliances.

There is likely to be some inherent reluctance to allowing external control of in-house programmes of activity that are based on wider considerations, and there will be a need to establish tariff incentives. Schemes that are in place around the world are generally based on energy use that is less sensitive to short-term interruption like air conditioning and heating systems. Schemes implemented in the USA include tariff incentives to encourage participation and have proved cost effective compared to other options. In the United Kingdom, one option that could be available providing flexibility, is managing the charging of electric vehicles as their use expands. **Demand side regulation can make a contribution to balancing intermittency locally but is not likely to prove a panacea for overall system balancing with large tranches of wind generation and other intermittent sources.**

The increasing cost of energy is affecting the economics of embedded generation. Local CHP schemes can realise high levels of efficiency where the waste heat can be utilised. They can also avoid the high costs associated with transmission and distribution network losses when they supply local demand. Small-scale wind farms are being built that can be connected at the distribution level and it may prove attractive to use demand side regulation to support balancing their intermittency.

The developments taking place on the distribution networks moves them from being largely passive networks to more active controlled networks. This requires a more interactive interface between the distribution and transmission system operators to manage the overall system security. It also introduces complication in the process of predicting future demand for generation capacity planning. The traditional forecasting approach used by the transmission system operator was based on historic demand profiles as recorded at the bulk supply points to the distribution systems but these records are now being distorted by control action at the distribution level affecting the demand and output of embedded generation. Use of system tariffs are often based on the peak demand, in the United Kingdom, the triad demands are the peaks as recorded over three separated days. The impact of the use of demand side management is to 'flatten' the peaks to reduce tariff costs but at the expense of any shortfall resulting in a higher level of lost MW h over a longer period.

10.9   Market mechanisms to realise capacity

In many market implementations, specific capacity payments are not included. The generators are expected to recover their fixed costs through the energy contracts with suppliers/distribution companies and end users. The economic theory is that market prices will rise as capacity margins become tighter and eventually reach new entry prices. This establishes a climate when investors will see profit from sponsoring new generation and plant margins will be restored. Because all investors see the same potential to make profits, there may be an overreaction leading to too much additional capacity with reduced market prices leading to investment cycling.

There is also a problem in predicting the potential utilisation and revenue through the long life cycle of generation. This is particularly the case where a large tranche of intermittent generation make up part of the plant mix. With the priority given to using renewable sources whenever available, the utilisation of conventional generation operating at the margin can be seriously undermined.

Another approach to securing availability, that can be used in a Pool Market, is to add a supplement to the selling price based on an assessment of the LOLP derived from analysis of the probable generation availability and the demand profile i.e.

$$\sum \text{LOLP} = \sum_{1,1}^{t,n} \left(H_{nt} \left(D > G_n^t \right) \right) \tag{10.4}$$

where H_{nt} is the number of hours when demand $D_t >$ generation; D_t is the demand at time t; G_n is the output of n generators at time t. The system marginal pool selling price PSP is then increased above the normal system marginal price (SMP) according to the value attributed to lost load and calculated as

$$\text{PSP} = \text{SMP} + \text{LOLP} * (\text{VLL} - \text{SMP}) \tag{10.5}$$

This option is also useful in encouraging generators to make their plant available in the short-term but is not useful in signalling the need for new capacity in the long-term taking account of construction timescales.

A third approach is to establish capacity auctions for several years ahead to allow time for development. Generators can bid their blocks of existing or planned capacity for 4 years ahead into the auction with a price. The demand side can also bid into their capacity to reduce demand 1 year ahead reflecting their flexibility and shorter term production requirements. Capacity below a 2-MW threshold has to aggregate with others to participate. A descending clock auction process can be used where the price is set and generators bid in the capacity they are prepared to supply at that price. The price is then reduced through successive rounds to iterate towards the target capacity with those bidders remaining being awarded contracts at the clearing price. The bids may be technology neutral although some utilities like Eskom have considered technology specific capacity incentives to better manage the development of their plant mix. The process offers a good opportunity for price discovery. It is claimed that having separate capacity payments will reduce the market wholesale price for energy offsetting some of the capacity costs. In practice, the marginal plant setting prices may not receive capacity payments and could be open cycle gas turbines with lower efficiency and higher prices.

10.10   Capacity market Ireland

The Irish power system market is operated as a single entity called the Single Electricity Market (SEM). It is organised as a pool with generators obliged to adhere to a bidding code of Practice by bidding their short run marginal cost (SRMC) with capacity payments to cover other costs. Bilateral trading takes part outside the pool through contracts for differences. Some direct contracts are put in place to manage market power as advised by a market modelling group.

The capacity market in Ireland is part of the joint regulatory arrangements administered by the SEM. All contributors to capacity are paid an annual fee through the capacity payment mechanism based on prices determined by the Regulatory Authority. The prices are based on a best new entry open cycle gas turbine e.g. Alstom GT13E2 costing typically €80/kW/year.

The requirement is based on analysis of some 15 demand scenarios with generation capacity deratings based on technology type and size. The object is to provide a derated MW requirement resulting in a target loss of load expectation (LOLE) of 8 h/year. Typical percentage derating factors are shown in the first row of Table 10.2 with the second row showing the scheduled outage time in days/year.

The target is set to establish the course of least regret comparing the cost of unserved energy priced at the VLL with the additional costs of providing more generation capacity.

In 2016, the capacity requirement was set at around 7,000 MW to meet standard security requirements resulted in an annual capacity payment sum of €560m. This pot of money is distributed to contributors based on the proportion of registered capacity adjusted for outages. Payments are made monthly with 30% year ahead, 40% month ahead and 30% month end. The capacity requirement is calculated using a mixed integer programme model managed by the market operator. The wind generation capacity contribution is based on a derived curve and varies from 16% down to around 10% at higher levels of wind capacity. The curve used is shown in Figure 10.5 with the lower line related to Northern Ireland, the middle line to the Republic and the top line to all Ireland.

Table 10.2   Capacity derating factors

GTs	Steam	Hydro	Demand	Storage
3.8%	7.2%	3.9%	24.7	7.1%
17.4	23.8	45	16.5	18.0

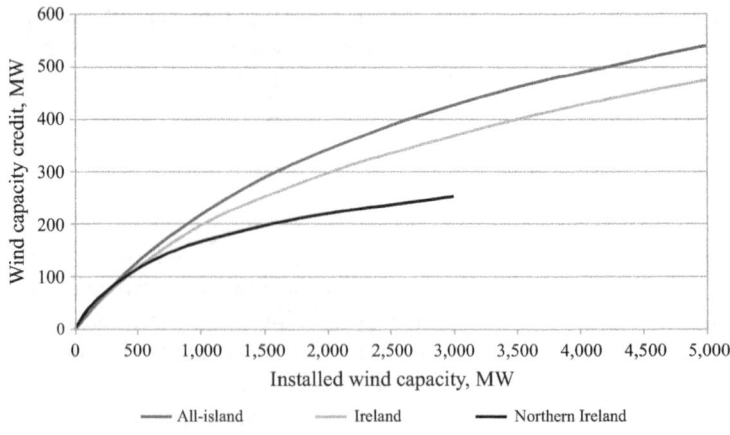

Figure 10.5   Wind capacity credit Ireland

An advantage of the scheme over that applied in the United Kingdom is that all generators with capacity available receive payments, whereas only those contracted through the UK auction process receive payments. The result is that some generators not receiving payments may choose to close having a negative impact on the plant margin. The Irish system also reacts to oversubscription as the amount distributed to each contributor will automatically reduce if the capacity target is exceeded and vice versa.

10.11   Capacity market design – United Kingdom

The UK auction process uses a capacity demand curve with a target capacity set to deliver a reliability standard initially set to a LOLE of 3 h/year. The cost of new entry (CONE) is based on open cycle gas turbine prices and is used as a benchmark initially set to £49,000/MW. A cap is set as a function of the net CONE with no capacity procured above it where

Net CONE = Gross CONE – Expected market revenue

A capacity demand curve of price against capacity is used to enable the trade-off between cost and reliability to be automatically determined through the auction process according to the gradient of the line as shown in Figure 10.6.

The eligible generation includes new and existing capacity, demand side response, storage and aggregated smaller capacity of less than 2 MW. Interconnection is not explicitly included because its availability cannot be guaranteed although a figure is included. Existing generation is expected to be a price taker bidding capacity up to the price threshold. New generation and the demand side can

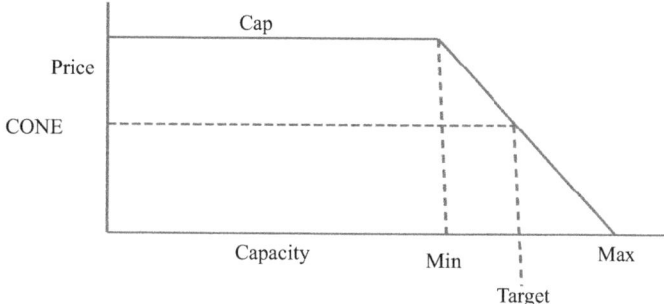

Figure 10.6   Auction price/capacity curve

bid capacity and prices up to the cap. A timetable includes pre-qualification 22 weeks ahead with capacity submissions 3 weeks ahead and prices 10 days ahead. The delivery body apply derating factors based on a 7-year history of winter availability. The capacity is based on that registered in the use of system agreement.

10.12   Capacity auction results

The capacity market auction, introduced in the United Kingdom, was an attempt to secure future capacity by awarding contracts each year to companies prepared to build reliable power plants to meet future demand. **The scheme was intended to encourage the build of new efficient combined cycle gas-fired generation but failed to do so with new small-scale diesel and gas generators bidding capacity at lower prices.** These technologies are highly polluting and contrary to realising reduced emissions and are also less efficient. These smaller generators connect directly into the distribution network and avoid paying the transmission network charges incurred by larger generators. They also get extra revenues from local energy suppliers by helping to reduce their peak triad demand and hence their network charges. Predictably, they were able to undercut offers from larger generators.

What makes distributed generation financially attractive is the potential to also avoid distribution network charges where energy can be supplied by direct wiring. At the lower voltages, the charge for use of the network is of the same order as the energy charges made by suppliers. The network tariffs are designed with a capacity element and a small energy element to reflect losses. The extension of the practice of bypassing the network would have serious implications to the financial viability of the network business. There is a case for rethinking the approach to charging for use of the network reflecting the backup security provided and its wider benefits in providing a trading hub and market liquidity.

The first outcome of the UK capacity market auction for delivery 4 years ahead (2018–19) showed 17 GW oversubscribed with just 49.3 GW contracted at

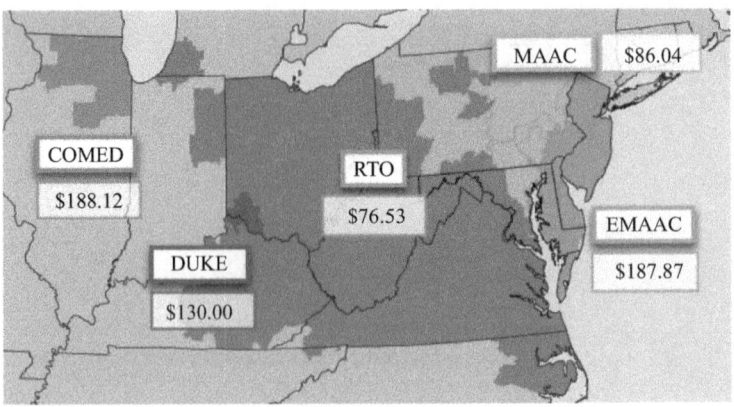

Figure 10.7   US capacity auction (source PJM)

£19.4/kW/year. **The price was much lower than the government estimate of £49/kW/year** resulting from existing capacity seeking to stay open and new low capital cost engines connected at distribution level. The contracted capacity included 57% of existing plant, 25% of refurbished generation and 15% new i.e. 7.4 GW with an average size of 61 MW and mostly connected to the distribution network. The initial submission included 67.4 GW of capacity with 8.4 GW dropping out with the intention to close with just 174 MW of demand side participation. **The net result was to end up with the lowest forecast plant margin on record.** This was illustrated by National Grid Company (NGC), the system operator issuing a capacity shortfall warning in May 2016 following a failure of an import cable and outages for maintenance at a time when there was no wind generation output. As a result of the Notice of Inadequate System Margin alert, one generator had to be paid £1,250/MW h.

10.13   US capacity auctions

The US electric companies in New England (Massachusetts Electric/Narragansett Electric) have been participating in forward capacity market Auctions since 2010. The Pennsylvania Jersey Maryland (PJM) area in the United States used a capacity auction in 2017 with the grid operator procuring a total 165 GW of capacity providing a 23% margin against a target margin of 16.6%. For the period, from June 2020 to May 2021, the prices averaged $76.5/MW/day. Higher prices were recorded in some areas due to transmission constraints and generation retirements with prices falling between 26% and 66% of the net CONE ($288/MW/day) as illustrated in Figure 10.7. **The average PJM price of $76.5/MW/day (£21.4/kW/year) is higher than that realised in the UK auction at £19.4/kW/year** that is equivalent to:

$$£19,400/\text{MW}/\text{year}/365 * 1.3 \ (\$/\pounds) = \mathbf{\$69.1/MW/day}$$

Table 10.3 PJM MW capacity procurement (source PJM)

Type	New gen.	Gen upgrades	Imports	Dem. resp.	Efficiency
20/21	2,389	434	3,997	7,820	1,710
19/20	5,373	155	3,875	10,348	1,515
18/19	2,954	587	4,687	11,084	1,246
17/18	5,927	339	4,525	10,974	1,338
16/17	4,281	1,181	7,482	12,408	1,117
15/16	4,898	447	3,935	14,832	922
14/15	415	341	3,016	14,118	822
151,492	26,237	3,484	31,517	81,584	8,670
%	17	2	21	54	6

The expected UK capacity price of £49/kW/year is much higher and equivalent to $174/MW/day. This compares to the net CONE occurring in the United States of $288/MW/day equivalent to:

$$\$288/MW/day * 365/1,000/1.3 = £80.8/kW/year$$

This is at a low exchange rate of 1.3 in 2017 and at 1.5 would be around £70/kW/year and closer to the UK target figure of £49/kW/year.

The contracted generation capacity has to be 100% available with any imports bid covered by reserved generation and transmission capacity. The procurement figures since the year 14/15 are shown in Table 10.3 together with the percentage mix. It can be seen to include a high proportion of demand response (54%) and imports from adjacent areas (21%) as well as new generation and efficiency improvements.

10.14 Conclusions

There is a fundamental limitation in fully liberalised markets in using short-term energy markets to guide the development of the optimum level of generation capacity to meet future security requirements. The simplistic economic approach is that market price rises will encourage new capacity. But, there are increasing levels of government intervention in the market in the form of subsidies, capacity payments and contracting for nuclear generation aimed at meeting wider environmental and security objectives. Understandably, these have the effect of undermining the market and general investor confidence leading to potential capacity shortfalls in future years. The expansion of intermittent energy sources, embedded generation, demand side regulation and interconnection as discussed in this book further add to the complication. There are also questions to be addressed related to the interaction of intermittent sources with less flexible generation like nuclear and CHP that affect the plant mix. At times of light load, wind output may need to be curtailed to accommodate inflexible generation and maintain sufficient conventional generation reserve in service for the system operator to maintain regulating capacity and secure system operation.

There is an urgent need for the development of an overall indicative energy plan that analyses and quantifies all the issues and enables investors to analyse their options and assess their risks. There are several examples of countries publishing medium-term energy plans providing a background to investment analysis. In the current situation, the subject area is too complex to assume that all will be resolved by market mechanisms and it is too open to influence from lobby groups with particular vested interests. The mixture of generation development also has significant implications to the future requirements of the transmission and distribution systems. The overall costs may run into hundreds of billion pounds and a mechanism is needed to encourage development that is close to the optimum to maintain business competitiveness.

There are a number of areas of research that are necessary to support future regulatory policy related to setting security standards:

- Establishing the capacity contribution to system security that can be expected from renewable sources;
- Facilitating demand side participation in energy and capacity markets through the application of smart meter installations;
- Determining the levels of security that should apply at the transmission and distribution levels consistent with consumer needs and overall system security;
- Ensuring access to the networks and markets by new local embedded generation sources and energy storage systems.
- Restructuring the regulatory price review processes to accommodate the changes in system infrastructure and facilitate innovation;
- Determine a regulatory regime appropriate to the application of smart meter technology to ensure consumers receive a share of any benefits.

As distribution networks become more active with embedded renewable generation, CHP schemes and domestic heat pumps, solar energy systems and electric vehicles, modelling capability will need to be developed to establish the collective impact on demand levels and profiles. The regulatory agencies will play an expanding role in building a framework to meet the changing environment whilst protecting consumer interests.

The probability analysis in this chapter indicates that wind generation contribution to firm capacity is equivalent to around 15% of the installed total capacity with load factors of 30% falling to 12% with 25% LF. This was confirmed by a full year half dispatch with and without wind to identify the conventional generation displaced. It is also noteworthy that on cold days in the United Kingdom, there is often no wind.

Analysis of plant margins will often assume support from interconnectors that may not be available from adjacent countries that are also experiencing adverse weather. Support from interconnectors can only be guaranteed when firm capacity contracts are in place for both generation and interconnector capacity with the neighbour.

The UK capacity scheme was intended to encourage the build of new efficient combined cycle gas-fired generation but failed to do so with new small-scale diesel and gas generators bidding capacity at lower prices. The outturn price of the first auction at £19.4/kW/year was much lower than the government estimate of £49/kW/year that was based on open cycle gas turbine prices. The net result has been to end up with the lowest forecast plant margin on record. The UK capacity prices compare with the average capacity prices realised in capacity auctions in the PJM auction of £21.4/kW/year in the USA.

The contribution to capacity provided by demand side response amounts to 54% of the total procured by PJM in the USA. This may in part be realised by contracting interruptible air conditioning load. The scope in the United Kingdom is currently limited but may change with the take up of electric vehicles and heat pumps.

Question 10.1 Assuming that the data in the table below can be represented by an average unit size of 61 MW and that the volume weighted average availability is 43% use equation 10.1 to calculate the probable availability profile of the embedded generation in hours/year.

Zone	Capacity MW	Biomass	CHP	Diesel	Gas	Hydro	Thermal	Waste	Wave	Wind	Total
1	Scott/South.	7.1	0	14	0.2	147	14.4	26.9	26.9	11.2	248
2	Southern	85	216	34	489	0	0	0	0	0	824
3	CE electric	215	227	60	365	5	276	77	0	204	1,429
4	North West	18	195	38	141	0	65	11	0	426	894
5	Eon central	189	162	35	185	0	76	134	0	321	1,102
6	UK Pow. SE	106	534	105	105	0	231	222	0	666	1,969
7	Scott. power	12	22	0	0	91	0	76	0	365	566
8	North West	3	29	0	159	111	14	35	20	330	701
9	North Wales	0	390	12	0	79	0	94	0	398	973
	Total	635	1,775	298	1,444	433	676	676	47	2,721	8,705
	Load factor	**0.5**	**0.6**	**0.29**	**0.5**	**0.5**	**0.35**	**0.46**	**0.2**	**0.286**	
	Total GW h	2,782	9,329	757	6,326	1,897	2,074	2,722	82	6,818	32,786
Emissions	ktCO$_2$	0	2,078	568	2,505	0	1,707	0	0	0	6,859

Chapter 11

Cross-border trading

11.1 Link benefits

There are three main reasons why investment in interconnection may prove an attractive proposition.

- Energy trading from a source with lower marginal incremental costs of production. This could be the export of nuclear energy as from France to the rest of Europe;
- Reserve sharing agreements where the provision of emergency support can be shared minimising the spare capacity each area has to maintain;
- Importing green energy from countries with excess renewable capacity like hydro or geothermal.

New links may be established on a commercial basis with the capacity auctioned though annual, monthly and daily periods. The United Kingdom uses a cap and floor arrangement that has reduced risks and encouraged investment. The available link capacity is usually much smaller than the physical capacity due to wider system constraints that can vary. Extra capacity may be made available as the expected future system conditions become clearer. The standard approach to determining the available transmission capacity (ATC) is to first determine the realisable total transmission capacity (TTC) from network studies taking into account security standards. There may be some capacity contracted long term referred to as already allocated capacity (AAC). It is also usual to provide a transmission reserve margin (TRM) to cater for unplanned incidents. The available capacity is then given by:

$$\text{ATC} = \text{TTC} - \text{AAC} - \text{TRM}$$

The AAC often relates to the provision of firm power supplies from one system to another and is distinct from opportunity traded energy exchanges in that the supply is backed by capacity and contributes to the plant margin in the receiving system. The transmission capacity available through auctions, the ATC, supports shorter term opportunity trading where marginal prices between the connected systems are sufficiently large to more than cover the cost of reserving the transmission capacity. The TRM is usually set by the TSOs managing each system so that the provision of reserve can be shared.

11.2   Trading evaluation

The evaluation of the potential benefit from trading requires a longer term comparison of system marginal prices. The evaluation needs to establish a view through a period consistent with the project financing arrangements and may be for 20 years ahead. The analysis requires a prediction of fuel prices, of future system demands and generation new entry and closures. Of particular importance is a comparison of load profiles as this affects short-term marginal pricing in each of the coupled systems. This in turn requires an estimate of the marginal price function with respect to demand profile. The maximum benefit from energy exchanges occurs when the transfer across the interconnector results in marginal prices being equal to each other, sometimes referred to as the 'equal lambda criterion'.

Figure 11.1 shows an example of two marginal price functions for two systems that could be coupled by interconnection. **It shows the spot generation level in each system when the two marginal prices equate when meeting a combined demand of 30 GW for a period in time.** In this example, the interconnection flow is not restricted and the optimum transfer is 3.3 GW as calculated below.

The marginal cost function for the two systems can be represented by polynomials as shown:

$$\text{System 'A' marginal price} = 0.02 * D_a^2 + 0.2 * D_a + 15$$
$$\text{System 'B' marginal price} = 0.03 * D_b^2 + 0.3 * D_b + 12$$
$$\text{Cost of production systems isolated} - \text{'A'} = 15{,}000\,(0.02 * 15^2 + 0.2 * 15 + 15)$$
$$= 337{,}500$$
$$\text{'B'} = 20{,}000\,(0.03 * 20^2 + 0.3 * 20 + 12)$$
$$= 600{,}000$$

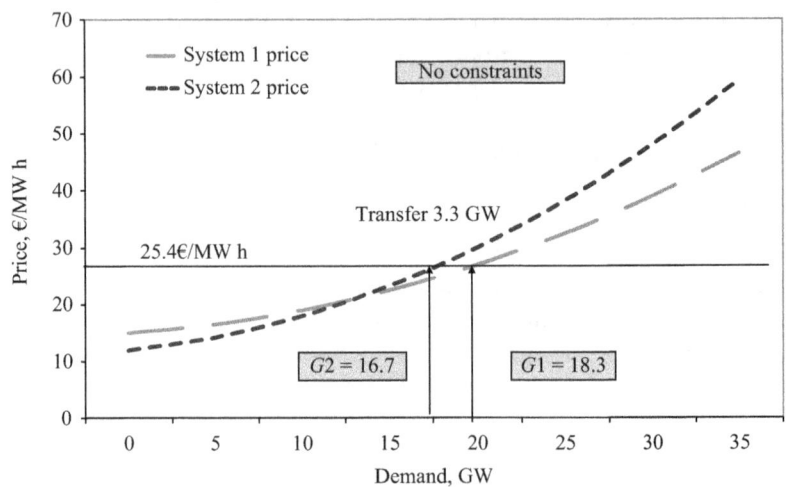

Figure 11.1   Equal lambda plot

Total production cost for 1 h = €937,500
  Minimum cost occurs when marginal prices are equal i.e.

$$0.02G_a^2 + 0.2Ga + 15 = 0.03G_b^2 + 0.2Gb + 12$$

Since Ga + Gb = 35 GW
  Substituting for Gb

$$0.02G_a^2 + 0.2Ga + 15 = 0.03\,(35 - Ga)^2 = 0.3\,(35 - Ga) + 12$$

Re-arranging

$$0.02G_a^2 - 0.03G_a^2 + 0.2Ga + 2.1Ga + 0.3Ga = 12 - 15 + 36.75 + 10.5$$

Re-arranging

$$0.01G_a^2 - 2.6Ga + 44.25 = 0$$

Solving for Ga = $(-b \pm$ root $(b2 - 4ac))/2a$

$$Ga = (2.6 - \text{root}\,(6.76 - 1.77))/0.02$$
$$Ga = (2.6 - 2.2338)/0.02 = 18.3 \text{ GW and transfer}$$
$$= 18.3 - 15 = 3.3 \text{ GW}$$

Marginal overall system price = $0.02 * 18.3^2 + 0.2 * 18.3 + 15$
$$= €25.36/\text{MW h}$$
New energy cost = $25.36 * 35 * 1,000 = €887,523$

It can be seen that enabling the transfer reduces the total production cost for the hour considered from £937,500 down to £887,523 i.e. by €50,000. The solution can also be found using a linear programme formulation with the same result. If the ATC is restricted to 2 GW then a new solution can be found when the two marginal prices are different as would occur with market splitting. In this case, system 'A' has a marginal price of €24.2/MW h and system 'B' of €27.2/MW h i.e. a difference of €3/MW h. The price difference is that charge for use of the link that would cause the minimum cost solution to result in a transfer of 2 GW. A range of solutions with different link capacity charges are shown in Figure 11.2.

It can be seen that the maximum revenue for the interconnector owner occurs when the charge is around €4/MW h with a link transfer around 1.5 GW. For the energy suppliers, the optimum transfer would be 3.3 GW with no charge for use of the link. There may also be other network security considerations that cause the TSOs to opt for a 2-GW link capacity. It will be necessary to evaluate a range of joint system demand levels to build up an annual assessment of potential revenues. The assessment will have to be projected through the life of the project.

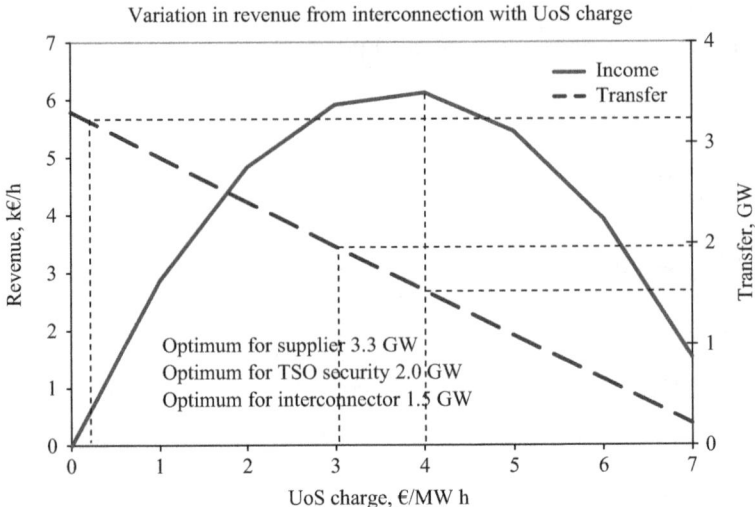

Figure 11.2   Interconnection revenue vs UoS charge

11.3   Renewable energy exchange

The general philosophy is that renewable energy should be fully utilised when it is available i.e. it has priority dispatch. This could apply to hydro power during a rainy period when the output exceeds what can easily be utilised locally. Hydro is also very flexible and could be used across a border to support intermittent generation in an adjacent utility. The hydro capacity in Scandinavia may be used to balance intermittent generation in NW Europe. **It could also apply to surplus wind generation and solar output at times of low load when it may be undesirable to de-load conventional generation to accommodate all the renewable output.** This could be because of the need to maintain system security against contingencies or the sudden reduction in renewable output. Scotland exports a lot of wind generation output to England. Ireland also has untapped wind generation potential. Where the local system is reaching the limit of its capacity to absorb more renewable energy, it could be exported to a neighbouring country rather than curtailing the output. It is also advantageous to maintain loading on nuclear generation because of its low operating costs and emissions. France exports a lot of energy from its nuclear stations around Europe to maintain their units at full load.

By way of example, Iceland has a lot of untapped geothermal energy that could be used to provide renewable energy to the United Kingdom. The length of the link would be about 1,000 km and a direct current (DC) link may be the cheapest option. A typical set of key parameters are as shown in Table 11.1 for a 1,000-MW link.

Table 11.1   Icelandic link parameters

Icelandic link		
Link length	1,000	km
Cost/km	2,500	k Euros
Capital cost	2,500	Million Euros
Capacity	1,000	MW
Project life	20	Years
Interest rate	0.085	Per unit
Utilisation	0.85	Per unit
Losses	6	%
Exchange rates		
1 US$	0.88	Euros
1 US$	0.71	Pounds
1 Euro	0.81	Pounds

The annual transfer is estimated based on an 85% utilisation allowing for forced and planned outages and is given by:

$$1,000 * 0.85 * 8,760 = 7,446,000 \text{ MW h}$$

Net of losses of 6% the delivered energy would be 6,999,240 MW h.

Based on a capital cost of €2.5m euros/km, the transmission capital cost per year is given by

$$2,500 * 0.085 * (1 + 0.085)\vee 20/((1 + 0.085)\vee 20 - 1) = €264\text{m}/\text{year}$$

Dividing by the delivered energy gives a transmission cost per unit of €37.74 or $42.89/MW h (based on the exchange rates shown in Table 11.1). Energy prices in Iceland of $43/MW h have been quoted based on a 20-year contract (as of 2016). The delivered energy cost would be $45.74/MW h taking account of line losses. Adding the transmission and energy costs gives a total delivered cost of $88.64/MW h or £62.93/MW h without a profit margin. This compares favourably with the United Kingdom subsidised onshore wind power feed-in tariffs of £95/MW h and nuclear at £92.5/MW h with offshore wind power at £155/MW h. The price also compares favourably with expected future CCGT generation costs at £50/MW h based on a CO_2 price of £30/t. The results of the analysis of this potential link are as shown in Table 11.2.

11.4   Limits on renewable generation

There is a limit on the proportion of renewable generation that can be managed in operation of a power system. This arises for a number of reasons:

- Alternative generation needs to be available for when the wind does not blow and there is little solar radiation;

Table 11.2   Delivered energy cost (2016)

Transfer	7,446,000	MW h/year	0.85 util
Transfer delivered	6,999,240	MW h/year	Net of losses
Annual cost	264	Million €/year	
Trans. cost/MW h €	37.74	€/MW h	Delivered
Trans. cost/MW h $	42.89	$/MW h	Delivered
Energy 20-year contract	43	$/MW h	At source
Energy cost delivered	45.74	$/MW h	Delivered
Total cost UK, $/MW h	88.64	$/MW h	Trans + energy
Total cost UK, £/MW h	**62.93**	£/MW h	Trans + energy

- The power system requires a proportion of synchronous generation providing inertia that supports frequency management within limits following disturbances;
- A proportion of generation able to provide reactive power is necessary to maintain voltage control and avoid voltage collapse;
- Fast acting backup generation is required to be available for dispatch to balance sudden changes in wind and solar generation;
- The gas grid needs to be able to respond to sudden changes in demand when a lot of gas-fired generation needs to start up to replace sudden falls in renewable output.

Scotland has suggested that it can operate with a very high proportion of renewable generation but only by relying on support through the interconnectors with England. A new £1bn high voltage direct current (HVDC) link is planned from the Clyde to North Wales as much to support Scotland at times of low renewable output as enable exports of renewable energy from Scotland to England. Apart from the high cost of subsidies that UK consumers bear, this would constrain the proportion of renewable generation that could be in operation in England. In turn, Great Britain (GB) relies on support from Europe in its assessment of its generation capacity needs. There is a danger of assuming support at times of low wind output when many countries across Europe may be experiencing light winds and low output. There is often low wind levels associated with very cold conditions lasting for several days and massive storage capacity would be required to begin to balance supply and demand. **In Ireland, the proportion of non-synchronous generation is up to around 50% and is considered just viable based on practical operating experience.** Germany also has a high proportion of wind generation but because of its extensive interconnection with its European neighbours, a lot of the intermittency is effectively exported causing disturbances to neighbouring networks.

11.5   European strategy

The EU has persistently pressed for a fully integrated internal market spanning Europe mitigating price spreads through cross-border trading. At the same time, the need to reduce emissions is recognised with a target of 27% of energy from renewable sources by 2030. To accommodate the intermittency of renewable sources, the need for flexible generation operating in parallel is recognised, and a 'within day' market is planned to manage its operation. The trend towards more decentralised generation is also taking place across Europe and raises new issues. There is a more active distribution and consumer side equipped with smart meters and networks wishing to participate rather than just be a passive user. A number of actions are necessary to facilitate these developments including provision of a new framework with

- Clear price signals for investors;
- Regional coordination on energy policy;
- Cooperation on the development and management of renewables;
- Managing security of supply from a European perspective.

Specific developments include

- An EU wide system to support 'within day' trading;
- A cross-border short-term market facilitating trading up to the event;
- To enable the exploitation of storage, demand side response and flexible generation through market coupling;
- To extend balancing markets to cover larger areas based on network zones rather than national borders.

There is also concern about investment with the market exhibiting long periods of zero-price with generation over-supply in part due to high renewable output across regions. This needs to be offset by allowing high prices at other times to reflect scarcity. It is expected that suppliers and producers should manage volatility on behalf of consumers exploiting demand side response. There is also a need to facilitate longer term contracting to support investment and identify the prime zones for renewable generation. There is a need to incentivise investment in demand side response to support balancing with aggregators playing a coordinating role. **The retail side needs to engage with customers with time varying tariffs linked to the wholesale markets.**

The EU sponsor projects of common interest for interconnection development where market coupling is enhanced. The target for interconnection capacity is that it should be equal to 10% of a country's generation capacity rising to 15% by 2030. The income from congestion charges for the use of links should be used to support new investment. The United Kingdom has plans to reach the 10% target and uses a cap and floor arrangement for new interconnection. The floor protects investors with a minimum return to de-risk the project. The cap avoids excess profits by sponsors to protect consumers. It is recognised that system operators need to coordinate at regional level to manage interconnection flows. There is also a need to harmonise capacity adequacy assessments and system security and review the need for a capacity mechanism. Given the changing role of distribution system operators, there is a need for a better interface to transmission system operators to manage interaction. **Progress on realising a fully integrated European wide single market has been slow and the transformation now taking place is likely to hamper harmonisation because of the large increase in the number of players that are largely autonomous.**

11.6 EU cross-border trading optimisation

To optimise energy exchange across Europe, the EU introduced a trading platform called EUPHEMIA, an Electricity Market Integration Algorithm. It is designed to effect market coupling of seven power exchanges and calculates the transfer energy allocation and prices to maximise the overall welfare rather than for a particular participant. Buyers and sellers of energy submit orders and the algorithm allocates exchanges to maximise the overall benefit. The allocation takes account of transmission limitations based on the ATC and any ramping limitations and can include flow tariffs.

The output includes details of orders accepted or declined and the net position together with the market clearing price for each bidding area and period. Bidders can introduce constraints and submit block orders. To facilitate computation, the orders in each area are aggregated to produce piecewise linear curves processed initially with integer constraints relaxed using a mixed integer quadratic algorithm. Where bids of equal prices occur, allocation is based on a merit order.

There is a problem with modelling trading based on assuming a single transfer limit between adjacent trading areas in that the physical flows follow routes determined by the network impedances rather than traded exchanges between the two areas. Loop flows will occur in adjacent networks that will use the ATC of other interconnectors as well as disturb the planned flows within the networks subject to the loop flows. This problem can be managed by formulating definitions of interconnected physical flows in a 'Flow model' that forms a constraint based on network analysis. This introduces complexity that may make some optimum solutions appear counter intuitive as illustrated in the example in Figure 11.3 that shows a three-area system with different market clearing prices with a flow from 'B' (with MCP of €63/MW h) to 'C' (with MCP (Marginal Cost of Production) of €60/MW h).

In this example, there are constraints on the area net exchanges (nex) and an overall flow constraint given by:

Constraint $0.25 * \text{nex } a - 0.5 * \text{nex } b - 0.25 \text{ nex } c \leq 125$

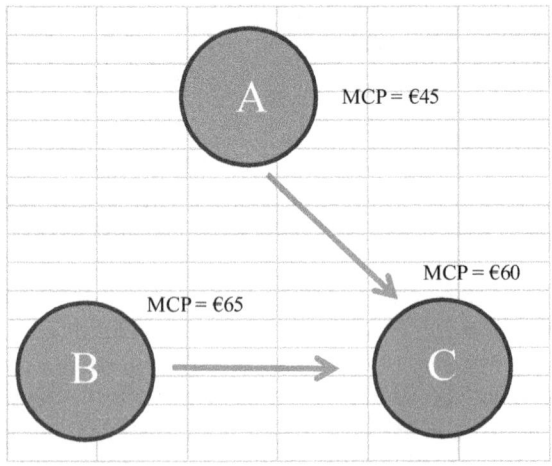

Figure 11.3   Area traded exchanges

Table 11.3   Optimum exchanges

Exchanges		€/MW h	Cost €/h
Nex *a*	250	45	11,250
Nex *b*	0	65	0
Nex *c*	−250	60	−15,000
	0		−3,750
Constraint 0.25 ∗ nex *a* − 0.5 ∗ nex *b* − 0.25 nex *c* ≤ 125			
Limit ≤	125		
Exchanges		€/MW h	Cost €/h
Nex *a*	400	45	18,000
Nex *b*	300	65	19,500
Nex *c*	−700	60	−42,000
	0		−4,500

Table 11.3 shows two solutions that both meet the constraints. The first is based on no exchange from the higher priced area 'B' with a cost saving of just €3,750/h, while the second solution enables a flow from area 'B' resulting in a higher cost saving of €4,500/h. The complex nature of the constraint in this example illustrates the broader issue of trading across borders with associated loop flows through meshed networks.

11.7   Predicting cross-border flows

In evaluating future investments in generation or interconnection, it is necessary to predict market prices. In an interconnected system like the UCTE (Union for the Coordination of the Transmission of Electricity) system, spanning Europe, this will be influenced by future energy exchanges that will affect price relativities between countries. The process involves

- predicting prices for each country node isolated based on a dispatch type simulation;
- estimating interconnector flows based on initial price differentials;
- revising nodal prices based on the change in the net demand altering the marginal plant and prices;
- repeating the process until stable conditions are realised.

The initial flows can be estimated using a simplified system model derived from analysis of the interconnected system using historic data. The recorded energy exchange can be reduced to an average flow for a year along each route and the associated nodal transfers. These can be used to determine the impedance matrix for all routes and hence the admittance matrix. By substituting the nodal transfers into each column, the set of determinants can be found. These are used to estimate a set of relative nodal voltages as illustrated in Table 11.4.

Table 11.4 *European flow model*

| | Determinates | Flow on 100 MVA base = voltage differential * admittance | | | | | | | | | | | | | | Voltages derived | Relative prices |
		Austria	Belgium	France	Germany	Italy	Lux	Netherl	Portugal	CZ	Spain	Switz	UK	Poland	DK		
Austria	−6.5E + 10	0	0	0	454	−180	0	0	0	551	0	−368	0	0	0	0.46	32
Belgium	−3E + 10	0	0	1,004	0	0	−41	−107	0	0	0	0	0	0	0	0.44	32
France	−2.9E + 10	0	−1,004	0	−2,191	−2,437	0	0	0	0	−1,008	−1,010	−1,001	0	0	2.85	20
Germany	−1.8E + 11	−454	0	2,191	0	0	−373	−1,761	0	998	0	814	0	114	240	0.93	30
Italy	−6E + 10	180	0	2,437	0	0	0	0	0	0	0	2,562	0	0	0	−1.75	43
Lux	1.14E + 11	0	41	0	373	0	0	0	0	0	0	0	0	0	0	−0.73	38
Netherl	4.75E + 10	0	107	0	1,761	0	0	0	0	0	0	0	0	0	0	0.32	33
Portugal	−2.1E + 10	0	0	0	0	0	0	0	0	0	203	0	0	0	0	−1.15	40
CZ	7.47E + 10	−551	0	0	−998	0	0	0	0	0	0	0	0	695	0	1.77	25
Spain	−1.1E + 11	0	0	1,008	0	0	0	0	−203	0	0	0	0	0	0	−0.15	35
Switz	9.85E + 09	368	0	1,010	814	−2,562	0	0	0	0	0	0	0	0	0	0.04	34
UK	−1.1E + 11	0	0	1,001	0	0	0	0	0	0	0	0	0	0	0	1.65	26
Poland	−1.6E + 11	0	0	0	−114	0	0	0	0	−695	0	0	0	0	0	2.50	22
Denmark	−1.3E + 11	0	0	0	−240	0	0	0	0	0	0	0	0	0	0	1.93	25
Nodal injections		−456	−856	8,650	−141	−5,179	−414	−1,868	−203	853	−805	371	−1,001	809	240	1.08E−12	

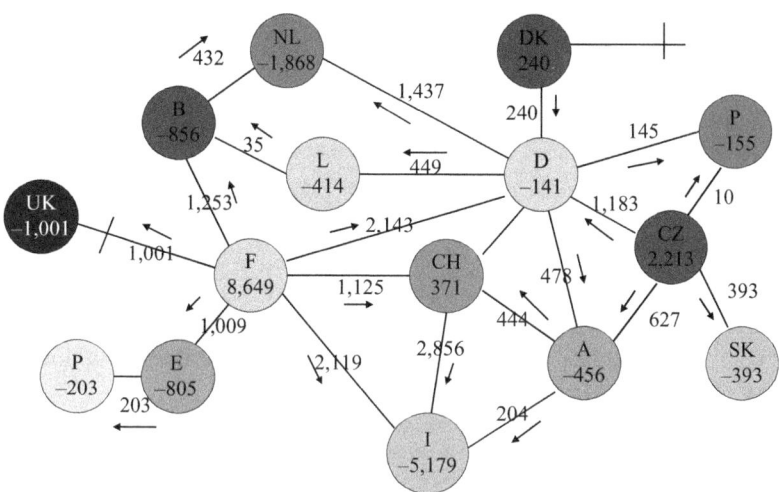

European interconnection – net MW flows

Figure 11.4   European typical average MW flows

The interconnector flows shown in the table are derived from the nodal voltage differences times the admittance. The nodal voltages can be translated into an equivalent nodal price shown in €/MW h. These representative nodal prices are based on a function of each nodal voltage with the highest voltage translated into the lowest price. Future flows can then be estimated by converting the predicted set of price differentials into equivalent nodal voltages and using the admittance matrix to translate these into flows. Each route has a transfer limit and if it becomes congested, a constraint price will be added to restrict the nodal transfer. In being based on recorded flows, the mechanism reflects the physical aspects of the inter-connected network.

The pattern of flows consistent with this model (and recorded flows) is shown in Figure 11.4 together with nodal transfers. It can be seen that France is the largest exporter in this example with an average flow of 8,650 MW. This enables high load factors to be maintained on the high volume of French nuclear capacity. In contrast, Italy is shown as a big importer that reduces the use of local expensive oil-based generation. **The process provides an indication of the average annual energy flows around Europe that is useful to support initial investment analysis and expected utilisation of proposed new generation.**

Ireland is not directly connected to Europe but links through the UK network. Its energy prices are amongst the highest in Europe when the capacity cost is included. Consideration is being given to establishing a direct link from Ireland to Europe to facilitate price convergence.

11.8   Cross-border trading auction

Transmission facilitates operation of the market in enabling wholesale trade between generators and suppliers largely independent of their location. This maximises the liquidity in the market and promotes competition. In general, the larger the consumption in the interconnected market the greater the liquidity and competition. It enables traders for the most part to regard the grid as infinite. In some circumstances, constraints on the network may become active when the difference between the generation and demand in an area of the system exceeds the interconnecting transmission capacity with the rest of the system (having allowed for security requirements). Several approaches are used to deal with this situation:

- Split market and let zonal prices separate until transfers match capacity available (the Nordpool trading area is designed to split when constrained);
- Ignore constraints in market operation – let TSO resolve constraints in real time and share costs through uplift – ex post;
- Explicitly charge for use of inter-connector to manage transfer through bilateral contracts or an auction (Europe).

The market would be split by the market operator when it is apparent that interconnecting flows would be exceeded. This has the advantage that those customers within a constrained zone bear the additional costs as opposed to the cost being spread amongst all users through uplift.

The second option for dealing with internal network constraints is to establish the total market solution and then, in the event, let the grid operator instruct generation so as to manage constraints. In this case, the generator may receive 'constrained off' payments related to lost profit or 'constrained on' payments based on the generators bid into the market.

A third option is to charge for use of interconnection if the demand for capacity by traders across the route exceeds the capacity. This can be realised through auctions where traders bid to reserve capacity a year or perhaps months ahead of the event. Prices are established competitively and have to be recovered by users from the proceeds of the trades. The volumes traded reduce nearer the event as positions become clearer and the ATC is confirmed. Table 11.5 shows the results of

Table 11.5   Trading auction

From	To	Available, MW	Obtained, MW	Price, €/MW h
Elia	Tennet	328	328	4.7
Tennet	Elia	328	327	0.11
RWE Transportnetz Strom	Tennet	356	356	7.14
Tennet	RWE Transportnetz Strom	356	356	0.07
E.On Netz	Tennet	216	216	7.02
Tennet	E.On Netz	216	216	0.01

a typical auction in NW Europe related to interconnectors between the Netherlands (Tennet), Belgium (Elia) and Germany (RWE and E.On).

It shows the MWs available for reservation in both directions, the capacity obtained and the prices paid in €/MW. It can be seen that interconnector capacity from RWE and E.On in Germany to the Netherlands attracts a premium of around €7/MW consistent with the Netherlands being a net importer from Germany as illustrated in Figure 11.4. The Netherlands is also shown to import from Belgium albeit at a lower premium of €4.7/MW of transmission capacity. The prices for export capacity from the Netherlands are a small fraction of a Euro and notional.

11.9 Conclusions

The expansion of interconnection capacity in Europe is targeted to be 10% of generation capacity rising to 15% by 2030. The objective is to facilitate cross-border trading resulting in price convergence. It will also provide benefit in enabling reserve sharing and in managing the intermittency of renewable generation through flexibility sharing.

There is a danger in crediting interconnection capacity with a contribution to the plant margin. There is no guarantee of availability of generation or link capacity unless it is backed by firm capacity contracts. Adverse weather may affect wide areas influencing the availability of generation in neighbouring countries at the same time reducing any spare capacity.

The evaluation of the potential benefit of trading across a proposed new link needs to be based on a comparison of expected marginal prices through the life of the link. This in turn requires an analysis for each system of future demand, fuel prices and generation additions and closures. The respective profiles of demand will also be relevant in evaluating potential within day trading opportunities. The process is made more complicated with interconnected networks with associated loop flows. In some instances where the plant mixture is very different, there may be clear opportunities e.g. exploiting Norwegian hydro to balance wind intermittency; using spare French nuclear capacity to displace the use of oil in Italy; a link from Iceland to the United Kingdom to export spare geothermal energy.

From an investment perspective, it's the long-term potential for energy trading rather than 'within day' trading that will determine the likely return for sponsors. This can be reviewed by an analysis of historic energy flows and an understanding of their rationale. The process provides an indication of the average annual energy flows around Europe that is useful to support initial investment analysis and expected utilisation of proposed new generation.

Interconnection can also support the wider use of renewable generation. There are technical constraints on the proportion of non-synchronous generation in operation at any one time. These relate to ensuring system reserve capability to cater for changes in renewable output and managing frequency changes following sudden loss of generation. The non-synchronous renewable generation does not contribute to the system inertia that slows the initial frequency change. The rapidly falling frequency may cause other generation to trip initiating a cascade in loss of capacity. In Ireland, a proportion of non-synchronous generation of up to around 50% is considered just viable based on practical operating experience. The availability of additional interconnection to England would enable more renewable generation to be installed in Ireland and Scotland with the excess energy exported.

The EU sponsorship of interconnection development is based on the perceived potential improvement in market coupling. But, progress on realising a fully integrated European-wide single market has been slow and the transformation now taking place in the industry is likely to hamper harmonisation because of the large increase in the number of players that are largely autonomous.

Question 11.1 To evaluate risks, calculate the impact on the price of energy delivered to the United Kingdom from Iceland if the link cost/km is €3,000k rather than €2,500k and also if the link availability reduces utilisation to 75% rather than 85%.

Icelandic link		
Link length	1,000	km
Cost/km	2,500	k Euros
Capital cost	2,500	Million Euros
Capacity	1,000	MW
Project life	20	Years
Interest rate	0.085	Per unit
Utilisation	0.85	Per unit
Losses	6	%
Exchange rates		
1 US$	0.88	Euros
1 US$	0.71	Pounds
1 Euro	0.81	Pounds

Chapter 12
Demand side management

12.1 Scope

The introduction of large tranches of renewable intermittent generation onto power systems has highlighted the need for mechanisms to profile demand to smooth and better match the renewable generation output to prevailing demand. It is envisaged that a proportion of the intermittency could be managed by regulating the demand side at the grid, distribution or end-user level. This chapter reviews the potential applications that may create the returns to finance developments and encourage participation. There are schemes that operate at the supergrid level often administered by the system operator and others that operate at the distribution level that may be managed by the supply companies. The system-wide schemes tend to be automatic or directly controlled and are designed to support primary system operation in adverse circumstances and include

- Tripping demand on low frequency relays following generation loss incidents;
- Reduce voltages to consumers by tap changing on bulk supply point transformers;
- Direct switching of demand contracted to accept interruption for a few hours each year;
- Pumped storage schemes that may be controlled by the SO or react to system prices;
- Contracted system reserve operating automatically or under instruction through various timescales.

At the supplier/distribution level, the demand control mechanisms tend to be based on schemes with voluntary participation or encouraged by price signals including

- Real-time pricing signals to encourage consumers to switch demand to lower price periods;
- Tariffs with variable time-of-day prices and a maximum demand element;
- Tariffs to suppliers for use of system charges with a maximum demand clause;
- Direct control of the demand of participating consumers linked to smart metering;
- Minimising peak demands to reduce network charges for suppliers and end users using embedded generation or direct control of local demand that can accept short interruptions;

- Providing short-term system reserve with typically 10-min notice and up to 2-h delivery using grid or embedded OCGTs and diesel engines to reduce demand;
- Providing capacity to the generation market from local generation sources or by demand reduction in response to instruction;
- Deferring capital investment in network transformers or circuits by managing peak periods.

It is possible that qualifying installations could participate in several market sectors establishing viable revenue streams to support investment. **But, it should be recognised that some market requirements are limited in scale like reserve provision or asset replacement deferral.** Also, savings in containing system peaks only saves money if generation capacity and capacity payments can be reduced.

The range of typical customer demands is illustrated by those defined in Eurostat as shown in Table 12.1. The table also shows how a total system demand can be synthesised with a typical number of consumers allocated to each category. In this example, the system has 26.6 million users with a total annual demand of 404 TW h. It provides a useful basis for analysing the scope for demand side control. The United

Table 12.1   Eurostat demand types

Sector	Type	Capacity, MW	Energy, MW h	Consumers	Energy, GW h
Domestic, TW h 109	Da	0.003	0.6	8,155,507	4,893
	Db	0.0035	1.2	6,265,797	7,519
	Dc	0.0065	3.5	5,063,969	17,724
	Dd	0.0075	7.5	3,911,478	29,336
	De	0.009	20	2,469,710	49,394
Total				25,866,462	108,866
Commercial and small sized industry TWh 79	la	0.03	30	427,406	12,822
	lb	0.05	50	240,501	12,025
	lc	0.1	160	98,413	15,746
	ld	0.5	1,250	31,041	38,802
Total				797,362	79,395
Medium-sized industry, TW h 81	le	0.5	2,000	8,269	16,538
	lf	2.5	10,000	1,811	18,110
	lg	4	24,000	1,930	46,320
Total				12,010	80,968
Large industry, TW h 135	lh	10	50,000	1,455	72,750
	li	10	70,000	889	62,230
Total				2,344	134,980
Year 2015		Total, all types		26,678,177	404,209

Kingdom identifies two classes of domestic consumer each with a low, medium and high demand. Class I includes a low of 1.9 MW h/year, a medium of 3.1 and a high of 4.6 MW h/year. Class II includes a low of 2.5 MW h/year, a medium of 4.2 and a high of 7.1 MW h/year. The range is smaller than that defined in Eurostat.

The infrastructure to manage demand control involves receiving a signal from a remote source or from a local device like a frequency monitor or energy meter and switching demand. The costs of the control facilities will not be radically influenced by the size of the load. Equally, the effort involved in managing and contracting for services and associated financial transactions will not be influenced much by the demand size. In consequence, it is generally cheaper to set up demand control with larger consumers who are able to offer control of more MWs, and this application area has been widely exploited in the past. It can be seen from Table 12.1 that the **domestic sector represents around 27% of total demand but over 96% of consumers**. The organisation of their involvement in demand side management represents a challenge technically and financially given the small contribution each consumer could make. The roll-out of smart meters to domestic consumers widens the scope and establishes a basis for their involvement. The applications would need to be largely automatic to be cost effective without consumer's intervention other than having the option to opt in or out and override the facility.

12.2   Impact of intermittency on needs

The developing need for demand control is characterised by the intermittent nature of renewable energy sources like wind and solar. One of the problems of high levels of wind capacity is the need to curtail output at times of low system load. This is necessary in order to maintain conventional generation in service either to provide dynamic reserve or because the generation is inflexible like nuclear and CHP. These are not usually designed to provide the fast regulation required to track rapid changes in renewable generation output. The amount depends on the demand level and proportion of inflexible generation. There is also a need to maintain system inertia to contain the initial frequency fall following a system loss incident. It would be valuable to be able to exercise some control over demand to increase it to use the wind generation rather than curtail it. A model, based on a half-hour dispatch, was used to analyse the need for curtailment through a year for a system with an equivalent firm capacity of 70 GW. The percentage of wind energy that needed to be curtailed is shown in Figure 12.1 against the installed wind capacity for two different system demand levels (333 and 376 TW h) and a system with 1.5 GW of more inflexible generation. The curtailment need is minimal with a wind capacity at about 35% of the total generation capacity (i.e. 25 GW) but rises rapidly towards 20% curtailed output with 70% of wind capacity included (50 GW). The current expectation, with priority dispatch, is that this curtailed energy would have to be paid for further adding to the costs of using intermittent renewable generation. The alternative option of using storage or demand side control to balance the intermittency is considered in this chapter.

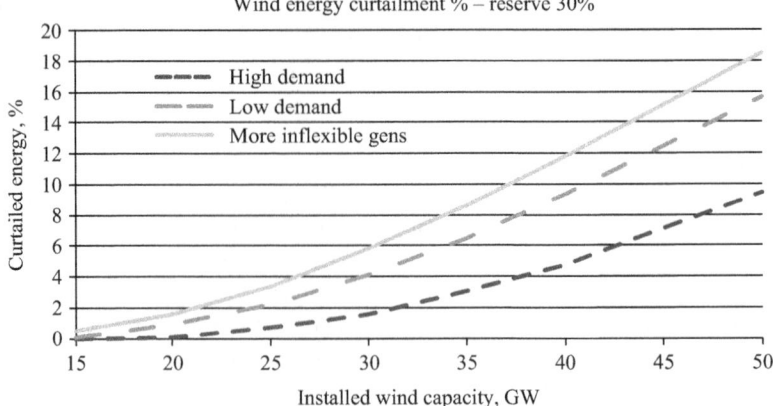

Figure 12.1   Curtailment of wind energy

An indication of the costs of curtailment can be estimated assuming 30 GW of wind capacity when the need for curtailment would be between 2% and 6% of wind annual output. Based on a 25% load factor, annual output would be 65,700 GW h (30*8,760*0.25) with curtailment between 1,314 and 3,942 GW h. Assuming a FIT of £95/MW h, the cost of full compensation for lost output would be between

£125m/year (1,314,000 MW h ∗ £95/MW h) and £375 m/year.

The potential costs increase as the wind capacity is increased in relation to system demand as there will be more periods when wind generation has to be curtailed. A detailed analysis shows that with 2.2% of energy curtailed, the average MW curtailment was 2,448 MW over 963 hhr with a peak of 10,196 MW. With a lower demand and higher curtailment of 4.1%, the average was 2,782 MW over 1,591 half hours with a peak of 11,687 MW. This could be managed by 10 million consumers adjusting their demand by 1 kW. **The cost saving would be around £300m/year or £30/consumer/year and may justify the infrastructure cost and incentives for consumer to participate.**

In addition to accommodating high proportions of intermittent generation during low demand periods, costs are also incurred in regulating conventional generation to track the changes in renewable output. This results in less efficient operation with more frequent stops and starts and ramping burning more fuel and increasing emissions.

12.3   Pumped storage

The use of a pumped storage scheme to smooth the wind output profile is an option worth investigating. This analysis was based on a dedicated 2,000-MW installation with a storage capacity assumed to be 5 h at full load i.e. 10,000 MW h. These parameters were set in as variables that could be readily changed. The overall

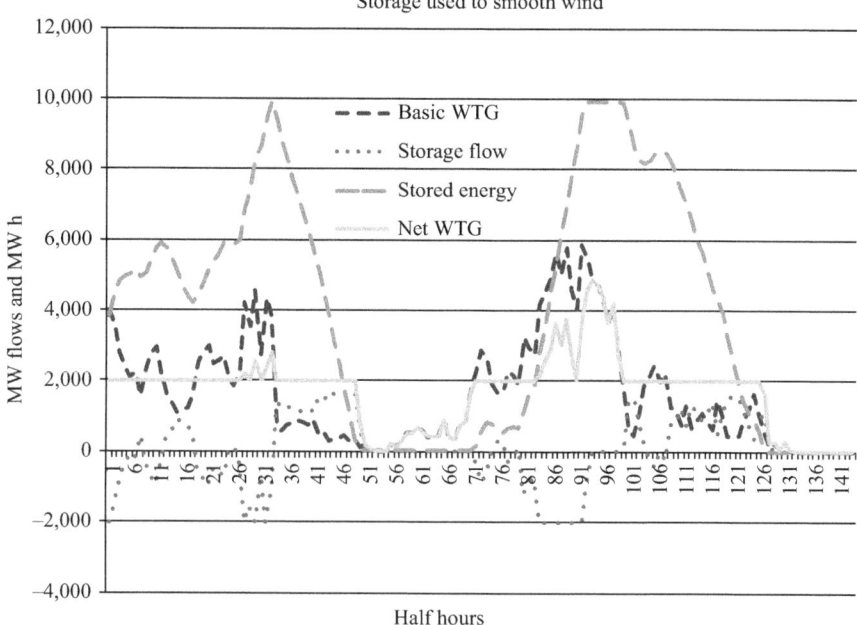

Figure 12.2   Smoothing wind output using storage

pumping/generating cycle efficiency is assumed to be 70% including 15% loss of energy in pumping with a further 15% in generating. A prototype model was built to test the principles applied to an installed wind capacity of 6 GW. The results of smoothing wind generation are illustrated in Figure 12.2.

The dashed line shows a typical wind output over a 3-day period and the solid line the smoothed net output that results when energy is taken in/out of storage. It can be seen to be flat for long periods of time. The flow in/out of storage can be seen to be limited by the size of the installation at 2,000 MW. The other constraint is the reservoir capacity assumed to have an upper limit of 10,000 MW h with a lower limit of zero. The upper line shows the stored energy profile and when the limits are reached. The bottom line shows the flow in/out of storage and can be seen to be a mirror image of the wind generation, other than when the capacity or reservoir limits are reached.

Given a wind capacity of 6 GW, the point of zero transfer into storage was chosen based on the expected load factor. In this example, a figure of 33% is used giving an average wind output of 2,000 MW. **It can be shown, using an linear program (LP) formulation, that choosing the zero transfer point in/out of storage to be equal to the average wind output maximised the balancing effect.** In this example, the standard deviation of the wind output was reduced from 1,540 to 1,064 MW or by about one-third. There is also a benefit where storage is located close to the wind installation in reducing the transmission requirement. The energy

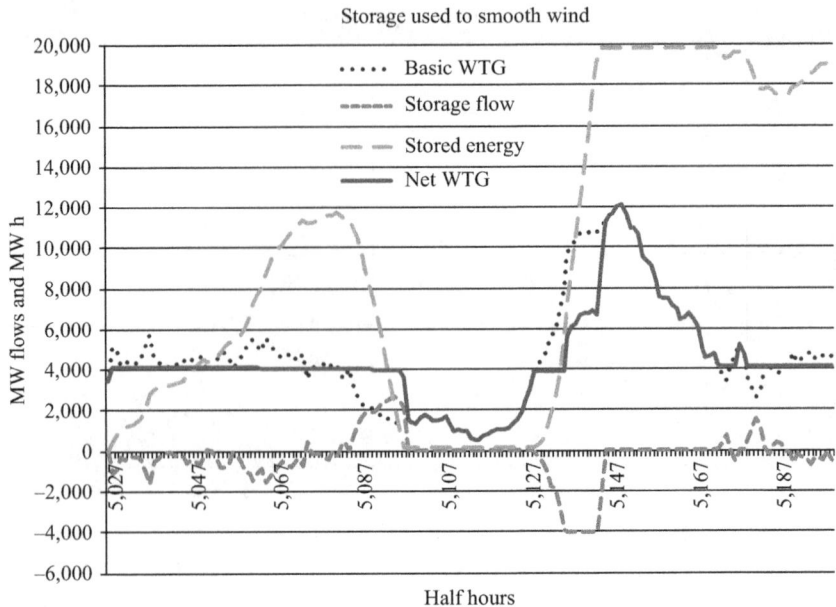

Figure 12.3   Storage used to smooth wind

losses incurred through the period analysed with occasional pumping/generation were 6%.

This approach was applied to a full year's data for 25 GW of wind operating on a system with 70 GW of capacity to establish the financial benefit. Because of the wide variation in wind output, it was found that better results were obtained using a monthly moving average zero transfer target. A section of the result is shown in Figure 12.3 where the smoothed net output is shown by the flat lines. A storage system with a capacity of 20,000 MW h and maximum flow of 4,000 MW was assumed and resulted in a reduction in the standard deviation of the output from 4,200 to 4,060 with wind energy losses of just 1% (compared to an overall pumped storage efficiency of 30% reflecting the low energy throughput of 3.4 TW h compared to wind energy of 53 TW h). The smoothed output was fed into a model to calculate the reduction in operating costs, principally in ramping. **This amounted to £26m/year, but this was largely offset by the cost of replacing the energy lost in the storage cycle.**

12.4   Storage market operation

An alternative approach was tested using storage to buy and sell energy in the market taking advantage of the normal price spread. The details of the installation evaluated are shown in Table 12.2 with capital costs of £800/kW for a 5-h capacity i.e. £160/kW h. In the installation evaluated, the maximum flow was 2,000 MW

Table 12.2   Storage characteristics

		Store, £/kW	800		
		£/kW h	160		
Storage	Capacity, MW		2,000	Storage	10,000
Characteristics	Total cost, £m/year		113	WACC	7
	Efficiency		70	Life, years	70
Optimal operating strategy					
Max pump price, £/MW h	70		Gen price, £/MW h		100
		Revenue, £m/year		115	

with a storage capacity of 10,000 MW h. The overall pump/generation efficiency was assumed to be 70% with a service life of 70 years.

The storage control was set to buy energy when the price dropped below a defined level and sell at prices above a higher level based on the efficiency loss. The approach was tested using the half-hour wholesale market prices derived from a dispatch study with 25 GW of wind capacity. These were based on calculations using the half-hour SRMC and LRMC with an exponential function estimating the market price depending on the level of market competition. The maximum pump price was found to be optimum when set at £70/MW h with generation at prices above £100/MW h. Taking account of the efficiency in storing energy and in generating, this led to a net revenue based on the half-hour flows of £115m/year. **To break even financially, the storage costs would have to be about £800/kW (£160/kW h)** with a WACC of 6% resulting in an annual capital charge of £113m. More revenue could be realised through the provision of balancing services at premium prices. From 2010 to 2015, the cost of lithium-ion battery cells fell from $1,000/kW h down to $400/kW h (£300/kW h). If the trend continues, as is expected, the storage option could become commercially viable by 2020 based on market wholesale prices. At the distribution domestic level energy, charges are around double those of large intensive users. There may be more potential to save money, when time-of-day tariffs become available, making use of storage to shift demand. The domestic storage may be based on the utilisation of electric vehicle batteries.

12.5   Battery storage for frequency control

There is potential to use fast response storage devices, capable of providing output within 1 s, to manage frequency deviations resulting from loss of generation. This option may provide a faster and cheaper solution than using conventional generation operating with free governor action in response to frequency. The provision of generation for regulation in the United Kingdom in 2015 was estimated at £170m/year. The solution using storage devices was expected to save around £50m/year or £200m over a 4-year contract period.

The service would provide a proportional response to frequency deviation on an automatic basis for both rising and falling frequency within 1 s. The delivery would need to be sustained at 100% of capacity for at least 15 min with discharge from maximum to minimum over 30 min. The automatic control system included a dead band typically from 49.95 to 50.05 Hz.

In response to an enquiry, nearly all the bids were based on lithium-ion battery technology with eight successful bids providing 201 MW of capacity at a cost of £66m i.e. around £328/kW of capacity; these included

- Electricity de France (EdF) £12m for 49 MW
- Vattenfall £5.7m for 22 MW
- Eon £3.8m for 10 MW
- Low carbon storage £15.3m for a 10-MW and a 40-MW installation
- Element power £10m for 25 MW
- Belectric £14.6 for 10 MW
- RES £4.2 for 10 MW (plus previous 20 MW installation)

This is for just 15 min of sustained output given an equivalent stored energy price of around £1,312/kW h. The payment for availability would be around £5/MW h e.g. for 15 MW available 20 h/day. The daily payments would be

$$15 * 5 * 20 = £1,500/day.$$

It is important to note that storage systems are economic in this specialist role where fast response is crucial. It would not be viable economically for balancing the intermittency of some generating installations like wind.

12.6 Desalination

In hot climates like the Middle East, there is often a shortage of water, and desalination of sea water is an attractive option that gives some flexibility in when electricity is used. **This demand provides an option to support balancing supply and demand by exploiting the option to store the desalinated water.** This could prove a particularly attractive option where a tranche of renewable generation is installed or planned.

After years of struggling with drought, Australia brought six desalination plants online from 2006 to 2012, investing more than $10 billion. The plants all use some renewable generation for power, mostly from nearby wind farms that feed energy into the grid. The Sydney Water desalination plant, which supplies about 15% of water to Australia's most populous city, is powered by offsets from the 67-turbine Capital Wind Farm about 170 miles to the south.

Solar energy is attractive for many countries that rely on desalination, particularly those in the Middle East and the Caribbean where solar energy is plentiful. In one of the more ambitious projects, the United Arab Emirates energy company Masdar announced in 2013 it's working on the world's largest solar-powered desalination plant, capable of producing more than 22 million gallons/day, with a planned launch in 2020.

The process is usually based on reverse osmosis with water pumped at high pressure through semipermeable membranes with energy making up some 50% of the variable cost. The energy use varies from around 2 to 4 kW h/m^3 or 7.6 to 15.2 kW h/kgal, assuming 10 kW h/kgal a plant producing 10 million gallons a day would consume

$$(10{,}000 \text{ kgal } * \; 10 \; (\text{kW h/kgal}))/1{,}000 \; = \; 100 \text{ MW h/day}$$

This would be equivalent to a standing load of 4.2 MW. Given an average usage of 50 gallons/person a day, to supply a city with a population of 1 million would require 50 million gallons a day requiring 500 MW h/day with a standing demand of over 20 MW. This level could readily be supplied by a mixture of local wind and solar generation given the right climatic conditions.

12.7 Domestic demand control

The deployment of smart meters offers the opportunity to profile domestic demand. ENEL the utility in Italy was one of the first to establish a smart meter programme. The Telegestore project rolled out 31 million meters linked by a telecommunications network with concentrators in every medium voltage sub-station enabling two-way communications. The total cost was reported at €2.1 billion or €70/meter with in-house links to devices at €20. The savings realised were in automatic meter reading as well as peak demand shaving. It was estimated that the information made available to consumers resulted in 57% changing their behaviour including

- 30% moving the use of 'white' appliances to evening hours;
- 12% altered their use of white goods;
- 8% switched off electronic appliances rather than leaving them on standby;
- 7% reduced usage of white goods.

The cost of the automatic reading scheme at €70/consumer compares favourably with other reported costs of €193 for the United Kingdom, €220 in Sweden and €220 in California. It is generally accepted that apart from the benefits of demand profiling, there is a reduction in energy use of around 5% from things like switching off electronic appliances.

Control of air conditioning is a popular application for demand side control in countries where it is widely used but less so in the more northerly countries. The Long Island Power Authority implemented a scheme controlling air conditioning load within 90 s including 20,000 residential consumers and 300 commercial/small industrials. The set-up costs were $515/customer or $487/kW with a one-off payment to domestic consumers of £25 and $50 for small commercial users.

Progress Energy of Florida used radio paging to control water and space heaters and pool pumps with participating residential consumers receiving $11.5/month from November to March or $57.5/year. This enabled a total of up to 5 h of continuous heating and hot water interruption or 16.5 min in any half hour. The domestic utilisation averaged 600 kW h during 5 winter months and 200 kW h during the 7 summer months, equivalent to 4.4 MW h/year or 366 kW h/month. Larger users were paid on a pro rata basis for winter energy above the 600-kW h/month base at $57.5/month. The expected energy deferred/month needs to be set to cater for all the peak periods during a month. The total scheme operating costs are shown in Table 12.3 at $44.8m/year. The initial installation cost was $182/kW.

It can be seen the potential reduction in peak demands is around 1.13 GW at a cost of $44.8m/year and is equivalent to $39.6/kW/year plus set-up costs. This compares favourably with the option of additional CCGT capacity at around $800/kW with annual capital costs of around $75/kW/year (7% interest rate). Assuming that the energy use is diverted to lower demand period, it is reasonable to assume that other fuel and operating costs remain the same. The average set-up costs of $182/kW would be spread over the lifetime of the facility at about $17/kW/year.

The Norwegian distribution company, Kraftnet, used radio and power line carrier to affect demand peak reduction. The scheme involved 11,000 mainly domestic consumers incentivised by a time-of-use tariff (TOUT) of 0.88 NOK/kW h (8 pence/kW h) at peak times (from 07.00 to 11.00 a.m. and 4.00 to 8.00 p.m.) and 0.02 NOK/kW h off-peak. The project set-up cost was 680 NOK/customer (about £60 or $71) and included automatic meter reading facilities. It also had the capability to invoke load control in reaction to spot prices.

The estimated domestic demand in the United Kingdom by 2020 is about 117 TW h with 26 million consumers each consuming on an average 4.1 MW h/year. The make-up of the energy is as shown in Table 12.4 with the estimated average loads assuming full diversity. This is the load that might be expected to be in service at peak times. **It can be seen that the loads that could potentially be controlled, at least for short periods, are space heating and cold appliances amounting to around 3 GW.** Not all consumers would want to participate and take up could be limited to 1–2 GW of peak lopping by domestic consumers. This situation will change if electric vehicle and domestic heat pumps are taken up.

Table 12.3 Demand management scheme – Florida

Consumer class	Residential	Comm./small ind.	Large industrial
Number participating	400,000	4,600	170
Average demand, kW	1.2	100	1,000
Winter, kW h/month	600	36,000	360,000
Total, MW	500	460	170
Incentive, $/year	57.5	3,450	34,500
Annual costs, $m	23	15.9	5.9

Table 12.4   Domestic load composition, United Kingdom

All appliances	Space	Lighting	Cold	Cooking	Electronic	ICT	Wet	Other	Total
GW h/year	9,704	17,214	15,821	13,114	17,014	10,971	14,226	18,386	116,449
Average MW loads	1,108	1,965	1,806	1,497	1,942	1,252	1,624	2,099	13,293
Percentage	8	15	14	11	15	9	12	16	

12.8   Aggregators

There is a role for aggregators to participate in demand side management although they are not usually recognised as direct market participants. It would not be financially or logistically viable for the SO function to directly interface to a large number of end users that have a wide range of potential capability. There is a role for aggregators to act as an interface to the SO and coordinate participation to deliver a viable coordinated service. The demand side participation in balancing has been limited to a few per cent by the limited availability of metering and control facilities. The advent of smart metering opens up new opportunities with an ambition to increase the demand side participation from around 6% up to 30%. In Germany and the United Kingdom, aggregators needed the agreement of registered suppliers to participate, whereas in France, free access was available. **In the United Kingdom in 2016, there were 19 aggregators registered with the SO** of which 11 were involved in DSR with generation, 7 were suppliers with 1 DSR not focussed on the domestic level. The characteristics of domestic demand side response place it in the category of short term operating reserve (STOR) contributing just 7% of the total contracted provision of 3.4 GW. To fully realise the demand side potential to supply services, the barriers to aggregators' participation need to be removed. Their role could be formally recognised, and they could be registered as a balancing market unit. Their involvement in the capacity market may be further enhanced by establishing longer contract periods over which to recover investment.

12.9   Electric vehicles

As more electric vehicles take to the road, their connection to the network introduces new opportunities to manage demand profiles. This may also be necessary to help contain the loads on the network. Figure 12.4 shows an estimate of a typical daily pattern of MW flows for charging 1 million vehicles each with a 50-kW h capacity both with and without a time of use tariff (TOUT). It also shows the percentage of vehicles connected during the day. It can be seen that the cheaper TOU overnight tariff encourages a higher level of charging during that period with more daytime charging without a TOUT. It is assumed that some 55% of the total energy capacity is used through the day and fully restored by the next morning.

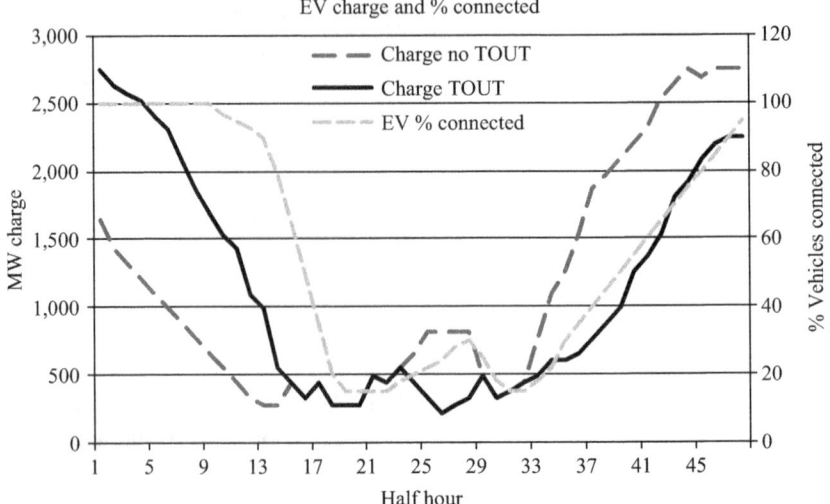

Figure 12.4   EV charging and energy

In terms of the capacity available for demand side regulation, the EV fleet can be considered more as a regulated load than a source of energy. **The scope to draw energy from the batteries is limited by the need to realise a full charge and the financial implications of the cycle efficiency.** The other complication is that the energy stored varies through the day as does the number of connected EVs restricting the potential flow. It is assumed that the total energy stored would not be depleted below a figure set in this example at 20 GW h for this fleet of 1 million vehicles with a combined capacity of 50 GW h. This leaves a potential charge capacity available for regulation varying from 2,800 MW (1,400 MW h/hhr) at midnight down to around 400 MW (200 MW h/hhr) during the day restricted in any half hour by the number of vehicles connected to the grid. There is potential to interrupt charging but limited scope to export battery energy. The charge on individual vehicles will vary widely, and their participation in any demand side regulation would need to be controlled so as not to fall below a set limit. There would need to be an option to deselect participation when long journeys are imminent.

The corresponding profile of energy status associated with charging vehicles is shown in Figure 12.5 based on the right hand scale. It shows the energy input to the vehicle fleet, mostly overnight, and the energy used by the vehicles during the day. The combined profile shows the energy in the fleet rising close to the capacity of 50 GW h by early morning and then reducing to its lowest level by early evening. The potential to receive energy or export energy to the grid will depend on the number of vehicles actually connected during any half hour as shown in Figure 12.4. Overnight, it is expected that most vehicles will be connected and available to be charged or discharged within limits. It is assumed that owners will want their

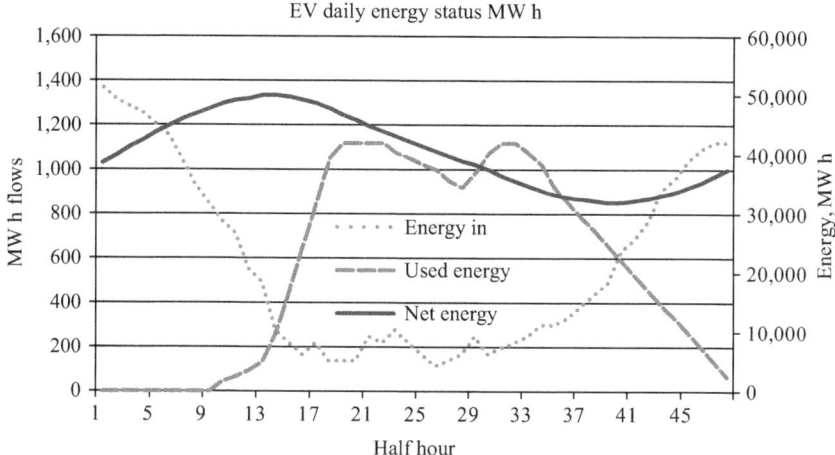

Figure 12.5　EV MW h energy status 1 million 50 kW h EVs

vehicles charged up to an average value of at least 85% of full charge by morning. There will be less concern during the evening about interruption of charging or exporting energy with time to recover the target charge. The potential for energy exchange during the working days will be limited by the number of vehicles connected. **However, there will be a lot of users without off-road parking who will want to charge their vehicles while at work in car parks.**

There is some scope to control the EV load to balance the wind output intermittency, but it is constrained during the day by the number of vehicles connected to the system during any one-half hour and the need to realise close to a full charge by the morning. Figure 12.6 shows a 3-day plot of using control of charging 1 million vehicles to balance 6 GW of wind generation. The charging is controlled to track the periods of high wind output whilst meeting close to daily charge energy requirements.

Whilst the highest peaks are reduced, it can be seen how the number of vehicles connected limits the scope during the day between about 8 a.m. (16 hhr) and 6 p.m. (36 hhr). The maximum flow to EVs was 3,000 MW with the minimum at 450 MW to ensure a satisfactory end charge in this example. Higher charging rates would increase scope but could overload the networks.

The wind generation output data for the United Kingdom has been processed to identify any time-of-day bias through the year. It can be seen that the variation is minimal with a slight preference to values some 7% higher during daylight hours from around 9 a.m. to 10 p.m. as shown in Figure 12.7. **The implication is that the ability of EVs to smooth wind intermittency will be largely limited to the 12-h a day outside working hours.** Based on recorded data, the estimated standard deviation of wind data in the United Kingdom during a year was 4,200 MW for an installed capacity of 25 GW.

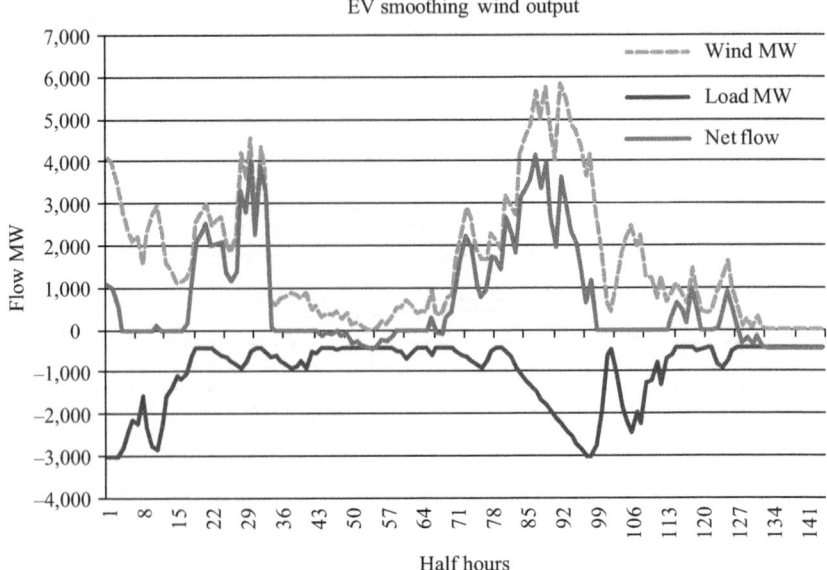

Figure 12.6   EV load control smoothing wind

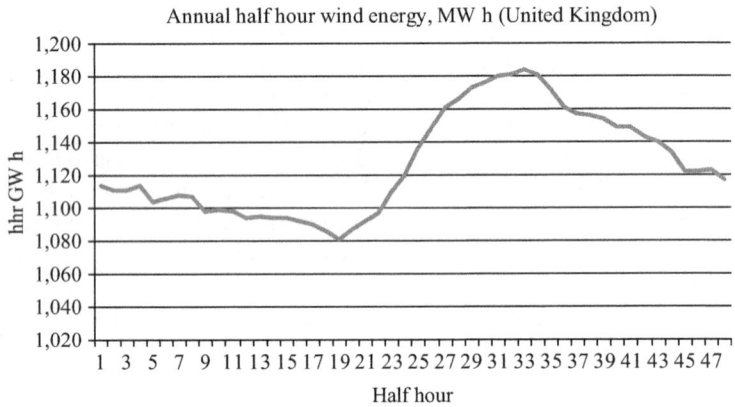

Figure 12.7   Wind energy by half hour (the United Kingdom)

12.10   Industrial and commercial demand control

The voluntary management of the demand of larger premises and factories is more established and is encouraged by the tariff structures as well as bilateral contracts. Tariffs will usually include a capacity element but the proportion varies between countries. Table 12.5 shows the **mixture of tariff power/capacity and energy elements with capacity varying from zero in Hungary up to 77% in Estonia,**

Table 12.5   European tariff structures (source ETSO)

	Power part	Energy part		Power part	Energy part
Austria	25%	75%	Ireland	34%	66%
Belgium	48%	52%	Italy	0%	100%
Czech Republic	11%	89%	Lithuania	72%	28%
Denmark	0%	100%	Netherlands	67%	33%
Estonia	77%	23%	Norway	22%	78%
Finland	0%	100%	Poland	25%	75%
France	37%	63%	Portugal	2%	98%
Germany	87%	13%	Romania	0%	100%
Great Britain	60%	40%	Slovak Republic	9%	91%
Greece	59%	41%	Slovenia	49%	51%
Hungary	0%	100%	Spain	31%	69%
			Sweden	56%	44%

Table 12.6   Large user tariffs

	RWE			**Netherlands**	
Tariffs			Tariffs		
Capacity tariff	€/kW/year	23.28	Capacity tariff	€/kW/year	10.27
Commodity	€c/kW h	0.19	Commodity	€/kW h	0.004
Monthly, kW			Monthly, kW	€/kW/month	1.03
Standing+meters	€/month	377	Standing+meters	€/year	7,800
Maintenance			Maintenance	€/year	
System service			System service	€/kW h	
Assumed load			**Assumed load**		
Capacity	MW	10	Capacity	MW	10
Energy	MW h	70,000	Energy	MW h	70,000
Monthly peak	h	7,000	Monthly peak	MW	1.6
Charges	Charges		**Charges**		
Monthly capacity	€/year	232,800	Monthly capacity	€/year	19,776
Annual capacity	€/year	133,000	Annual capacity	€/year	102,700
Commodity	€/year	365,800	Commodity	€/year	280,000
Total			Total	€/year	402,476
Other fixed charges	€/year	4,524	Other fixed charges	€/year	7,800
Total charges	€/year	370,324	Total charges	€/year	410,276
Charges	**€/MW h**	**5.3**	**Charges**	**€/MW h**	**5.9**

while the energy part varies from 100% in Italy down to 11% in the Czech Republic. There are also variations across the different customer types.

It would be expected that cheap hydro energy, often remotely sited, would lead to a higher proportion of network charges to cover the network capacity costs and losses, whereas dependence on imported oil and gas would lead to higher energy charges but there are variations between different customer types and apparent inconsistencies. **High tariff capacity charges will encourage large system users to minimise their peak demand using their own demand control.** There are also specific contracts between large users and system operators that are usually based on reducing demand for several hours during a year under instruction through advance notices from the SO.

Two typical tariffs for large consumers in Germany and the Netherlands are shown in Table 12.6 with similar total costs/MW h. It can be seen that the capacity charge predominates in Germany representing 63% of the average charge, whereas in the Netherlands, the energy/commodity charge is larger at 70%. The consequence is that there would be more incentive to reduce the peak demand and capacity requirement in Germany, whereas in the Netherlands, the focus would be on energy efficiency. **Tariffs do not currently directly value flexibility other than through time-of-day energy pricing.** This is feasible for large users, but they will have limited scope to adjust their planned production programmes at short notice.

12.11   Conclusions

Demand control has always been used in the power sector primarily to reduce short-duration demand peaks. At the system-wide level, techniques include tripping demand on low frequency relays, direct control of contracted consumers demand, pumped storage schemes and voltage reduction. These applications have generally been restricted to larger consumers where the control infrastructure costs/MW is small compared to the domestic sector. Applications at the distribution level have generally focussed on local peak demand reduction to defer the need for reinforcement. The United Kingdom domestic demand represents around 30% of the total demand but with some 95% of consumers indicating the scale of the problem.

In the current situation, the wider application of demand side control is driven by the need to manage the intermittency of increasing levels of renewable generation. In its wake will follow the expected increase in network and generation capacity requirements to support decarbonisation of heat and transport. The roll-out of smart meters offers the potential to engage the smaller consumers in this demand control through time-of-day tariffs or through participation in direct control of selected loads. The current scope would be limited to a few GWs and would not be sufficient to smooth the output of a large tranche of wind farms. As the wind capacity reaches around 50% of the conventional capacity, there are periods of low system demand when wind output has to be curtailed. Detailed modelling showed curtailment average levels of around 3 GW over 1,600 hhrs/year costing over

£300m/year to compensate for lost FIT payments. This could be managed by 10 million consumers increasing their demand by just 1 kW to avoid the need for any compensation payments worth £30/consumer/year. The potential in the domestic sector is expected to increase further as and when electric vehicles and domestic heat pumps are deployed. A fleet of a million vehicles would, depending on the time of day, have a charge demand varying from around 400 up to 2,800 MW that could be regulated. There is currently scope to influence the design of these schemes to include sufficient energy capacity to support DSR.

Storage schemes based principally on pumped hydro have featured in power system operation as a mechanism to lop peaks. The wider application is limited by the cycle efficiency that may be as low as 70% necessitating a wide spread in peak/trough prices for arbitrage to be viable. Lithium battery technology is being deployed in small volumes to provide fast response to arrest frequency changes following incidents. These applications are viable in specialist areas where the speed of response is critical but would not, at current price projections, be viable smoothing the output of large tranches of wind generation.

Current tariff design is very varied and does not specifically encourage flexibility. There are wide variations in the proportion of charges related to capacity compared to energy. In Hungary, there was no capacity charge with Estonia at 77%, while the energy part varies from 100% in Italy down to 11% in the Czech Republic. A high-capacity charge encourages consumers to manage their own peaks but, in balancing renewable output, the requirement may be to increase demand to use up excess wind/solar generation. A variable time-of-day tariff is required to encourage demand profiling but consumers will need some revenue certainty to justify their investment in flexibility. This could in part be covered by a participation fee linked to the range of flexibility offered.

Question 12.1 Compare the total annual cost of realising a reduction in peak demand using DSR with the characteristics as illustrated in the table below with the alternative of installing OCGTs at a capital cost of $400/kW with a 7% interest rate to meet the same level of demand. The DSR set-up cost is assumed to be $182/kW. Calculate the extra fuel costs in using OCGTs instead of CCGTs assuming that CCGT fuel costs are $26.43/MW h compared to OCGT fuel costs of $36.81/MW h with peak lopping for 500 h/year.

Consumer class	Residential	Comm./small ind.	Large industrial
Number participating	400,000	4,600	170
Average demand, kW	1.2	100	1,000
Winter, kW h/month	600	36,000	360,000
Total, MW	500	460	170
Incentive, $/year	57.5	3,450	34,500
Annual costs, $m	23	15.9	5.9

Chapter 13

Emissions and interaction with heat, transport and gas

13.1 Global energy utilisation and emissions

Figure 13.1 shows the global energy consumption by fuel type for 2016. It can be seen that fossil fuels remain dominant with oil at 33%, gas at 24% and coal at 28% with renewable accounting for only 3%. Nuclear adds 4% and hydro 3% of the total of 154,272 TW h/year. To manage a reduction in emissions would require a massive shift in energy utilisation that is unlikely to be realised in the short term.

The statistics of energy use by region are shown in Figure 13.2. It can be seen that the utilisation in the Asia-Pacific region approaches that from the rest of the world with China and India making the largest contribution with a dependency for 50% of energy on coal.

Table 13.1 shows the global annual energy utilisation expressed in the TW h of fuel used (rather than the converted useful energy) and represents an increase of 1% over the previous year. The table also shows estimates of the associated emissions, based on fuel utilisation and carbon content, with a total of 33,700 MtCO$_2$/year. **Figure 13.3 shows the distribution of the emissions by region with those from**

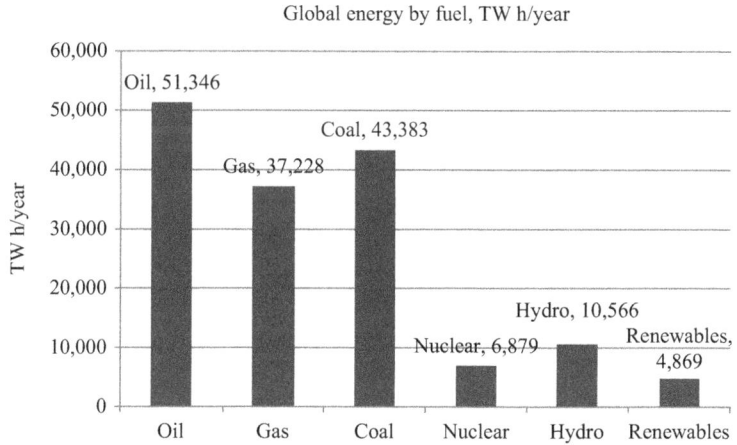

Figure 13.1   Global energy consumption by fuel type (source BP)

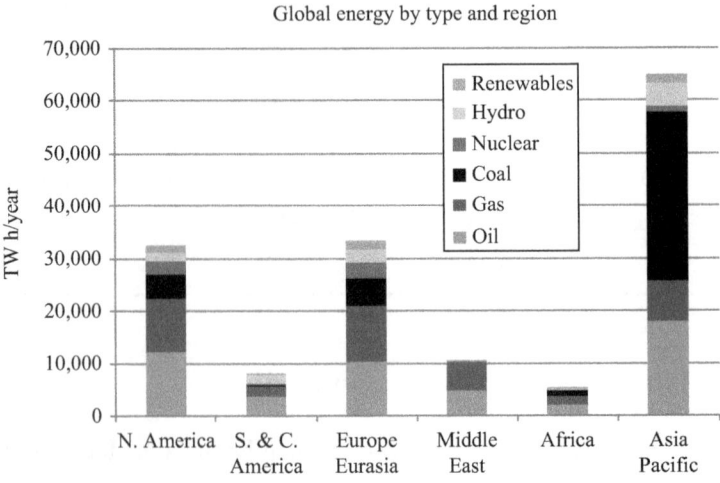

Figure 13.2   Fuel use by region (source BP)

the Asia Pacific region almost equalling those from the rest of the world. This is consistent with most of its energy being derived from coal (49%) followed by oil and gas. Although emissions in North America and South and Central America fell by 2% and 2.4% in 2016, those in the Asia Pacific region increased by 0.9% as did those in the Middle East 1.6% and Africa 1.1%. This resulted in a net global increase of 0.1%. Given the world-wide dependence on coal and other fossil fuels, any significant reduction in the short term looks difficult. One viable option is to retro fit carbon capture and storage to coal plant initially and subsequently gas. The decarbonisation of transport and heating will be helped by a shift to electric vehicles and heat pumps. This will require the production of significantly more electricity with low emissions and efficiency savings. It is reported that in 2016, EVs took 29% of the market in Norway, with the Netherlands at 6.4% and Sweden at 3.4% against a target of 30% and illustrates the potential pace of transport decarbonisation.

13.2   Target emission reduction by sector

The mixture of energy use is examined in more detail using figures for the United Kingdom. To realise the target reduction in overall UK emissions of 15% by 2020, targets were distributed amongst the main sectors as shown in Table 13.2 with electricity generation given the highest reduction of 34%. As well as the distribution of energy the table shows the equivalent renewable TW h required. In practice, some reductions can be realised by choosing a fuel with lower emissions like gas and improving efficiency in energy use. Progress is being made to reduce emissions from **electricity generation against the target of 34% reduction from a base case figure of 247 Mt/CO$_2$ with levels around 112 MtCO$_2$ that already exceed**

Table 13.1 Global fuel and emissions (source BP)

Primary energy, TW h	Oil	Gas	Coal	Nuclear	Hydro	Renewables	Total	CO_2
N. America	12,165	10,304	4,501	2,524	1,779	1,128	32,401	6,441.195
S. & C. America	3,791	1,791	395	64	1,814	326	8,182	1,439.238
Europe Eurasia	10,281	10,769	5,245	3,001	2,338	1,675	33,308	6,261.816
Middle East	4,850	5,361	108	16	55	8	10,398	2,263.747
Africa	2,152	1,442	1,116	42	300	58	5,110	1,171.693
Asia Pacific	18,108	7,560	32,017	1,233	4,280	1,675	64,872	16,122.18
Total	51,346	37,228	43,383	6,879	10,566	4,869	154,272	33,700
Percentage	33	24	28	4	7	3	100	
Asia Pacific, %	28	12	49	2	7	3		
Total fuel burn, TW h	51,346	37,228	43,383					
Emission tCO_2/GJ	0.073	0.0495	0.0869					
Emission tCO_2/MW h	0.2628	0.1782	0.31284					
Emissions by fuel	13,494	6,634	13,572					
Total $MtCO_2$/year			33,700					

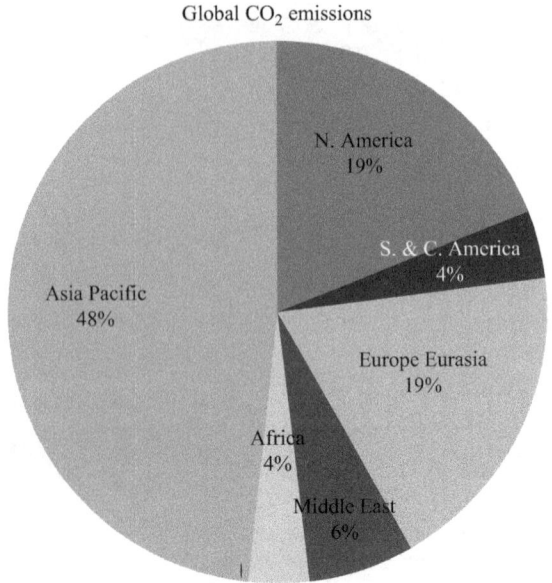

Figure 13.3   Global CO₂ emissions

Table 13.2   Emission targets United Kingdom

	Electricity	Transport	Heating	Total
Total energy, TW h	330	628	675	1,633
Target reduction, %	34	10	10	15
Renewable, TW h	112	63	68	245
Emissions base, MTCO₂/year	247	171	182	600
Emission target, MTCO₂/year	163	154	164	481

the target. This has been largely realised by phasing out coal-fired generation in favour of gas, substituting renewable generation and improving efficiency. In contrast, progress in heating and transport has been slow.

To promote reductions in emissions from heating the UK government introduced a scheme called the domestic renewable heat incentive, issuing certificates to compliant installations through the micro-generation certificate scheme (MCS). Incentives were offered to encourage adoption of the new technologies including biomass 12 p/kW h, heat pumps 7.3 p/kW h and solar 19.2 p/kW h. These developments were intended to encourage the use of technologies with lower emissions and higher efficiencies. In transport, the main incentive scheme has been the introduction of subsidies to encourage the take-up of electric vehicles. Progress has been slow but is being given incentive by the recognition of the impact on health of emissions in cities from diesel vehicles leading to added charges and in some areas a complete ban on use in cities.

Table 13.3   Emissions USA (source IEA)

CO$_2$ Mt/year	Residential	Commercial	Industrial	Transport	Electricity	Total
Coal	0	6	129	0	1,364	1,499
Natural gas	252	175	474	49	530	1,480
Petroleum	67	40	338	1,816	24	2,285
Other	0	0	0	0	7	7
Electricity	723	702	495	4		
Sector total	1,042	923	1,436	1,869	1,925	5,271

Table 13.3 shows the source of emissions in the **USA at about 10 times the targets for the United Kingdom**. It shows electricity generation and transport as the largest sources of emissions. The UK transport contribution is around 28%, whereas that in the USA approaches 36% that may reflect the relative size of the countries. Electricity production is responsible for 37% of the emissions in the USA compared to 34% in the United Kingdom. This reflects a continuing reliance on coal for electricity production in the USA that the government proposes to maintain. However, **US electricity emissions at 1,925 MtCO$_2$/year is less than China at 2,943 MtCO$_2$/year** with an even higher proportion of coal use, whereas the USA has advanced the use of shale gas.

The reduction in emissions across all areas needs to take account of the interaction between the sectors and end-user costs. Solutions advanced to reduce emissions in transport, and heating are often based on using more electricity, effectively loading the problem onto the electricity sector. It needs to be established that this can be managed without adding to emissions from generation and excessive costs. A key factor that will influence the outcome is the time of use. Implementations that use generation capacity at off peak times will offer capital cost savings as will the use of waste heat from generation in CHP schemes. The deployment of storage can also offer benefits if the cost and cycle efficiency can be improved. This chapter discusses some of the technology available and how a total solution may be realised and its implications.

13.3   Heat pumps

Some 20% of total emissions result from domestic use for space and water heating. The energy required can be reduced to a third of that from conventional sources using heat pumps. They operate like a refrigerator in reverse using a compressor to drive refrigerant through a condenser and expansion valve into an evaporator as illustrated in Figure 13.4.

They extract energy from the environment with a seasonal performance factor (SPF) of 3.0 to 4.0 that means that the output energy is three to four times the input energy. Figure 13.5 shows the energy flows of a typical installation based on an air source heat pump with an SPF of 4.0. Higher SPFs can be realised using ground source heat pumps.

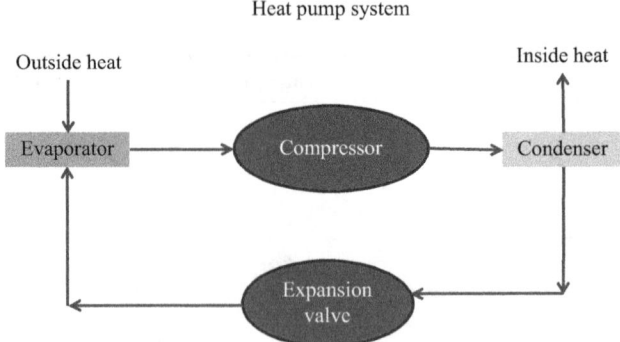

Figure 13.4   Heat pump system

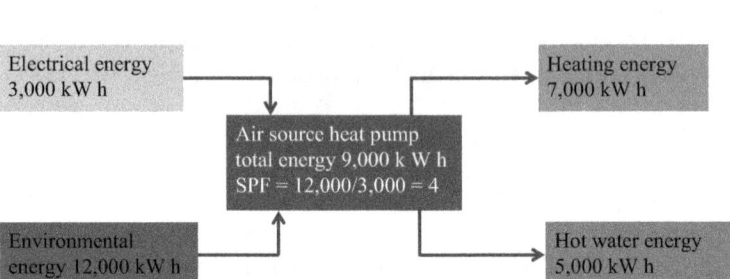

Figure 13.5   Air source heat pump

In this example, the heat pump uses 3,000 kW h/year of electrical energy that extracts 9,000 kW h from the ambient air based on an SPF of 4 providing a total of 12,000 kW h of which 7,000 kW h might be used for heating and 5,000 kW h for hot water. A typical dual fuel domestic user annual demand of 15,100 kW h made up of 3,100 kW h of electricity at 12 p/kW h and 12,000 kW h of gas[1] at 4.3 p/kW h costs

$$(3,100 * 12 + 12,000 * 4.3)/100 = £888/year.$$

Using a heat pump to supply heating and hot water and assuming that the electrical load for other appliances remains the same costs

$$(3,100 * 12 + 3,000 * 12)/100 = £732/year.$$

---

[1] The United Kingdom uses a domestic demand for gas with a low of 8,000, medium of 12,000 and high of 17,000 kW h/year.

Table 13.4 Domestic heating comparisons

Heating type	Air source	Convent.
Capital costs installed	9,700	3,000
Interest rate	0.06	0.06
Life	15	15
Annuity, £/year	999	309
Electricity 3,000 kW h at 12 p/kW h	360	0
Gas 12,000 kW h at 4.3 p/kW h	0	516
Total, /year	1,359	825

The savings in annual operating costs from using a heat pump are £156/year. The installed capital costs are in the region of £9,700 for a heat pump installation (excluding the internal heating system) as opposed to around £3,000 for a conventional boiler. The annual capital and running costs are compared in Table 13.4.

Although the energy costs are £156/year cheaper, this is more than offset by the higher capital charges of the heat pump installation at around £690/year. **The unit price for heat is around 9 p/kW h from the heat pump compared to 5 p/kW h from the conventional boiler.** Although the net energy output is improved, this is offset by gas prices currently being around 1/3 those of electricity per unit of energy. This balance could change if emission costs escalate.

13.4 District heating

This can be fuel efficient when the heat is derived from a local power station facility. It can embrace a broad range of generation technologies, and **because of its inherent storage potential, the energy can also be derived from renewable intermittent sources like wind and solar**. High efficiencies can be realised with relatively low labour and maintenance costs. The local users can also avoid high electrical transmission and distribution costs by direct wiring for electricity supply as well as heat offsetting some of the heat distribution costs.

The temperature difference across the network may be from 100 down to 50°C and influences investment costs;

$$\Delta t = T_{out} - T_{return}; \text{ heat capacity} = m\,C_p\Delta t;$$
$$\text{Pumping energy } P\,\alpha\,k\,Q^3; \text{ heat loss } \alpha\,\Delta t - T_{ambient}$$

The control of the system is optimum when the temperature change Δt is at a maximum and is realised by variable speed pumping. Heat network costs are high and have been estimated at £1.5 billion for 270,000 households (Powry) or £5.5k/household apportioned equally between the infrastructure, the local branch connection and user receiver and metering. This is around twice the cost of a domestic boiler installation. **For a scheme to be viable, a cheap source of heat is required like the residual heat from a local power station.** The economics are also improved with a high-density consumer base, like a block of flats, reducing the distribution costs. The costs associated with CO_2 emissions are also avoided where the heat is derived from a power station that has already accounted for its emissions.

In Denmark, district heating accounts for about 50% of heat supply with 98% in Copenhagen and in Austrian cities 36%, whereas in the United Kingdom, it accounts for only 2%. The barriers to deployment include a lack of understanding and local expertise as well as the high upfront costs and perceived risks. There are also issues when the station providing the source of heat falls out of merit. The waste heat from a typical 200 MWt generator could supply around 50,000 domestic consumers within a 15-km radius. The costs of heat pipes vary with those in the United Kingdom being about twice those seen across Europe partially due to lack of knowledge and management of the risks. Costs of £100/MW h (10 p/kW h) have been estimated and compare favourably with domestic energy rates when network costs are taken into account. The other key risk is the take-up by potential custo-mers, and installations are more easily managed if integrated into new town or district developments.

The schemes can be designed to be most efficient when the station can be operated in a baseload mode coupled with a gas boiler and heat storage facility to enable the troughs and peaks in heat demand to be managed. The scheme total costs vary from around 10 to 17 p/kW h of heat depending on the extent of the trans-mission and distribution needed and can compete with air source heat pumps at around 9 p/kW h. A review of existing sites in the United Kingdom has identified up to 5 TW h of district heating capacity. The installation on the Greenwich Peninsula costs £87m and supplies 87 MWt from a gas CHP engine to 15,700 homes through 10 miles of pipework and provides heat for 3.5 m sq. ft of com-mercial space. The unit has to be sized to meet the winter peak demands resulting in lower overall load factors unless other uses can be found for the output like refrigeration. Surveys of potential new sites identified several hundred with capa-city sized to supply 15,000 people in 6,000 households. Supply of waste heat from nuclear stations is claimed to be competitive when located up to 60 km away.

13.5   Electric vehicles

Electric vehicles provide the opportunity to reduce emissions from transport and also to support balancing the intermittency of renewable sources through demand management and exploiting their storage potential. Table 13.5 shows the key parameters of an all-electric vehicle including the average energy use per mile.

Table 13.5   Electric vehicle parameters

Tesla car Battery capacity, kW h	Range, mi	Charge/discharge Efficiency	kW h/mi	Gross, kW h/m
80	215	0.85	0.37	0.4378

Taking account of the charge/discharge efficiency, the gross kW h/miles can be calculated. Although there are no emissions from the vehicle, there are emissions from the power station used to charge its battery. In this calculation, it is assumed that a combined cycle gas turbine is used with the parameters shown in Table 13.6.

This shows the calculation of the emissions in gCO_2/mi resulting from charging a typical electric vehicle from a CCGT. The CO_2 emission per kW h of generation is a function of the CO_2 content/GJ and the efficiency of the generation and is given by

$$1,000 * 0.0495 * 3.6 * 55/100 = 324 \text{ g/kW h}.$$

Given that the EV uses 0.4378 kW h/mi then the emissions from the generation in providing electricity to motor, the vehicle through a mile is given by

$$324 * 0.4378 = 141.8 \text{ gCO}_2/\text{mi}.$$

This can be compared to the typical emissions from a petrol car at 220 gCO_2/mi with a saving in using the EV of 78 gCO_2/mi or 35%. The target 10% reduction in transport emissions could be realised by converting around 35% of the transport fleet to electricity from CCGT generation. The savings would be higher if some of the energy used to charge the vehicle is derived from renewable sources like wind generation.

The comparable assessment of emissions for charging the battery from a coal-fired generator would be above 500 gCO_2/kW h or 220 gCO_2/mi i.e. about the same as that from an equivalent petrol engine with no net savings in emissions. However, a key advantage of EVs is that the pollution in the congested cities is reduced, whereas **the emissions from the power station are likely to be from a remote site where there is the potential to manage them in future using carbon capture and storage technology.**

Table 13.6   CCGT emissions

Conventional CCGT	Units	Value	CO_2 emission included	
Net capacity	MW	377	Output, TW h	2.81
Load factor	%	85	Fuel cost, £/MW h	27.53
Gas cost	Pence/therm	40	Capital, £/MW h	10.09
Gas CV	kW h/therm	29.32	Fixed O&M, £/MW h	0.60
Gas grossCV/netCV	Ratio	1.11	Var O&M, £/MW h	1.54
Efficiency	%	55	Total, £/MW h	39.76
Interest 'i'	Per unit	0.07	CO_2 cost, £/MW h	22.68
Capital cost	£M	300	**Total + CO$_2$, £/MW h**	**62.44**
Fixed O&M	£/MW h	0.6	Energy cost, p/kWh	6.24
Variable O&M	£/MW h	1.54	Network, p/kW h	3.6
Project life	Years	20	**Total cost, p/kW h**	9.84
Gas	tCO$_2$/GJ	0.0495	Emission, g/kW h	324.00
CO_2 cost	£/tCO$_2$	70	**gCO$_2$/mi**	**141.83**

The CCGT costs shown in Table 13.6 can be calculated as shown below. The output in TW h is given by

$$377 * 8{,}760 * 85/100/1{,}000{,}000 = 2.81.$$

The fuel cost is given by

$$10 * 1.11 * 40/29.32 * 55/100 = 27.53 \, \pounds/\text{MW h}.$$

The capital cost is given by

$$(300 * 0.07 * (1 + 0.07)^{20}/((1 + 0.07)^{20} - 1))/2.81 = \pounds 10.09/\text{MW h}.$$

The fixed and variable costs are 0.6 and 1.54 £/MW h that when added to the fuel and capital costs gives a total of £39.76/MW h.

The cost of CO_2 emissions is calculated at an assumed price of £70/t and is given by

$$70 * 0.0495 * 3.6 * 55/100 = \pounds 22.68/\text{MW h}$$

Added together, the total energy costs are £62.45/MW h or 6.25 p/kW h. Adding a network charge of 3.6 p/kW h to supply the energy to the user premises, the total cost is 9.84 p/kW h. This is equivalent to a cost of 4.31 p/mile based on the EV consumption of 0.4378 kW h/mi. The cost for a petrol car is based on a fuel price of £1/l and consumption of 4.9 l/100 km and equals 7.84 p/mile i.e. the EV running cost/mile is 55% of the cost of the petrol-fuelled car. The results are summarised in Table 13.7.

If the CO_2 cost was less at £35/t, then the electric car costs would be lower at 3.8 p/mi compared to a petrol car at 7.84 p/mi i.e. about half the cost. The comparisons do not include an assessment of emissions associated with vehicle production that are assumed to be of a similar order. There is also the cost of the infrastructure required to supply fuel and electricity that should be compared. There is an existing network of petrol stations with large underground storage tanks and pumping facilities. In contrast, access to the electricity network is already widely available with limited work to engineer connection points for charging. The main difference is the time taken to charge an EV compared to filling with petrol. One option that could be considered is battery exchange where the energy supplier owns the battery. In the longer term, as the use of EVs increases, reinforcement to the distribution networks will be required. If the distribution network is designed with

Table 13.7   Comparison of EV and petrol car emissions and costs

Emissions associated with a Tesla car using electricity are gCO_2/mi	141.83	gCO_2/mi
The emissions from a BMW 3 series are	220	gCO_2/mi
Using a CCGT to run an EV saves emissions equal to	**78.17**	gCO_2/mi
The electrical energy cost p/mi for the Tesla	**4.31**	p/mi
The BMW petrol consumption is 4.9 l/100 km	0.0784	l/mi
If petrol is £1/l calculate the cost per mile	**7.84**	p/mi

an after diversity maximum demand of 1 kW/consumer, then a 500-kV-A transformer could supply up to 500 households. If 30% have EVs, all on charge at the time at 3.5 kW each, then the peak demand could be

$$500 * 0.3 * 3.5 = 525 \, \text{kW}.$$

Not all transformers and circuits will be fully utilised and will have some spare capacity but, at around the 30% level of take-up of EVs, reinforcement of a similar proportion of the network could be required. The rate of take-up of EVs is likely to be faster than the industries capability to reinforce the network, and some advanced work will be necessary.

13.6 Power to gas

There is considerable capacity to store energy in gaseous form in the gas grids. In Germany, the gas network storage capacity is estimated at 200,000 GW h. The UK national gas grid has a linepack capacity of 4.5 TW h with 41 TW h of explicit storage capacity with the capability to recover 10% a day. **This capacity dwarfs the UK daily electrical energy consumption of around 1 TW h a day and has the potential to solve the management of the variations in demand and generation due to intermittent renewable sources.** It could also be used in conjunction with managing variations in transport and heating demand by integrating their use with production of electricity. The objective would be to use renewable energy whenever it is available, irrespective of current demand, and to smooth the total demand to fully exploit the available capacity.

Surplus renewable electricity could be used to produce gas, if the process can be made efficient. One possible process involves splitting water into hydrogen and oxygen using electricity and injecting the hydrogen into the gas grid. There is an energy loss of 30% in producing the gas and further losses in the use of the gas depending on the application. The losses in transporting through the network are low but compression uses about 2.1% of the energy. The conversion efficiency to electricity is around 55%. **This gives an overall cycle efficiency of around 38.5% and comparable with conventional coal-fired generation but some emissions still result from the use of electricity to produce the gas.**

One option would be to use renewable energy that would otherwise have to be curtailed, to produce hydrogen. This would enable higher levels of renewable capacity to be installed, and the option could also be used to smooth the wind output when other regulating options may be more expensive. In this situation, the electricity is effectively free as it would otherwise be curtailed.

13.7 Combined heat and power

There is scope to improve overall efficiency of energy use through the application of **combined heat and power schemes that can operate at efficiencies of 85%.**

Table 13.8   Small-scale CHP

Micro generation

Cost installed (low)	Euro, 000	115	Running hours	h/year		5,890.0
Cost installed (high)	Euro, 000	140	Electrical production	MW h		589.0
Power production	kW	100	Hot water production	MW h		1,035.4
Electrical efficiency	%	30	Hot air recov. prod.	MW h		117.8
Hot water	kW	167	Avoided fuel in boiler	MW h		1,218.1
Hot air (recovered)	kW	20	MT fuel consumption	MW h		1,963.3
Boiler efficiency	%	85				
			Income with avoided elec.	Euro, 000		51.2
Electricity price	€/MW h	87	Income with hot water	Euro, 000		30.5
Boiler gas price	€/MW h	25	Income with hot air	Euro, 000		2.4
CHP gas price	€/MW h	25	MT gas consumption	Euro, 000		−49.1
Value of hot air	€/MW h	20	MT maintenance	Euro, 000		−8.8
MT maintenance	€/MW h	15	**Total income/year**	Euro, 000		**26.1**
			WACC %			8.0
Operating hours	h/year	6,200	Project life, years			15.0
Availability	%	95	Payback range from	5.4		4.4
Electricity price			Production	MW h	Costs	Euros
Avoided energy	E/MW h	31.5	Electricity	589	Capital	16,356
CCL tax on energy	E/MW h	3.5	Water	1,035.4	Fuel	49,083
Dx charge	E/MW h	42	Air	117.8	Maintenance	8,835
Tx charge	E/MW h	10	Total	1,742.2	Total	74,274
Total	E/MW h	87	Cost, E/MW h			42.6

The analysis of a smaller 100-kW embedded CHP scheme is shown in Table 13.8. The left hand side of the table shows the basic plant parameters. At the bottom, the make-up of the normal electricity price is shown including the energy and network charges. It can be seen that the network charges constitute 60% of the total €87/MW h. The right hand side of the table shows the electricity production, based on the estimated annual running hours of 5,890 h, and the avoided electricity costs. The unit also provides heat output of 167 kW providing 1,035 MW h of heat over a year. This is valued at the avoided cost of producing heat in the boiler using gas operating at 85% efficiency. The unit also enables recovery of 20 kW of hot air providing 117 MW h prices at €20/MW h. The cost of gas is based on the electricity production divided by the efficiency times the gas price of €25/MW h with an allowance for maintenance. At the bottom of the table, the total energy production in MW h is shown alongside the total costs of capital, fuel and maintenance. The equivalent CHP combined cost of energy is calculated at €42.6/MW h for comparison with the cost of electrical energy from the network of €87/MW h. Although the basic energy charge is a little higher, because of the comparatively low 30% electrical efficiency of the CHP generator, **the main difference in cost is due to the network charges that are avoided**. The installation is particularly viable if the CHP heat and hot air can be used throughout the year. In this type of installation, the network affords security and a backup, and the charging arrangements need to better reflect this.

The United Kingdom has around 2 GW of CHP schemes capable of generating 1,800 MW of electricity. There are around 200 schemes that are mostly in the range 1–10 MW. The 100-kW unit described in Table 13.8 produces 589 MW h/year and 1,035 MW h of heat plus 117 MW h/year of hot air. Scaling the figures up to 2 GW installed capacity results in 11.78 TW h of electricity production and 23 TW h of heat.

13.8 Implications to electricity demand of meeting emission targets

The decarbonisation of heat and transport could have a profound effect on the requirements from the electricity sector for generation and its distribution. If it is assumed that 5 GW of air source heat pumps could be installed, then the electricity consumption at a 35% load factor is given by

$$5\,GW * 8,760\,h * 35\%/1,000 = 15.3\,TW\,h.$$

At an SPF of 4.0, this would displace 61.3 TW h of heating demand that is close to meeting the initial target reduction of 10% or 68 TW h/year of heating demand by 2020. If we can assume diversity in demand for electricity similar to the heating load factor, then up to 40% (2 GW) of generation capacity would be required based on the assumed 5-GW installed capacity. This may not be extra capacity if the load can be controlled to occur at time of low load exploiting the natural thermal storage capacity of buildings.

To meet the requirement to reduce transport demand by 10% requires a reduction of 63 TW h/year. The Digest of UK Energy Statistics (DUKES) shows the use for transport at 53.413 Mtoe that is equivalent to 621 TW h/year:

$$53.413\,Mtoe * 11.63 = 621\,TW\,h/year.$$

Of this, road transport consumes 39.5 Mtoe equivalent to 459.4 TW h/year. One barrel of oil contains 35 gallons of fuel, and it is assumed that 1 toe contains 6.61 barrels. Assuming a typical annual mileage of 10,000 mi at 33 mi/gallon, each vehicle consumes

$$10,000/33\,gallons\,equivalent\,to\,10,000/33/35/6.61\,toe = 1.309\,toe/year.$$

This is equivalent to 15.2 MW h/year for each notional car (i.e. 1.309 * 11.63) or 15,200 kW h/year. This compares to the energy utilisation of the Tesla car of 0.4378 kW h/mile that for 10,000 mi is equivalent to 4,378 kW h/year reflecting the efficiency compared to the internal combustion engine. Given there are 30.3 million cars in the United Kingdom, the total fuel utilisation would be 39.6 Mtoe/year (1.309 * 30.3) and is consistent with the DUKES figure for road transport for the year.

To reduce transport fuel utilisation by 10% would require 3 million conventional vehicles to be replaced by electric vehicles. This would increase electricity demand by

$$3{,}000{,}000 * 4{,}378/1{,}000{,}000{,}000 \text{ TWh} = 13.1 \text{ TW h}.$$

The demand can be estimated assuming that vehicles are charged at 3.5 kW with a 33% diversity factor as

$$3{,}000{,}000 * 0.33 * 3.5/1{,}000{,}000 = 3.5 \text{ GW}.$$

The generation would operate at a load factor given by

$$13.1/(3.5 * (8{,}760/1{,}000)) = 42.7\%.$$

In practice, the need for extra generation capacity could be limited with a lot of charging off-peak.

The combined impact of meeting just the 10% reduction in heat demand and transport on electricity demand would be

Heating $-$ 5.0 GW at 35% load factor = 15.3 TW h
Transport $-$ 3.5 GW at 43% load factor = 13.1 TW h.

The combined effect would be a requirement for up to 8.5 GW of generation with an annual load of 28.4 TW h. These figures are based on an annual load factor of around 1/3 and assume not all cars will be fully discharged and charging at the same time and that heating requirements will exhibit some diversity. There may be occasions, during winter evenings, when a lot of cars are being recharged and heating demand is higher, but it is assumed that there would be some scope to defer the transport demand until the overnight period.

An alternative approach is to install more local CHP schemes that can operate at efficiencies of 85% producing electricity and heat. Based on the 100-kW CHP scheme described in Section 12.6 and assuming 2 GW of installed capacity, electricity production would be 11.78 TW h together with 23 TW h of heat. If more sites can be identified that can use the heat locally, this is a viable option towards decarbonising heat. It is likely that a range of options will need to be deployed to realise full decarbonisation of heat and transport. **Based on a 10% reduction, requiring 28.4 TW h/year to realise 100% reduction would require 284 TW h and virtually double the UK electricity demand.**

Across Europe, although coal utilisation has fallen to reduce emissions, gas use continues to rise. According to Gazprom, the Russian energy giant supplied Europe with 42.3 bcm in the first quarter of 2017 against 37 bcm in the first quarter of 2016. They also supplied a record 615 MCM on a cold January day. Sales have been further boosted by low prices dragged down by the collapse of oil prices. There has been traditionally a link between oil and gas prices market prices. Gazprom already supplies a third of European gas and plans to extend its capacity and influence with a second major gas pipeline through the Baltic sea to its major customer Germany that as well as reducing its dependence on coal is closing its nuclear stations.

13.9 Integrated control of transport, heat and electricity

Given the requirement to decarbonise heat and transport, there is an opportunity to employ a more integrated approach to their management to realise a more optimal provision of energy that exploits the inherent capabilities of each sector. There is no scope to store electricity, and it has to be converted into some other form through a process that uses energy both in transformation and in conversion back to electricity. This is the case with pumped storage schemes where the cycle efficiency may be around 70%. For it to be economically **viable, the wholesale price needs to 30% higher**, when the energy is used, than the cost at the time of pumping. The financial case for pumped storage is generally based on peak lopping and providing fast response to rebalance the system following generation losses or sudden demand changes. Other options include using the electricity to produce hydrogen gas that can be stored or used to charge batteries or run flywheels.

In contrast, heat energy can be stored relatively cheaply as in lagged household hot water tanks or systems based on phase change materials with little heat loss. Domestic storage tanks have been developed that can take energy input from a variety of intermittent sources like solar panels. The energy is stored as hot water that is released on demand. In the case of transport energy, storage is essential either as liquid fuel or in the batteries of electric vehicles. As the levels of renewable electricity generation increase, there will be times when the supply exceeds the available demand necessitating some curtailment. The true market price for the energy would be zero, and it could be used to drive heat pumps or/and produce gas for storage in the network. **The principals related to integrated operation can be appraised by analysing typical domestic energy profiles for heat, electricity and transport and identifying how they can be met at least cost.** This will be realised by flattening the profile to reduce the network capacity requirements and minimise the need for wide-scale reinforcement as EV and heating loads increase. The subsequent issue is how the realisation of the optimum can be achieved in a market environment based on time of day pricing mechanisms.

Demand profiles for a range of customer types have been constructed to establish domestic consumer daily demand patterns. It is based on the Eurostat domestic consumer types each with a defined profile. The total demand is calculated based on the number of consumers in each category and the demand profile that matches NGC national consumption levels. Figure 13.6 shows the average profile for domestic electricity demand on a winter day in January together with a typical pattern of demand for charging 10 million EVs supplied on time of use tariffs (TOUT).

It can be seen that the combined load shows a similar range of variation from minimum to peak demand as the basic electricity demand with the overnight trough reduced but the evening peak increased. There is limited scope to shift the EV demand to the late overnight period to reduce the peak but at the risk of not realising a full charge in the morning.

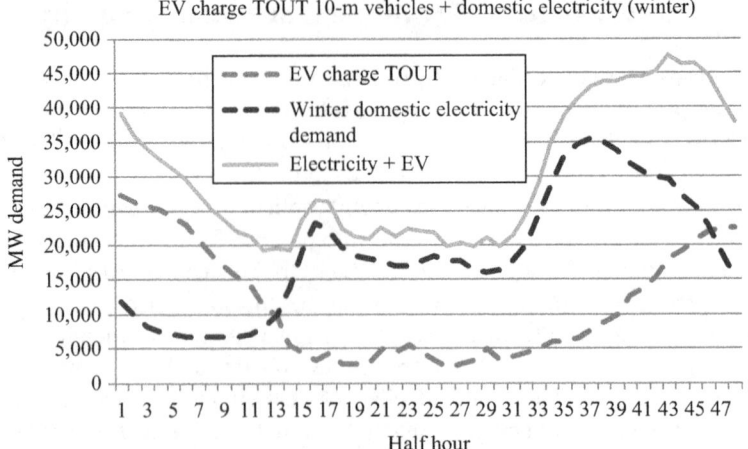

Figure 13.6   Domestic winter demand profile

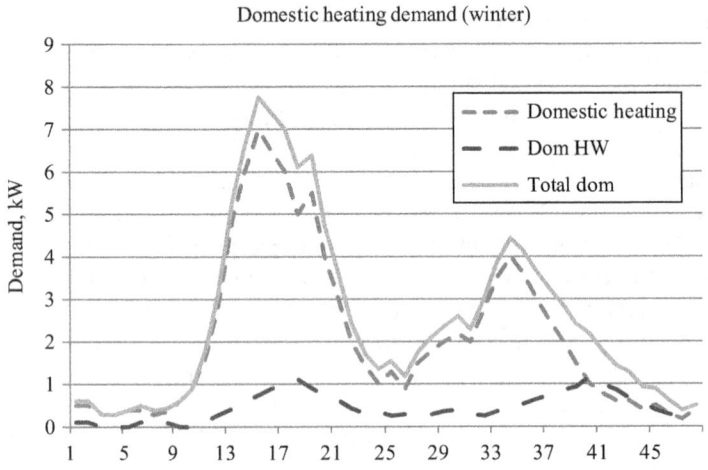

Figure 13.7   Domestic winter average heating demand

A profile of average winter domestic heat demand has been developed based on published data (Yao and Steemers energy and Buildings). It includes the heating and hot water demand and the combined total as shown in Figure 13.7. It shows that the combined heat profile exhibits high peaks during the early morning with a winter day consumption of 60 kW h.

On the assumption that 2.8-m domestic heating installations are based on heat pumps using electricity, the total domestic winter heat demand profile can be established as shown in Figure 13.8. The electrical energy consumption through 91 winter days would be 16 TW h displacing 64 TW h of heat demand with heat pump

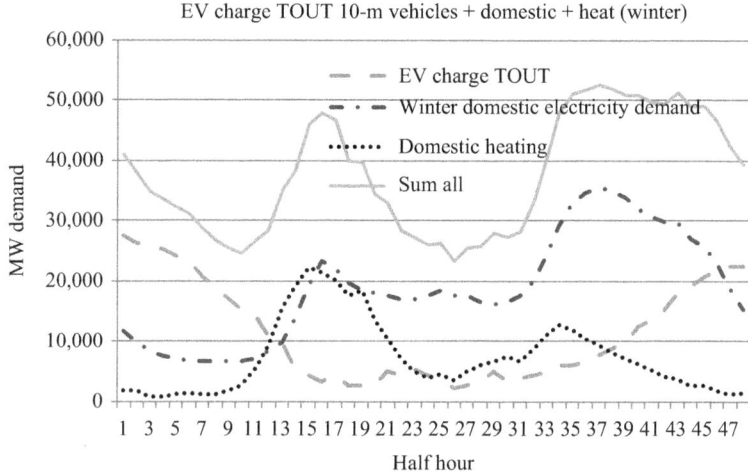

Figure 13.8   Combined domestic electricity demand (winter)

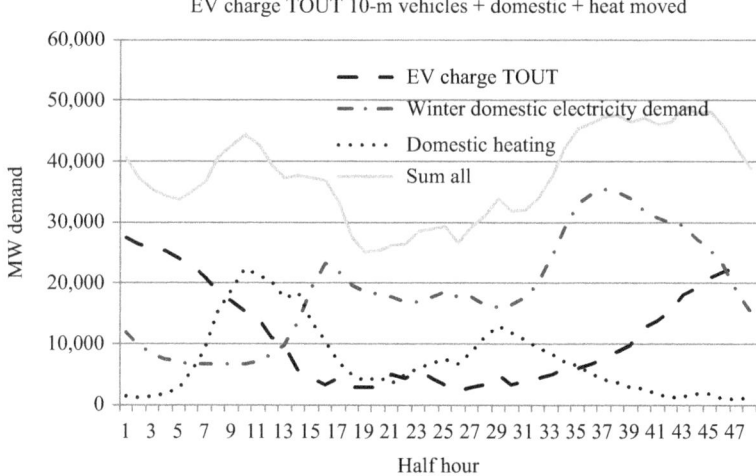

Figure 13.9   Combined winter domestic demand with heat demand advanced 2.5 h

SPFs of 4.0. This equates to the 10% target for heat demand reduction during winter months of 160 TW h. The heat consumption during the other seasons is of a similar order with an annual domestic total of 320 TW h. The total daily heat pump demand profile is added to the normal domestic electricity demand and the EV demand to establish the total profile for winter as shown in Figure 13.8.

It can be seen that the combined profile exhibits significant variations during the day of around **30,000 MW** from a low of 23,354 up to 52,444 MW. **The summation has been repeated advancing the heat demand by 2.5 h** resulting in a new combined profile as shown in Figure 13.9. It can be seen that the new profile is flattened with a range of demand variation of around 20 GW with a reduction in the peak. This is important in reducing the distribution network loading requirements. There is also scope to smooth the national demand profile to reduce the total generation capacity requirement as discussed further in Chapter 18. There is also some scope to adjust the short-term EV charging requirements to support the balancing process.

It is considered that the provision of heat storage offers the potential to smooth domestic demand profiles at the lowest cost with little loss of energy. Conventional hot water storage in lagged tanks is an established cheap option as are storage heaters. A range of domestic thermal storage systems has been developed to take input from a variety of sources including solar panels, wood burning stoves and heat pumps. The stores are able to manage the intermittency of energy sources, and buffer stores are also available that offer the potential to shift demand between periods of the day (e.g. Ecostat thermal storage). The shift to electrification of the winter domestic heat demand offers the potential to facilitate normal demand smoothing as well as balance intermittent generation.

The commercial arrangement for engaging with the wider system could be similar to existing schemes applied to demand management. The arrangements widely used in the USA and other hot climates for loads like domestic air conditioning include the end user receiving a payment for participation and call off. The availability of smart meters would facilitate monitoring with control affected through broadcast telephony signals. It is not expected that domestic consumers would participate directly in managing their loads but that they would contract with an aggregator. They would monitor wholesale prices and manage loads collectively on behalf of consumers.

13.10   Conclusions

Although emissions in North America and South and Central America fell by 2% and 2.4% in 2016, those in the Asia Pacific region increased by 0.9% as did those in the Middle East 1.6% and Africa 1.1%. This resulted in a net global increase of 0.1%. The emissions from the Asia Pacific region were almost equal to those of the rest of the world reflecting a dependency on coal. The US emissions at 1,925 $MtCO_2$/year were less than those of China at 2,943 $MtCO_2$/year. The United Kingdom had a target to reduce emissions from 247 $MtCO_2$/year by 34% and has already exceeded this with levels of 112 $MtCO_2$/year i.e. about 4% of those of China. Electricity production is responsible for 37% of the emissions in the USA compared to 34% in the United Kingdom. This reflects a continuing reliance on coal for electricity production in the USA.

Some 20% of the total emissions result from domestic use for space and water heating. To decarbonise heat, one solution is to use heat pumps that extract heat from the environment increasing the overall efficiency by a factor of three. In operation, the heat pump is cheaper than a conventional boiler but the overall costs are much higher because of the capital costs and the high price of electricity compared to gas. District heating schemes can be viable if there is a local source of waste heat and a dense group of potential users, and they are widely used in Austria, Denmark and Copenhagen.

Electric vehicles provide the opportunity to reduce emissions from transport and also to support balancing the intermittency of renewable sources through demand management and exploiting their storage potential. The reduction in emissions depends on the generation used to charge the EV batteries. Where conventional CCGTs are used, the net reduction compared to a petrol car is about 38% with half the operating costs. The use of local solar panels for charging results in no emissions or extra costs for those that have them installed.

The gas grid has massive potential to store energy and could be used to store excess renewable energy. The limiting problem is the overall cycle efficiency of around 38% of producing gas that is then used to generate electricity. It could be a viable option to use excess renewable energy that would otherwise have to be curtailed.

Combined heat and power schemes can operate at high efficiencies if the waste heat can be utilised. This generally results in small schemes with a local requirement for heat that are relatively more expensive than larger generating units. Their cost effectiveness is made more attractive by network charges being avoided when compared to grid prices for electricity.

The requirement to decarbonise heat and transport will have a significant impact on demand for electricity. Based on a 10% reduction in heat and transport, requiring 28.4 TW h/year to realise 100% reduction would require 284 TW h/year and virtually double the UK electricity demand. However, there is an opportunity to employ a more integrated approach to their management to realise a more optimal provision of energy that exploits the inherent capabilities of each sector. It is considered that the provision of heat storage offers the potential to smooth domestic demand profiles at the lowest cost with little loss of energy. A longer term view is necessary to determine what needs to be in place to facilitate development.

Question 13.1 Using the data in the table below. Calculate the impact on global emissions if Europe/Eurasia replaced the use of coal with gas. Also calculate the impact on emissions if North America doubled coal use at the expense of gas utilisation.

Primary energy, TW h	Oil	Gas	Coal	Nuclear	Hydro	Renewables	Total	CO_2
N. America	12,165	10,304	4,501	2,524	1,779	1,128	32,401	6,441.195
S. & C. America	3,791	1,791	395	64	1,814	326	8,182	1,439.238
Europe Eurasia	10,281	10,769	5,245	3,001	2,338	1,675	33,308	6,261.816
Middle East	4,850	5,361	108	16	55	8	10,398	2,263.747
Africa	2,152	1,442	1,116	42	300	58	5,110	1,171.693
Asia Pacific	18,108	7,560	32,017	1,233	4,280	1,675	64,872	16,122.18
Total	51,346	37,228	43,383	6,879	10,566	4,869	154,272	33,700
Percentage	33	24	28	4	7	3	100	
Asia Pacific, %	28	12	49	2	7	3		
Total fuel burn, TW h	51,346	37,228	43,383					
Emission tCO$_2$/GJ	0.073	0.0495	0.0869					
Emission tCO$_2$/MW h	0.2628	0.1782	0.31284					
Emissions by fuel	13,494	6,634	13,572					
Total MtCO$_2$/year			33,700					

Chapter 14
Network issues and tariffs

14.1 Introduction

Transmission facilitates operation of the market in enabling wholesale trade between generators and suppliers largely independent of their location. This maximises the liquidity in the market and promotes competition. In general, the larger the consumption is in the interconnected market, the higher the liquidity and competition. It enables traders for the most part to regard the grid as infinite. To ensure fair open access to all system users, there is a requirement to demonstrate independence from any participant involved in trading energy. This has not been realised where transmission ownership was coupled with generation so that the initial step was for transmission to be unbundled from generation.

In the organisation of state industries, it was usual for the operation of the power system to be integral with the wires business with common ownership. During 2007/08, the EU pressed for ownership to be unbundled to establish truly independent SOs able to manage operation of the system without any prejudice. In a statement in 2008, the European Regulators reiterated their support for the Parliament's position on "ownership unbundling" of transmission system operators. The key is to deliver effective solutions as soon as possible. Fair and equal access to the networks and appropriate investment incentives on transmission system operators are essential to Europe-wide security of supply. **Ownership unbundling is the most effective means of avoiding discrimination.** Moreover, the greater the degree of unbundling, the less additional, intrusive regulation is needed. In its advice to the European Commission, the European Regulators for Electricity and Gas (ERGEG) advocated that the model required in the third package is in principle ownership unbundling and that there is no justification for less unbundling in gas than in electricity, as the potential for discrimination does not differ.

The Independent System Operator is also responsible in most market models for identifying the need for development of the network based on operational experience. **Similar initiatives have taken place in the United States where the FERC (Federal Energy Regulatory Council) order 2000 required utilities to hand control of their networks to an independent entity**. This led to the establishment of Regional Transmission Organisations (RTO) exercising coordinating control over several transmission networks. The Pennsylvania Jersey, Maryland (PJM) RTO for example coordinates energy transfers across some

15 US states. As well as managing the network assets, they also continuously schedule and dispatch generation to keep the system in balance.

The distribution networks were fed from the transmission system via bulk supply transformers. These networks were largely passive with little generation and did not require active management. They were essentially radial in structure without loop flows. The networks are generally considered to be a natural monopoly and as such are subject to regulation. The objectives of the Regulator are to limit prices close to costs and drive them down through efficiency improvements. This chapter discusses the drivers leading to a shift in the role and security of distribution networks resulting from the connection of more generation.

14.2   Change drivers

The changes taking place in the industry are having an impact on the requirements of networks, their management and financing. The key drivers are the desire to realise sustainability by reducing emissions from electricity generation, heating and transport whilst containing costs and maintaining security of supply, the so called trilemma. The impact of these drivers includes

- A growth in renewable generation including the connection of remote wind farms that may be offshore and solar farms that may be connected to distribution networks at HV;
- The growth of generation embedded within the distribution network where there are efficiency gains from exploiting waste heat in CHP schemes or district heating;
- Drives to improve efficiency in generation and energy utilisation to reduce fuel use and associated emissions;
- More end-user engagement in local energy generation from solar panels and heat pumps facilitated by smart metering and networks;
- More interconnection between systems to exercise arbitrage in energy prices and exchange services to manage security and effect balancing;
- A growth in regulation of the sector with actions to foster more competition in the provision of network services.

These developments are having a profound effect on the operation of networks and their governance and financing. The distribution systems are becoming more active requiring system control, and there is an increase in customer engagement to minimise costs through efficiency savings and self-generation facilitated by smart metering. The rising cost of energy and system losses and the costs of providing transmission and distribution have a significant impact on end-user costs. **The use of system costs for the provision of networks has become an important factor in determining the choice of local vs remote generation.** The arrangement for charging for services is in need of review recognising the changing circumstances.

14.3 Distribution loading

The typical capacity of renewable generation like wind generators or arrays of solar panels can be very small. Wind generators are generally from 1 to 5 MW. Solar panels may range from a few kW for domestic roof top installations up to 50 MW for a large solar farm. This means that these installations can often be connected directly to the local distribution network. The 11-kV network is generally used to supply distribution transformers of 500 to 1,000 kV A and can accommodate small wind generators, but it was not designed to be an actively managed network with changes in the direction of energy flows. The distribution networks have traditionally been developed based on well-established diversity factors for consumers. As a consequence, the after diversity maximum demands (ADMD)/consumer are quite low. **For the average domestic consumers, the ADMD may be below 2 kW.** The natural diversity is assumed in connecting consumers to distribution transformers.

Figure 14.1 shows a typical domestic demand profile through a winter day normalised to 1 per unit with a peak around 6 p.m. The dashed line shows the potential reaction to prices as may be realised by an aggregator for a 10% price differential. Based on the Eurostat customer types and the number of consumers in each category, the annual domestic energy can be calculated in this example at 109 TW h as shown in Table 14.1. The proportion of the MW h for each consumer is assumed to split with 30% during 91 winter days, 20% during 91 summer days and 50% during the spring and autumn days. Based on the normalised profile, the average domestic ADMD can be calculated for each category at the peak consistent with the total annual energy. The customer-weighted value is 1.07 kW in this example rising by around 20% on the coldest winter days to 1.3 kW. The relatively low ADMD enables the distribution networks to support a lot of consumers off each transformer.

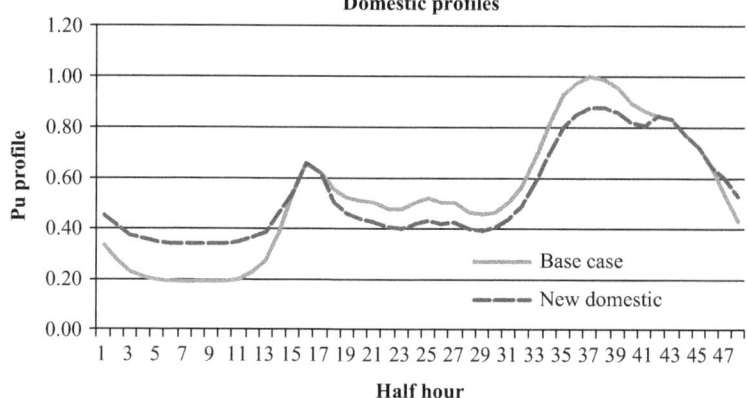

Figure 14.1 Domestic demand profile

Table 14.1   Domestic ADMD

Sector	Type	Peak, kW	MW h/year	Consumers	GW h/year	Winter	Summer	Rest	ADMD, kW
Domestic, 109 TW h	Da	3	0.6	8,155,507	4,893	0.18	0.12	0.3	0.155
	Db	3.5	1.2	6,265,797	7,519	0.36	0.24	0.6	0.306
	Dc	6.5	3.5	5,063,969	17,724	1.05	0.7	1.75	0.893
	Dd	7.5	7.5	3,911,478	29,336	2.25	1.5	3.75	1.9
	De	9	20	2,469,710	49,394	6	4	10	5.1

This natural diversity may be destroyed by a number of factors:

- Time-of-use tariffs facilitated by smart meters enabling end-user aggregators to take advantage of low market prices switching on consumer demands at the same time;
- The wide application of solar panels by end users, in the same area, will lead to common surges in output during hot sunny periods when other demands are low;
- Locally connected wind generators will also exhibit surges in output at the same time that will often not coincide with periods of high customer use.

These factors will have a significant impact on the distribution network loads and voltages and quickly lead to saturation of the networks as has been reported by some distribution owners. A 500-kV-A transformer may have 300 domestic consumers with an ADMD of 1.3 kW totalling 390 kV A. In future, if just 50 consumers have EVs charging at 3 kW, this could add 150 kV A during the early evening peak period resulting in a peak demand of 540 kV A.

Installing new distribution points in largely underground networks would be a significant and costly undertaking. Exercising control over the timing of some demands will enable some extension of the transformer asset life. The lower the voltage, the lower the connection cost for small installations, but there is an argument for the use of higher connection voltages of 22 kV for larger embedded generators to minimise the impact on the network and other consumers coupled to the same local network. For offshore wind farms, it is cheaper to collect the output from the generators offshore and connect them to the mainland grid at typically 132 kV. For the connection of several arrays of wind generators, multiterminal DC is an option being considered particularly where the windiest site is some distance from the mainland network.

The connection of more remote offshore wind farms has created the opportunity to introduce competition in the provision of the network extensions. These can and are put out to tender in the United Kingdom and has created sufficient interest to make the process worthwhile. Long-term contracts are used to support financing with payments linked to availability.

14.4   Security issues

The connection of generating sources to the distribution network alters the assessment of its security. The standards for distribution systems are usually based

Table 14.2 Distribution security standards

Class of supply	First outage	Second outage	Tm switching	Other	Maint. Tm
A ≤ 1 MW	Repair time	Nil	n/a	n/a	n/a
B 1–12 MW	3 h D – 1 MW	Nil	3 h	24 h	2
C 12–60 MW	D – 20 MW, Dt 3 h	Nil	3 h	15 days	18
D 60–300 MW	D – 20, Dt 3 h	Restore outage	3 h	9 days	24
E 300–1,500 MW	Dt immediately	Restore outage	n/a	9 days	24

on previous fault statistics and risk analysis compared to reinforcement costs. The usual approach is based on the assessment of a discrete demand area fed by bulk supply transformers. The security criteria will specify that it should be possible to meet a proportion of the demand following a forced outage of one of the main feeds to the group or a forced outage on top of a planned outage that can be restored. The proportion of demand met depends on the size of the group as illustrated in Table 14.2 where D refers to the group demand and 'Tm switching' the restoration time based on switching.

The introduction of embedded generation alters the network security assessment. For consistency, the loss of an embedded generator should not be more onerous than the loss of a circuit. The time 'Tm maintenance' is the minimum time that the generation source has to be available to contribute to securing the system following an outage. There is also a need to distinguish between controlled generation and random intermittent generation. The estimated contribution can be related to a proportion of the declared net capacity (DNC) and typical patterns of available output. For generation installations based on landfill, waste and CHP contributions of around 50% of DNC are estimated, whereas a CCGT installation with a controllable fuel supply a figure of 70% could be assumed. For intermittent generation, the availability can be represented as a probability of it being available at the time of the outage. This was analysed in Chapter 4 for wind generation and shown to be around 15% of the DNC. For small hydro, some control may be possible and up to 30% of capacity is estimated for a short period or more where storage is available as part of the hydro scheme.

The assessment of network security is made more dynamic by the introduction of embedded generation. **The other key variable is the demand that can be expected to increase with the take up of electric vehicles and heat pumps that will have less diversity than more conventional loads.** Normal domestic demand like kettles, dish washers and washing machines will have an after diversity demand of around 1.5 kW/dwelling. In contrast, EVs with a capacity of 80 kW h would require 24 h to fully charge at 3 kW, or with a level two charger operating at 80 amp and 20 kW, it would take 4 h. Although not all vehicles will be fully discharged, most owners will seek to maintain a full charge by topping up daily. The other key variable is the demand that can be expected as discussed in Chapter 3 where a full take up of EVs could generate an extra demand of 90 TW h/year, based on typical annual mileage rates, doubling the current domestic demand.

To support both the embedded generation and the increased demand, it would require a massive expenditure on network reinforcement or the adoption of smart systems to manage the impact through control of demand and generation coupled with local storage. The new investment required could be contained by establishing an integrated system of monitoring and control. At the transmission level, real-time security assessment is the norm with continuous assessment of the security of the system against network outages. Facilities are used to predict demand, dispatch generation and manage imports/exports. Similar developments will be required at the distribution level. Demand control exploiting smart meters offers the potential to manage forced outages for short periods and alleviate the peaks. Incentives will be required to encourage customer participation.

14.5   System development

It is suggested that changes in the use of networks will require new functionality that operators need to react to. There are many more participants in the sector that will bring new requirements. Some of the developments that are expected to have an impact are as follows:

- A shift from connecting large generation at transmission voltages to connecting more smaller generators at distribution voltages;
- The small embedded generators are not normally dispatched and may be wind or solar-powered exhibiting intermittency;
- Smart metering offers the opportunity for real-time pricing of energy encouraging changes in the profile of demand and peak lopping;
- New players in the form of aggregators will wish to establish interactive arrangements for managing load;
- There could be a significant increase in demand resulting from the take up of electric vehicles and heat pumps with less diversity than normally assumed.

The implication to the distribution networks is that they will move from being essentially passive to requiring active management along the lines of the transmission system. The difference is the degree of control that can be exercised over the many potential active participants that could run into several hundred on one distribution group. This in turn will make it very difficult to plan for network development. End users will not expect to seek permission as when to run their embedded generation or when to charge their electric vehicle or alter their demand. It would also be impractical to manage. **Management of the network utilisation would have to be fostered by a process of real-time pricing that reflects both energy prices and network limitations.** This type of arrangement has been successfully applied at transmission levels with zonal pricing. Time-of-day pricing has also been applied to encourage night-time use of energy to charge heat storage devices. When the network is stretched, prices would be high to discourage use and encourage the use of own generation. At lower demands, prices would be low encouraging energy import and less self-generation. The import/export prices would be determined by conditions on the supergrid. The distribution system would need a monitoring and control infrastructure interfaced to the national control centre as well as system users. Some of the costs should be levied on those users that are creating the change in requirements including embedded generators and storage operators.

Network planning in this environment will be challenging with difficulty in predicting the many small-scale developments that can be expected. One scenario advanced by NGC for the United Kingdom suggests that by 2030, there could be 3.3 million EVs, 6.6 million heat pumps and 71 GW of solar and wind capacity generating 39% of total electricity. Given this potential rate of change, the distribution companies will need to actively control development and define connection agreements that will meet longer term requirements for the exchange of data with active participants. This in turn will need to be facilitated by a regulatory and governance framework that embraces research and takes a whole system view. There will also be technical issues that need to be analysed and resolved at the planning stage related to voltage control, flicker and harmonics and the network protection taking account of the effect of rate of change of frequency on embedded generation. All these issues are similar to those that affect transmission system development and will inevitably lead to a step change in distribution charges.

14.6 Retail total charges

Figure 14.2 shows the makeup of a retail customer bill with a total price of 11.6 p/kW h. It shows that the wholesale energy costs are just 32% of the total with network costs slightly less at 27%. While energy prices have fallen, environmental costs have increased over the last decade with higher levels of subsidy that by 2016 added 20% to the retail bill with operations adding a further 16%. **It is not**

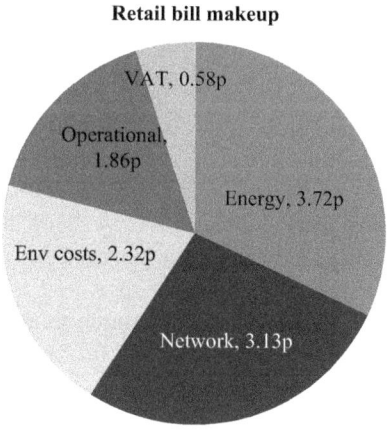

Figure 14.2   Retail bill makeup

surprising that avoiding the network charges by direct wiring from local generation to demand can appear financially attractive.

The consequence is that if one user pays less towards the service, others have to pay more. It could be argued that users with self-generation have the benefit of having a system connection providing backup all year and should contribute to its upkeep irrespective of its use. Since 2009, the Netherlands charges households and small industrial users a **fixed charge** based on the capacity of their grid connection. Since 2010, the United Kingdom has used a common approach to establish distribution use of system charges.

14.7 Distribution network charging methodology

The distribution company charges are regulated, and they have to establish with the regulator how much they are allowed to recover. The assessment in the United Kingdom covers an 8-year period and is based on a RIIO concept;

$$\text{Revenue} = \text{incentives} + \text{innovation} + \text{outputs (RIIO)}$$

This is designed to promote efficiency and innovation and improved performance in terms of reliability, the quality of service and environmental and social issues. The assessment is based on network and user-related costs based on forecasts of consumer numbers and consumption. It typically constitutes around 15%–25% of consumer energy bills.

A common distribution charging methodology (CDCM) is used to establish a set of tariffs that apportion costs across the customer base supplied at voltages below 22 kV. Charges for customers supplied at 22 kV or above are based on a forward cost pricing, extra HV charging methodology. The cost allocation between consumers takes account of diversity, volumes, losses and load characteristics and includes

Unit rates in p/kW h – in proportion to volumes;
Fixed charge in p/day – proportional to operating costs;
Capacity p/kV A/day – based on declaration in connection agreement;
Reactive pf – based on impact of low power factors.

Some tariffs may include time of day rates with special rates to cover green energy costs. **The tariffs for embedded generators reflect the fact that their operation may reduce demands at peaks and losses and hence the transmission use of system charge, and they may be negative.** A distinction is drawn between generation that is intermittent like wind, solar and tidal and non-intermittent sources like gas generators, CHP, landfill and waste-powered schemes. An intermittent source may be paid around 1 p/kW h and a non-intermittent source, 3 p/kW h. Positive charges apply to cover their impact on power factor and reactive requirements.

The CDCM process includes the following:

1. Estimate of impact of meeting a 500-MW increment in capacity base on asset and operating costs;

2. Allocate costs to each network level based on asset replacement costs;
3. Derive yardstick cost of load at each network level in £/kW/year;
4. For each user, derive unit and standing charge based on contribution to peak and capacity;
5. Aggregate for each user the unit and standing charges across the network levels used;
6. Adjust the results by scaling to match the revenue allowed by the regulator.

14.8   Distribution UoS tariff structures

A typical distribution tariff is shown in Table 14.3 for NW Europe; it is designed to recover the cost of the equipment used and is based on the declared capacity of the installation and operating costs and losses that reflect the energy delivered. The calculations shown are for a medium-sized commercial or industrial Eurostat type 'la' customer (0.03 MW peak demand and energy requirement of 30 MW h/year). All the charges are reduced to a price per kW h to establish an effective charge per unit. It includes a capacity charge of €1,207/year calculated from the product of the capacity tariff and the capacity (41.47 * 30 * 0.97) scaled by the 'E1' factor. The 'Y' factor is multiplied by the annual energy to

Table 14.3   Distribution tariff

Tariffs		
Capacity tariff	€/kW/year	41.47
Y factor	€/kW	0.003186
System service tariff	€/kW h	0.00024
Metering tariff	€/year	728
Losses tariff	€/kW h	0.00074
Pensions	€/kW h	0.0010368
Assumed load		
Capacity	MW	0.03
Energy	MW h	30
Load factor	%	11
Capacity charges		
E1 factor		0.97
Capacity charges	€/year	1,207.39
Proportional charge	€/year	95.58
Total capacity charge	€/year	1,302.97
Total capacity charge	€/kW h	0.04
Max allowed by CREG	€/kW h	0.07
Price used	€/kW h	0.04
Other charges	€/year	788.50
Total charges+Tx	€/year	2,415.47
Charges	€/MW h	80.5

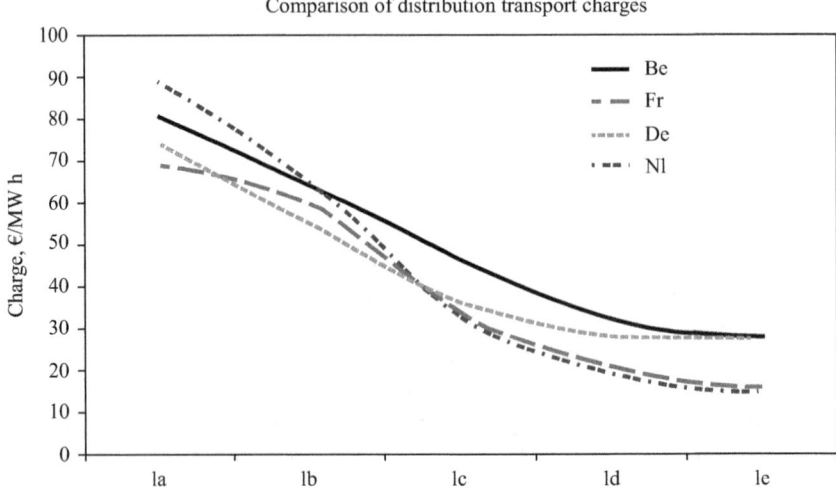

Figure 14.3   Distribution tariffs

calculate the proportional charge (0.003186 * 30 * 1,000 = 95.58). The 'other charges' includes a 'system service charge' covering the overall operation of the system and specifically identifies the cost of losses and pensions provisions all multiplied by the energy supplied and added to the annual metering charge of €728/year to total €788.5/year. A transmission charge of €324/year is added to the capacity and other charges based on 10.8 times the energy transferred (10.8 * 30 = 328). This is to cover the transmission use of system charge. These total costs per year are divided by the energy supplied to give an effective charge per MW h of €80.5/MW h. Similar calculations can be undertaken for the other distributors and other Eurostat consumer types to establish an overall pattern as shown in Figure 14.3.

It can be seen how the charges reduce for the larger customers with higher load factors. Type 'la' with a load factor of 11.4% pay around €68.6/MW h, while type 'le' with a load factor of 45.6% pay some €19.7/MW h.

The network owner is responsible for proposing a set of tariffs for review by the regulator who will compare the absolute level and the distribution of charges amongst customer types. It can be seen, in Figure 14.3, that similar charges apply across the countries in NW Europe in this example.

Table 14.4 shows the total annual utilisation of the different Eurostat customer types. The network costs for type 'le' users averaging around €80/MW h are around a quarter of those of 'la' consumers at around €20/MW h. This is consistent with the network fixed costs varying according to utilisation with 1,000 h for type 'la' up to 4,000 h for type 'le' i.e. four times as many units.

Table 14.4   Eurostat customer types

Category	Capacity, MW	Energy, MW h	Utilisation, h/year
Domestic			
Da	0.003	0.6	200
Db	0.0035	1.2	343
Dc	0.0065	3.5	538
Dd	0.0075	7.5	1,000
De	0.009	20	2,222
Commercial and small industry			
Ia	0.03	30	1,000
Ib	0.05	50	1,000
Ic	0.1	160	1,600
Id	0.5	1,250	2,500
Medium-size industry			
Ie	0.5	2,000	4,000
If	2.5	10,000	4,000
Ig	4	24,000	6,000
Large industry			
Ih	10	50,000	5,000
Ii	10	70,000	7,000

14.9   End-user charges

The end-user tariffs embrace the network charges and the energy charges. The energy charges depend on the pattern of use and reflect the load profile. Base load demand energy is cheaper than peaking energy that is influenced by start-up and shutdown costs with limited generator utilisation from which to recover their fixed capital costs. Table 14.5 illustrates the makeup of end-user tariffs in euros/MW h including the network charge and energy based on a base load price of €40/MW h with peak prices 24% higher and super peak 75% higher.

Both the network and energy charges reflect the level of utilisation of the facilities as illustrated in Figure 14.4. The utility costs include a high capital component and the more units these fixed costs are spread over, the lower the charge. This fact could have an important impact on the way the integration of heat and transport demand is managed. **Significant savings could result from managing the increased demand to provide a smoother profile that better exploits the assets.**

14.10   Transmission UoS tariffs

Charging for use of the transmission network is usually based on the assets utilised and the losses incurred in energy transfers across the network. They will therefore

Table 14.5   End-user tariffs

	MW	MW h	Util, h/year	Base %	Peak %	Super peak	Per unit price	Energy €/MW h	Tx/Dx cost €/MW h	End user €/MW h
Domestic										
Da	0.003	0.6	200	0.0	0.0	100.0	1.75	70.0	92.6	162.6
Db	0.0035	1.2	343	0.0	0.0	100.0	1.75	70.0	73.3	143.3
Dc	0.0065	3.5	538	0.0	0.0	100.0	1.75	70.0	56.0	126.0
Dd	0.0075	7.5	1,000	0.0	0.0	100.0	1.75	70.0	50.5	120.5
De	0.009	20	2,222	0.0	100.0	0.0	1.24	49.6	45.0	94.6
Commercial and small industry										
la	0.03	30	1,000	0.0	0.0	100.0	1.75	70.0	68.6	138.6
lb	0.05	50	1,000	0.0	0.0	100.0	1.75	70.0	55.1	125.1
lc	0.1	160	1,600	0.0	100.0	0.0	1.24	49.6	34.3	83.9
ld	0.5	1,250	2,500	0.0	100.0	0.0	1.24	49.6	23.3	72.9
Medium-size industry										
le	0.5	2,000	4,000	15.6	84.4	0.0	1.20	48.1	13.4	61.5
lf	2.5	10,000	4,000	15.6	84.4	0.0	1.20	48.1	12.6	60.7
lg	4	24,000	6,000	51.1	48.9	0.0	1.12	44.7	11.2	55.9
Large industry										
lh	10	50,000	5,000	33.3	66.7	0.0	1.16	46.4	8.6	55.0
li	10	70,000	7,000	68.8	31.2	0.0	1.07	43.0	6.6	49.6

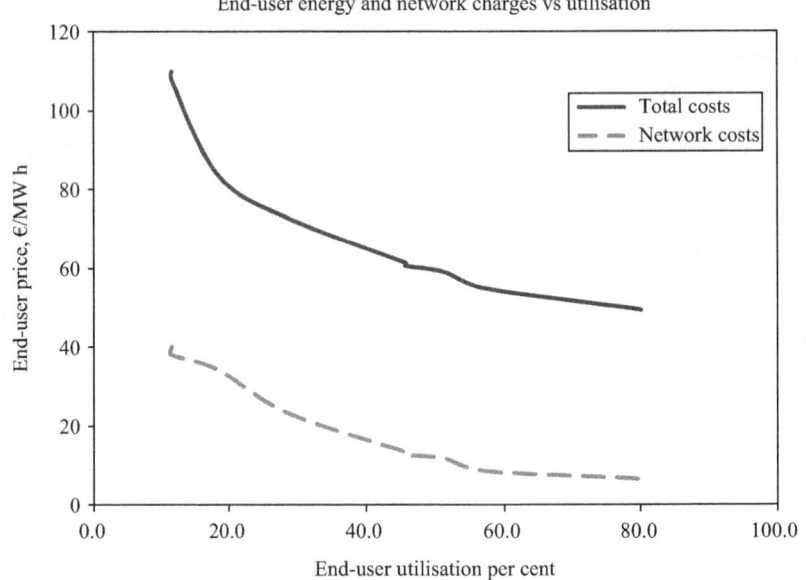

Figure 14.4   End-user tariffs vs utilisation

include a charge related to the capacity of the user's installation. This will affect the proportion of the transmission assets that has to be in place to meet the user needs at times of peak. They will also be a charge related to the energy transfer as this will affect the losses incurred on the network that have to be paid for. Typical transmission tariffs are shown in Table 14.6 for the Netherlands and Germany for a large 'li' consumer having a capacity of 10 MW with energy of 70,000 MW h/year.

It can be seen that the tariffs include a capacity payment and what's called a commodity payment for the energy. Principally, there are also other charges for metering either on a monthly or annual basis. Overall, the charges/MW h are about the same and relatively small for this user with a high utilisation as shown in Table 14.6. The tariff calculation above is for the largest consumer category with the highest load factor with an effective full load utilisation of 7,000 h. As the charges are related to the installed capacity or peak demand, the charges for consumers with lower levels of utilisation are higher in that **the fixed capital cost recovery has to be shared amongst fewer units of delivered energy**. This is illustrated by the graph of Figure 14.5 that shows how the transmission charge per unit increases for the lower utilisation type le customers.

The higher capacity consumers will be fed from higher system voltage levels and will therefore use less of the network with fewer transformation levels giving lower system losses. It can be seen from the graph that charges reduce with

Table 14.6   Transmission tariffs

Netherlands			RWE		
Tariffs			**Tariffs**		
Capacity tariff	€/kW/year	10.27	Capacity tariff	€/kW/year	23.28
Commodity	€/kW h	0.004	Commodity	€c/kW h	0.19
Monthly/kW	€/kW/month	1.03			
Standing+meters	€/year	7,800.00	Metering	€/month	377
Maintenance	€/year				
System service	€/kW h				
Assumed load			**Assumed load**		
Capacity	MW	10	Capacity	MW	10
Energy	MW h	70,000	Energy	MW h	70,000
Monthly peak	MW	1.6	Utilisation	h	7,000
Charges			**Charges**		
Monthly capacity	€/year	19,776			
Annual capacity	€/year	102,700	Capacity	€/year	232,800
Commodity	€/year	280,000	Commodity	€/year	133,000
Total	€/year	402,476	Total	€/year	365,800
Other fixed charges	€/year	7,800	**Other fixed charges**	€/year	4,524
Total charges	€/year	410,276	Total charges	€/year	370,324
Charges	€/MW h	5.9	Charges	€/MW h	5.3

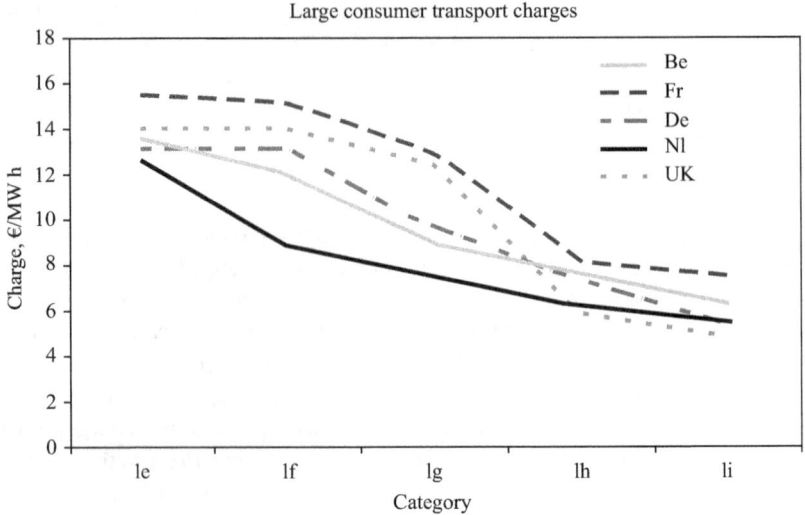

Figure 14.5   Transmission charges

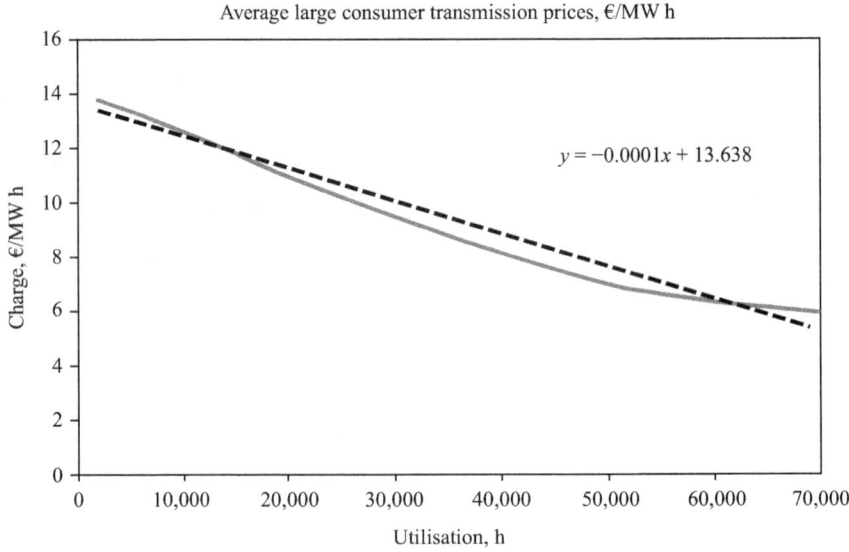

Figure 14.6   Transmission charge vs utilisation

utilisation. The charges also vary between the countries of NW Europe sometimes leading to complaints from industrialists who are in competition. A trend line of the average charges may be established based on the utilisation or MW h consumption as shown in Figure 14.6 where the trend is given by Cost = 13.64-MW h/10,000.

14.11   Transmission charging United Kingdom

The methodology employed in the United Kingdom for transmission charging is based on investment cost-related pricing. The charges vary between zones with 21 generation zones and 14 grid supply points. They are based on the impact on investment, maintenance and incremental costs to maintain defined security standards based on peak conditions. The tariff includes a location element based on utilisation and a non-location element to recover the residual revenue.

A DC load flow is used to assess the network MWkm utilisation at peak conditions for a MW injection at each node. The demand charges are equal and opposite but the revenue requirement is based on a split of 27% being recovered from generation and 73% from the demand side. The charges are adjusted for circuit type and voltage from a network reference point and weighted according to the required capacity. The expansion constant in £/MW/km is based on an annuity to cover the capital cost of transporting 1 MW over 1 km at 400 kV. A multiplier is used to cover the costs of different types of circuits and voltages. An additional factor is added to cater for maintaining security at each node. Charging zones are established based on groups of generators and supply points with charges within ±£1/kW.

The charging technique does not recover all the costs and the non-locational residual element has increased in recent years with super-grid generation closing and more units being connected to the distribution system. They are used to reduce the peak (triad demand) at the supply points further reducing the transmission revenue and increasing the residual element. It is necessary to remove this distortion to restore the competitive balance between supergrid and distribution-connected generation.

14.12   Distribution microgrids

The EU dictated that 80% of households should have smart meters by 2020. In the United Kingdom, it was proposed that suppliers rather than the distribution companies should be responsible for the installation programme. It has been argued by some academics studying the area that this was a mistake and the responsibility should have been placed on the distribution companies. Problems have been reported with equipment compatibility when switching suppliers. **The complete roll out cost has escalated and is expected to reach £11bn that will feed through to customer's bills.** Not all UK customers have been offered smart meters as of August 2017, but, of those that have, the take up has been below 50% with just 20% of customers with installations. Some protagonists see the smart meters as laying the foundation for 'smarter' management of household energy use. However, not all countries have accepted that there is a compelling business case and in Germany it is planned that only some 23% of consumers, with high loads, should have the installation.

The concept of enabling embedded distribution network operators to compete with conventional distribution network operators is accepted in the United Kingdom.

There is a utopian vision of smart microgrids embracing the control of local generation like solar and storage facilities to the point of being largely self-sufficient and capable of 'island' operation. A perceived advantage is that organisations like hospitals and factories are able to maintain essential services when the main grid fails, but usually they have to start diesel generators to maintain continuous supply. It is believed by some observers that microgrids could be adopted by many more users including domestic households sharing the output from solar panels. **It has been suggested that the inclusion of storage capability would enable microgrids to defect from the main grid.** The IREN2 project in the Wildpoldsried area of Germany has demonstrated island operation of a microgrid serving 2,500 consumers. The region has a high proportion of renewable generation and storage capability based on lithium-ion batteries and uses intelligent control to demonstrate decoupling and reconnection to the grid without loss of supply. It is less clear that microgrids would be cheaper to build and operate than conventional distribution systems with centralised generation. There are a number of reasons to question the viability of the concept:

- It is not clear who will sponsor, finance, operate and maintain the microgrid network that is serving a disparate group of users. The situation would be unlike that of a hospital with an engineering department managing the provision of services;
- How would consumers pay for the provision of the microgrid service and network losses?
- In the event of outages of the local generation facilities due to maintenance or faults supplies would need to revert to the main network and its availability would cost;
- How would the operation of the microgrid be coordinated to maintain normal supply voltages in conjunction with the main network?
- How would supply and demand be balanced and who would be responsible? The accepted concept is that balancing is most easily managed centrally to avoid the situation where each area is ramping supply or the demand side often in opposite directions, whereas the net system requirement may be zero;
- The costs of producing electricity from larger scale generators is usually cheaper than from small local generation. If there is part reliance on diesel generation, the high level of emissions would have an adverse effect on the local environment;
- Will microgrid developers adopt the same design standards as the distribution companies? The roll-out of smart meters by suppliers has demonstrated the dangers of fractioning supply leading to incompatibility;
- What happens if the microgrid management organisation goes into liquidation?

In effect, the microgrid would need a similar support infrastructure and skill set as the distribution company plus the ability and facilities to manage its generation in conjunction with the network. This is unlikely to be cheaper than the local distribution company or large supplier providing the services and taking advantage of economies of scale. On the positive side, the customer avoids paying for distribution services that can be half of consumer bills. **Currently, the distribution services are bundled into a simple tariff that does not specifically identify the provision of security and backup.** The distributor's costs would not change in the short term by the defection of microgrids. Longer term less grid reinforcement would be required and would mean that the option to change back to distribution supplies may well not be available to those fed by microgrids. The scope to engender competition is recognised and new entrants will see opportunities but not necessarily the scope of what's required. It may be like the roads on new estates that are eventually taken over by the council if they meet their design standards.

Embedded generation and the take-up of electric vehicles is having a major impact on the distribution network requirements and loading that will create the need for a radical rethink of the approach to its development. An alternative entry point for new network players will be in the provision of electric car charging facilities that will not be easily accommodated on the local domestic distribution network.

The situation in Spain highlights the difficulties that are arising from some 3.5 GW of solar generation resulting from the promotion of self-generation. The network revenues no longer cover the costs leading to reported deficits of $34bn in 2016. **The response has been to introduce laws requiring owners of self-generating PV systems of less than 100 kW to pay the same grid fees as all other system users (the so-called Sun Tax).** The law also prohibits them from selling surplus electricity that instead they have to donate free to the grid. Systems over 100 kW have to register before they can sell on the market and need permission from the local supplier and government before they can connect to the grid. Community ownership is prohibited for all sizes of systems designed for self-consumption. The new laws are retroactive and associated with high fines.

14.13 Conclusions

During 2007/08, the EU pressed for ownership of the networks to be unbundled to establish truly independent SOs able to manage operation of the system without any prejudice. Ownership unbundling is seen as the most effective means of avoiding discrimination. Similar initiatives have taken place in the United States where the FERC order 2000 required utilities to hand control of their networks to an independent entity. These initiatives have been coupled with the introduction of competition in the provision of networks as well as engaging the SO in the network-development process. The high use of system costs for the provision of networks to system users has become an important factor in determining their choice of local vs remote generation.

Distribution networks have traditionally been based on consumer after diversity maximum demand. For the average domestic consumers, the ADMD may be below 2 kW but this natural diversity may be destroyed by a number of factors; time-of-use tariffs; local solar panels all exporting; local wind turbines all generating. The management of the network utilisation would have to be fostered by a process of real-time pricing that reflects both energy prices and network limitations.

Use of system charges have traditionally been based on the level of utilisation with both the network and energy charges reducing as it increases to the benefit of larger users. This basis of charging does not cover the requirements of embedded generation. The current tariffs for embedded generators reflect the fact that their operation may reduce network demands at peaks and hence the supplier transmission use of system charge and also network losses and they may even be negative. The introduction of embedded generation also alters the network security assessment and management.

The other key change expected is the increase in demand resulting from the take up of electric vehicles and heat pumps that will have less diversity than more conventional loads. Significant savings could result from managing the increased demand to provide a smoother profile that better exploits both the network and generation assets.

The implication to the distribution networks is that they will move from being essentially passive to requiring active management along the lines of the transmission system. It has further been suggested that the inclusion of storage capability would enable microgrids to defect from the main grid. The IREN2 project in the Wildpoldsried area of Germany has demonstrated island operation of a microgrid serving 2,500 consumers. The additional costs incurred in managing a more active network will need to be reflected in the use of system charges to embedded generators and system users.

Question 14.1 Using the data in the table, calculate the impact on 'le' and 'li' end-user prices if 20% of the peaking energy comes from renewable sources at €90/MW h displacing conventional generation.

	MW	MW h	Util (h/year)	Base %	Peak %	Super	Per unit price	Energy	Tx/Dx cost	End user
Domestic						Peak		€/MW h	€/MW h	€/MW h
Da	0.003	0.6	200	0.0	0.0	100.0	1.75	70.0	92.6	162.6
Db	0.0035	1.2	343	0.0	0.0	100.0	1.75	70.0	73.3	143.3
Dc	0.0065	3.5	538	0.0	0.0	100.0	1.75	70.0	56.0	126.0
Dd	0.0075	7.5	1,000	0.0	0.0	100.0	1.75	70.0	50.5	120.5
De	0.009	20	2,222	0.0	100.0	0.0	1.24	49.6	45.0	94.6
Commercial and small industry										
la	0.03	30	1,000	0.0	0.0	100.0	1.75	70.0	68.6	138.6
lb	0.05	50	1,000	0.0	0.0	100.0	1.75	70.0	55.1	125.1
lc	0.1	160	1,600	0.0	100.0	0.0	1.24	49.6	34.3	83.9
ld	0.5	1,250	2,500	0.0	100.0	0.0	1.24	49.6	23.3	72.9
Medium-size industry										
le	0.5	2,000	4,000	15.6	84.4	0.0	1.20	48.1	13.4	61.5
lf	2.5	10,000	4,000	15.6	84.4	0.0	1.20	48.1	12.6	60.7
lg	4	24,000	6,000	51.1	48.9	0.0	1.12	44.7	11.2	55.9
Large industry										
lh	10	50,000	5,000	33.3	66.7	0.0	1.16	46.4	8.6	55.0
li	10	70,000	7,000	68.8	31.2	0.0	1.07	43.0	6.6	49.6

Part IV

How do we get to where we want to be?

The fourth part focuses on the analysis of options to meet the three concurrent objectives of security, cost and sustainability through curtailing emissions. It discusses the shortcomings of some of the market mechanisms and the regulation that has been applied to meet them. A set of scenarios is developed to provide a framework for analysis based on the use of an overview model developed to embrace the technical operational aspects of system management. It will also draw comparisons with other approaches like the 2050 model. The work concludes with a review of the governance and market arrangements to identify the need for any changes to meet the future requirements resulting from the impending transformation.

Chapter 15 Future scenarios – This chapter discusses the drivers of change and their impact on market players. It discusses the growth in embedded generation at the expense of large supergrid connected stations. It highlights the need for flexibility to manage renewable generation and their impact on system security. It references smart meter roll-out and end-user engagement.

Chapter 16 Scenario evaluation – This chapter illustrates an approach to establishing a development strategy that meets emission target reductions at minimum cost. It also demonstrates the analysis of the risks to the outturn due to erroneous data assumptions. It advances a robust scenario and illustrates the assessment of the risks and factors that could influence the outturn.

Chapter 17 Trading in an uncertain environment – This chapter discusses the difficulties faced by traders in predicting demand and prices given the impact of intermittent and embedded generation. The appraisal of investment options is described and the assessment of the values at risk. The management of fuel price risk is described and the assessment of the link to marginal prices.

Chapter 18 Smart grids development – The transition of the distribution networks from passive to active is discussed with the expansion of embedded generation. The deployment of smart metering is described to facilitate developments embracing network and demand-side management with storage. The potential impact on distribution system management requirements and security are analysed together with future market arrangements.

Chapter 19 Optimum development strategy – This chapter identifies some of the key measures that will be necessary to realise a favourable outcome from the transition. It discusses the need to establish a plant mix tailored to match the demand requirements at least cost and provide flexibility. It advocates a single buyer/seller market model as that most likely arrangement to coordinate the changes incident upon the sector transition. It reviews various modelling approaches to support the transition process.

Chapter 20 Key findings – This chapter includes a summary of the key findings from the book.

Chapter 15

Developing future scenarios

15.1 Objectives

The key objectives of the energy sector are usually framed by government policy and regulation. This will be formulated in consultation with energy users and take account of wider issues. This chapter critically reviews the power system development process. The three key current objectives in the developed world are

- Affordability – to ensure that all members of society can afford a supply and that local industry is not disadvantaged in competing with imports;
- Security – to achieve a balance between the costs of loss of supply with the costs of maintaining security standards;
- Sustainability – meeting environmental targets for emissions as developed by government and apportioned across the energy sectors.

The emphasis placed on these objectives will vary with the developing countries more focussed on realising electrification and developing their production capability. There will be emphasis on containing costs, but there are constraints in moving from an existing plant mix to realise new targets that will influence the development path. Where coal has been a major source of energy it will not be possible to suddenly phase it out putting miners out of work and destroying mining communities. A phased transition will be needed that takes account of the wider political issues. Prices in the developing countries may be set low to foster electrification. In the developed counties, there may be a need to protect energy intensive users, like the steel industry, from high prices to retain their competitiveness.

There will also be different attitudes to setting security standards affecting both network requirements and generation capacity margins that are usually specified in terms of the number of hours of potential shortfall during a year. In many developing countries, any available generation runs and demand is controlled to match its capacity. The emission targets set for each country will be apportioned across sectors reflecting the perceived capability to respond. The electricity sector is often expected to produce results in the short term in contrast to heating and transport energy use where the transition options are less readily available.

To identify needs and plan system development, there is a requirement to predict the likely future demand across the domestic, commercial and industrial sectors. The key determinant of future demand is the expected progression of the country's GDP and usually a good correlation can be established from previous statistics. Longer term improvements in efficiency will shift the relationships as will specific policies or developments. There is some demand sensitivity to price but it is often limited. Specific policies related to environmental issues will have more impact on future demand trends and incentives may be in place to encourage more efficient use of energy.

The realisation of environmental targets may cause a shift from one energy source to another e.g. electric vehicles or heat pumps that will increase electricity demand significantly. Specific measures may be in place to exploit demand-side response to contain peak demands and limit the total generation capacity required. The distribution networks are usually designed based on an after diversity maximum consumer demand. Electric vehicles and heat pumps not only constitute large loads at the domestic level but also may not exhibit much diversity because of the longer period of time they may potentially be taking energy from the network. **As their use increases the capacity, limits of the distribution networks will be reached.** Either high costs will be incurred for network reinforcement or access to capacity may need to be restricted. This could be through selective time of use tariffs or acceptance of controlled access to capacity. Schemes are already in place across the world where for a participation fee consumers with time insensitive loads like air conditioning are remotely controlled to manage network capacity restrictions. **In general, new tariff structures will be needed with more value placed on flexibility and pricing access to capacity rather than just energy.** There may also need to be payments or discounts to accept the interruption of less sensitive loads for short periods like heating appliances and freezers.

15.2 Policy options

The European Commission has outlined priorities for regulation in the electricity markets including

- Maintain a flexible approach to regulation appropriate to a fast moving market;
- Enable innovation with balanced regulation;
- Facilitate the operation of the markets and ensure they work;
- Secure supplies by restoring appropriate price signals;
- Develop regional markets.

The EU aims to strengthen regulation in the area of network operation and adequacy. They seek to establish a clear and effective framework for regulatory oversight.

The US National Renewable Energy Laboratory has provided policy guidance appropriate to their perception of the sector transformation including

- Avoid committing to conventional approaches with long lead times like nuclear to leave room for innovation;
- Take account of disruptive technologies like shale gas to assess their full development impact;
- Maintain an energy services focus rather than just electricity (perhaps need to assess heat and transport and potential interaction);

- Expect the breakup of the normal vertical supply chain through generation to transmission to distribution;
- Expect more local distributed generation rather than large-scale centralised generation parks.

A key aspect of the impact of the transformation is the risks perceived by potential investors with no clear direction and frequent regulatory intervention. This coupled with the reduced utilisation of conventional generation is leading to unprecedented low-generation capacity margins. **The shift from large centralised generation to small embedded is also leading to lower utilisation of the transmission network whilst stretching the capacity of the distribution networks to manage the transition to a more active network.**

15.3   Key variables

There are a lot of largely independent variables that will affect the outcome of the developing energy market for electricity and gas including developments on the distribution networks with embedded generation and a more active demand side. General factors affecting the energy markets and demand levels include

- The development of the GDP and that of neighbouring countries;
- The price of energy and tariff arrangements compared to other commodities;
- Consumer choice affecting the proportion of EVs and low carbon heating;
- Policy on efficiency improvements and insulation;
- Consumer expectations on the security of supply and the required plant margin;
- Domestic demand mix development for cooking, heating, cold, wet, computing;
- Demand mix for industry and energy intensive industry development;
- Regulations affecting supply, e.g. proportion of energy from renewable sources;
- Policy on heat and transport emissions and incentives for electrification.

The consumer choices can be expected to be strongly influenced by the cost of facilities and ongoing energy prices. The government interventions may be to meet wider environmental targets or support industries that may otherwise be uncompetitive e.g. steel production facing international competition.

Given the expected demand and its profile there are other factors that will influence the choice of generation to meet it at the lowest cost whilst meeting environmental constraints including

- Fuel prices – for coal, gas, oil and associated CO_2 prices and uranium fuel;
- The cost of capital, the availability of suitable generation sites;
- The market for embedded generation e.g. CHP;
- The environmental regulations, constraints and subsidy mechanisms;
- The opportunities for import and export;
- The network charges for use of the system and interconnectors;

- The availability of storage systems;
- The market and regulatory arrangements.

These factors affect the perception of risk to realising a profit by potential investors. If perceived to be too high investment in new generation may be discouraged or at best delayed leading to lower than normal plant margins. The analysis of risk will embrace consideration of options to hedge the risk.

15.4   Typical scenarios

In conjunction with stakeholders, the UK National Grid Company developed four scenarios that attempt to reflect energy user sentiments including

- Gone green – based on ambitious policy intervention to move to a low carbon energy sector that is enabled by high levels of prosperity;
- Slow progress – where economic conditions limit the ability to move to a low carbon world;
- No progression – business as usual prevails with the focus on short-term affordability not dominated by emission reduction;
- Consumer power – where developments are driven by the market and consumers focussing on costs rather than emissions.

These scenarios are very diverse and of limited value to a potential investor who has to manage risk. Two scenarios of the plant mix for 2030 are shown in Table 15.1 and compare a 'slow' development scenario with an ambitious 'green' scenario.

Table 15.1   Plant mix scenarios for 2030

Technology Mix 2030	MW Slow	MW Green
AGR	1,200	1,200
PWR	8,051	11,510
BIT	0	1,987
CCGT	44,265	29,469
OCGT		5,579
CHP	4,253	4,982
OIL	748	782
CCS	0	2,000
CCS	0	2,588
Interconnectors	7,200	7,600
Offshore wind	20,832	35,956
Onshore wind	13,590	20,985
Solar PV	6,106	15,813
Biomass	4,806	6,736
ROR	1,952	2,176
PMST	2,744	3,356
Marine	40	854
Total, MW	**115,787**	**153,573**

The main scenario difference is in the proportion of renewable generation in the mix with twice as much generation capacity in the green scenario in the form of wind, solar and biomass at the expense of gas and nuclear capacity. It should be noted that the 'green' scenario includes an additional 38 GW of capacity reflecting the low-energy output of renewable sources coupled with the need to retain the availability of conventional generation for when renewable generation output is low. **This conundrum could be overcome with more flexible low carbon emission generation like CCS without increasing emissions.** The CCS capacity in the 'green' scenario is assumed to be limited by slow development timescales. Given that large-scale systems were commissioned in the USA in 2016 a faster CCS development could be envisaged based on retro-fitted to existing coal and gas units. Given the high cost of renewable subsidies, this option could offer a lower cost solution to meeting emission reduction targets more quickly. An LP formulation was constructed to identify a least cost optimum plant mix as shown in Table 15.2. The result showed a higher proportion of CCS generation ('fast CCS') at the expense of less wind and solar generation whilst meeting the same emissions as the 'green' scenario of 22 Mt/year but at a cost of some £6bn/year lower (Table 15.2).

Table 15.2   Fast CCS scenario

Scenario Technology	2030 slow MW	2030 green MW	Fast CCS 1 MW	Fast CCS 2 MW
AGR	1,200	1,200	1,200	1,200
PWR	8,051	11,510	11,510	11,510
BIT	0	1,987	1,987	1,987
CCGT	38,686	29,469	14,935	11,930
OCGT	5,579	5,579	12,375	12,375
CHP	4,253	4,982	4,982	4,982
OIL	748	782	782	782
CCS gas	0	2,000	9,952	12,957
CCS coal	0	2,588	5,518	5,518
Interconnectors	7,200	7,600	7,600	7,600
Offshore wind	20,832	35,956	12,600	12,600
Onshore wind	13,590	20,985	7,400	7,400
Solar PV	6,106	15,813	6,106	6,106
Biomass	4,806	6,736	4,806	4,806
ROR	1,952	2,176	2,176	2,176
PMST	2,744	3,356	3,356	3,356
Marine	40	854	854	854
Total, MW	115,787	153,573	108,139	108,139
Maximum demand		59,812		
Less wind/PV, MW	68,019	72,365	73,579	73,579
Costs, £b/year	**34**	**41**	**34**	**34**
Emissions, mtCO$_2$/year	**49**	**22**	**29**	**22**

15.5   Risk assessment

Prices in the electricity sector are largely fixed by the prevailing generation plant mix and the associated fuel prices and subsidies. There is little scope to reduce prices in the short term, and the focus needs to be on optimising long-term investment. The latest round of UK capacity auctions illustrated how things can go wrong with undesirable consequences with a lot of small inefficient and dirty generators winning capacity contracts. The current use of system tariffs has also distorted the capacity market in favour of small embedded generators. **The overall uncertainty has led to a dearth in investment in large-scale generation leading to unprecedented low plant margins.**

During a period of significant transition, the market cannot be entirely relied on to realise the optimal outcome because of the complex interactions. Generally, there is a recognised need for a more holistic approach but not how it may be realised. The four scenarios produced by National Grid are very diverse with no clear direction. It would be of interest to compare the transmission requirements associated with each of the scenarios.

There is a need to identify a central case for development that can be illustrated to be robust against variations in key parameters like fuel prices. Regulation and market mechanisms then need to be designed to promote this outcome. Some countries have opted for capacity auctions by technology type to support the process. Government intervention is already happening in underwriting new nuclear and subsidies for renewable generation but without an overall plan coupled with a risk assessment to establish the course of least regret.

To illustrate the point Table 15.3 compares the cost of the 'green' scenario with one with a more rapid development of CCS. The model used is based on a half hourly optimal dispatch and takes account of the dynamics of tracking variable wind and solar output with conventional generation fitted with CCS. It shows a robust net annual benefit of a 'fast CCS' scenario through a range of variation in key variables whilst still meeting emission targets. The CCS scenario would also require less transmission development. A large-scale CCS scheme with EOR (enhanced oil recovery) has recently been commissioned in the USA, and the

Table 15.3   Sensitivity of impact of CCS on costs

Variable	Units	Fast CCS 2	Green	Difference
Base case costs	Cost £m/year	34,248	40,641	6,393
Capital at 12% WACC	Cost £m/year	36,425	42,284	5,859
CO_2 £200/t	Cost £m/year	37,937	44,012	6,075
Off shore FIT £120/MW h	Cost £m/year	33,434	37,973	4,539
Off shore FIT £100/MW h	Cost £m/year	32,783	36,402	3,619
CCS cap. £1,422/2,708/kW	Cost £m/year	34,532	40,719	6,187
12%, CO_2 200, FIT 100, cap	Cost £m/year	38,491	41,443	2,952
CCS capture 90% base	Emiss. Bt/year	22	22	0

worldwide market for CCS technology has been estimated at $31bn by 2021 offering significant export potential.

The base case assumptions in this study are shown below together with the extreme values used in the sensitivity studies shown in brackets:

- Gas prices £7/GJ
- CO_2 costs £70/t (sensitivity tested at £200/t)
- Gas CCS capital £1,236/kW; coal 2,475/kW (sensitivity gas at £1,422/kW coal £2,708)
- CCS carbon capture rate 90%
- CCS transport and storage £19.6/t
- Nuclear capital £3,885/kW; FIT nuclear FIT £92.5/MW h
- Wind FIT onshore £95/MW h, offshore £155/MW h; PV (photo voltaic) £120/MW h; sensitivity study at FIT of £100 or £120/MW h for offshore wind.

The emission levels for the two future scenarios are the same at 22 Mt/year. The last scenario result combines the extremes of a high WACC (12%), a high CO_2 price (£200/t), a low CfD (contract for differences) for offshore wind (£100/MW h) and high CCS capital costs (£2,708/kW) but still results in the 'fast CCS 2' scenario costs being lower than 'Gone Green' by £3b/year.

The sensitivity of the two scenarios to changes in gas prices is shown in Figure 15.1. The annual costs rise more quickly in the fast CCS scenario that includes more gas-fired generation but is still significantly cheaper. The rise in emissions reflects the negative impact on the cost comparison with coal as gas prices rise resulting in more coal being dispatched. Given the development of shale gas, there is reason to expect gas prices to stabilise if not to fall. This trend could be accelerated if gas use for heating is displaced by heat pumps.

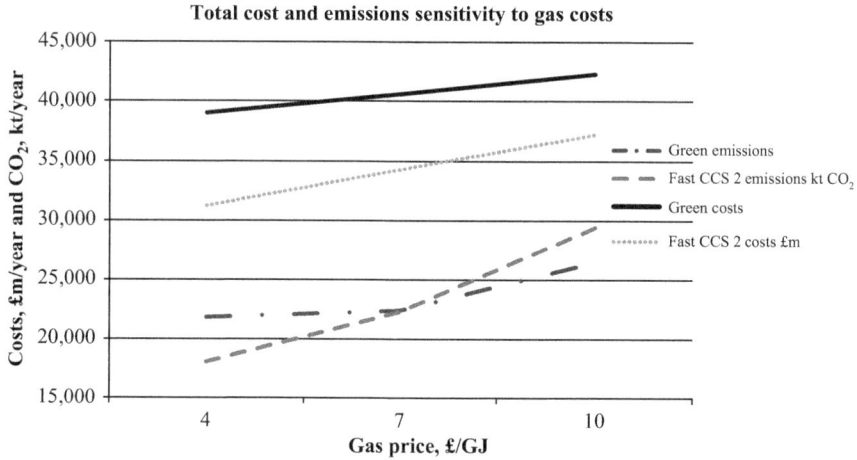

Figure 15.1   Scenario sensitivity to gas prices

In current market structures, there is no agency forward looking with responsibility for taking a holistic view of the developing industry and assessing risks and it is glibly assumed that the market will solve all. This presumes a perfect market can be structured to give a robust optimal plant mix at a time when there are many more new players being encouraged to operate in the market. In practice, sponsors and investors often assume high levels of availability and utilisation in their project assessments that cannot be realised in practice. The market needs to be structured to recognise and reflect the technical constraints associated with some development paths. In a situation of rapid change, the single buyer/seller market model can be shown to offer the potential to realise a holistic outcome. The buyer acts on behalf of all consumers to contract for tranches of generation capacity with longer term power purchase agreements for nuclear and hydro coupled with shorter term procurement from existing players. Through a series of auctions for different plant types across various timescales ahead, it is possible to tailor the plant mix to match the emerging demand profile in a competitive environment whilst managing risk. The current use of capacity markets is a step in this direction but does not afford influence over the plant types to match their likely operating role.

Governments should consider the establishment of a single buyer/seller market model to realise a holistic outcome. It should be responsible for appraising developing options and risks and managing the competitive provision of grid connected generation. It would also underwrite key developments like nuclear generation and sponsorship of CCS trial installations, initially focused on retro fit options to enable a gradual phase out of coal-fired generation.

15.6   Development constraints

There are a number of potential technical issues that could constrain the development paths open to the industry in the medium term as discussed in Chapter 2. They include system management technical and operational issues related to

- Managing intermittency;
- Maintaining frequency and voltage control;
- Predicting demand net of intermittent sources;
- Loading on the distribution networks;
- Managing interconnection loop flows.

These can limit the ability of the system to accommodate intermittent and embedded generation at acceptable costs. There are also financial constraints on companies and consumers including

- Reduced utilisation of conventional generation to the point of non-viability;
- An increasing proportion of consumers unable to afford energy costs and subsidies;
- The decimation of energy intensive industries faced with international competition;

- The risks perceived by potential investors in promoting developments;
- The cost constraints on consumers limiting the takeup of EVs and heat pumps by consumers.

The scenarios promoted for discussion across the industry have been based on axes related to prosperity and the extent to which government and consumers are focused on reducing emissions and air pollution. There are also a number of political issues concerning the competitive position of domestic industries in relation to countries that are more interested in developing industries and jobs than sustainability. **There are also a number of external potential developments that could drive development paths including**

- The full realisation of the impact of emissions on health;
- The development of cheaper reliable sources of energy from fusion power, geothermal, CCS or modular nuclear;
- The availability of cheap energy storage;
- A reversal of trends in global warming and the impact of emissions;
- A sudden increase in global warming with increased demand for air conditioning;
- Consumer concerns over the security and quality of supply.

These need to be considered in establishing a development path of least regret and the timing of key milestones. It should be demonstrated how the development path could change in the event of sudden changes in the key drivers. Commitment to ongoing subsidies and large-scale projects can limit future changes in direction. There is also a question of timing developments of networks to avoid delaying system user aspirations. This could particularly apply to distribution networks that are faced with increasing amounts of embedded generation, EVs and heat pumps.

In the emerging situation of many more participants in the market engaged in supply of self-generation, it is difficult to engender a coordinated development strategy through just market mechanisms, and there is a case for a different model to manage the transition.

15.7   External potential developments

There is increasing awareness of the impact of emissions on public health by politicians and the public at large. There is now open talk about tens of thousands of premature deaths due to pollution in cities. In China, restrictions have been introduced on the use of cars in cities like Beijing based on alternate days. The public have found it necessary to use protective masks to restrict inhalation. In London's busy shopping areas, emission levels reached their annual limit in January and consideration is being given to raising congestion charges on polluting vehicles or outright bans on entry.

Car manufacturers are aware of the health hazards and potential impact on their business and are actively pursuing the development of hybrid and all electric vehicles. Governments and manufactures are also **promoting schemes to scrape**

polluting vehicles and accelerate the move to cleaner transport. Given these initiatives, the transition to vehicles using electricity is likely to accelerate quickly. End users will expect to charge their car at will and not be aware of any network restrictions. The industry has a tradition of supplying electricity and not curtailing its use but widespread charging of EVs with little diversity could create network loading problems as well as additional generation capacity requirements.

There is a need to maintain a proportion of flexible generation in the plant mix to track demand and intermittent renewable generation output. There will be a requirement for more generation as the use of EVs and heat pumps increases and consideration needs to be given to their potential operating role. It was shown in Section 13.8 that the combined impact of meeting just the 10% reduction in heat demand and transport on UK demand for electricity would be

Heating − 5.0 GW at 35% load factor = 15.3 TW h
Transport − 3.5 GW at 43% load factor = 13.1 TW h.

The combined effect would be a requirement for up to 8.5 GW of generation with an annual load of 28.4 TW h. The increase in capacity required will depend on the timing and diversity of charging. Claims have been made of the reduction in generation capacity required that can be realised by engaging customers in demand management but without an explanation of how it can be realised or the costs. The potential to track demand and intermittent generation output variations will require a significant tranche of generation like gas with CCS that can be dispatched. There will be occasions when demand is falling while renewable generation output is rising that will require rapid changes in conventional generation output or curtailment of renewable generation. Gas generation with CCS looks the most likely new generation option while the prospect for fusion power looks remote, but there may be scope for small modular reactors that may offer more flexibility.

The availability of cheap storage would transform the ability to smooth the demand profile. The problem to be overcome is the overall cycle efficiency between storing the electricity, by transforming into some new medium like compressed air or water pumped to a high level, and subsequently using the stored energy to generate electricity. The overall efficiency may be as low as 70% and a significant price differential is required for the process to be financially viable and to recover capital costs. The prospect of large-scale efficient cheap storage becoming available looks remote. There are exceptions when storage schemes can be viable; when there is a premium for a service to provide fast response to manage frequency; and to use excess renewable energy that would otherwise have to be curtailed. A more viable proposal is to use EV batteries that are already paid for by the vehicle owner. This would be a particularly viable option to use surplus solar and wind energy. With a charge/discharge efficiency of around 80% price differentials would still need to be significant.

Any perceived acceleration of global warming could have a marked effect on the speed with which progress is made to low carbon electricity generation. Given the level of dependence on coal-fired generation across the world retro-fitting CCS may be the only viable short-term option.

15.8 Security analysis

There is a need to assess the implications to system security of increasing proportions of intermittent renewable generation in the plant mix. Security standards are usually defined as a loss of load probability (LOLP) and are typically around 4–8 h/year in the developed economies. This means that there will be a potential shortfall in generation meeting demand 4–8 h during a year. The analysis needs to be extended to accommodate the potential contribution from intermittent renewable generation. From statistical theory, the probability of r generators being unavailable from a population of n is given by

$$P_0^r = \frac{n!}{r!(n-r)!} \cdot P_0^{(n-r)} \cdot (1 - P_0)^r$$

where P_0 is the unit average availability; r is the number of equivalent generators; P_0^r is the probability of r units being unavailable. It has been found that a good approximation to an annual demand profile is given by a normal distribution of the form

$$D_n^h = \frac{K}{\sqrt{(2*3.124)}} * \exp - \left(\left(\frac{D_n - D_m}{std} \right)^2 / 2 \right)$$

where D_n^h is the number of hours during a year that the demand falls within the band n; K is a constant; D_n is the nth demand; D_m is the mean demand and 'std' the standard deviation. The parameters are adjusted to match the shape of actual records of the demand with some exhibiting a narrower range.

The formulation can be used to illustrate the relationship between the plant margin and the LOLP as well as the associated MW h of demand not met during the year as shown in Figure 15.2. It can be seen that the un-served energy increases exponentially as the plant margin is reduced and LOLP increases approximately in proportion to $10*(1{,}000*LOLP)^2$ GW h/year for this 325 TW h/year system. The formulation is simplified by assuming a set of units of similar size and availability that can be based on average actual data. The result shown is for a system with a 60-GW peak with 137 units of 500 MW. An average unit availability of 0.9 per unit is assumed in this simulation covering forced and planned outages. In practice, planned generation outages can result in shortfalls when they run over into colder weather periods or a sudden cold spell occurs towards the end of the outage programme season.

The assessment of security is complicated by the presence of a tranche of intermittent generation like wind or solar that also has an availability probability that determines its likely contribution to capacity. Using the model described above, the process described in Section 10.4 can be used to make an estimate of the contribution of wind generation to security in comparison with conventional generation.

The total system wind output profile is represented as blocks of generation with a probability of occurrence as shown in Figure 15.3 for 25 GW of installed

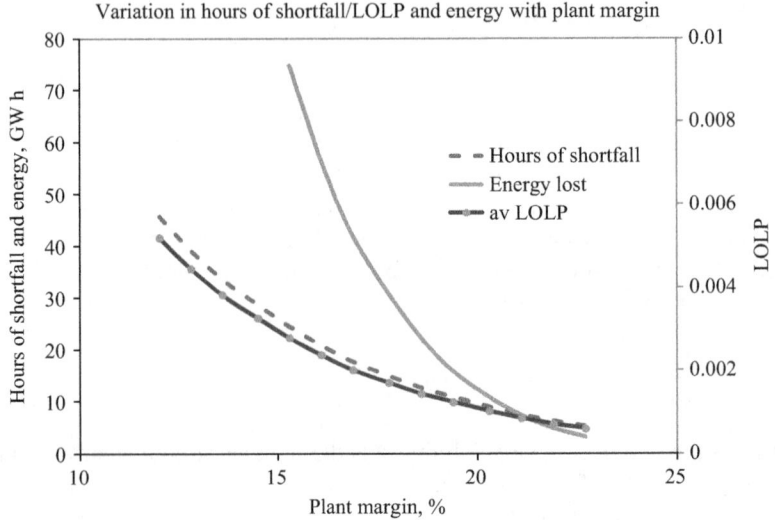

Figure 15.2   LOLP and shortfall hours vs plant margin

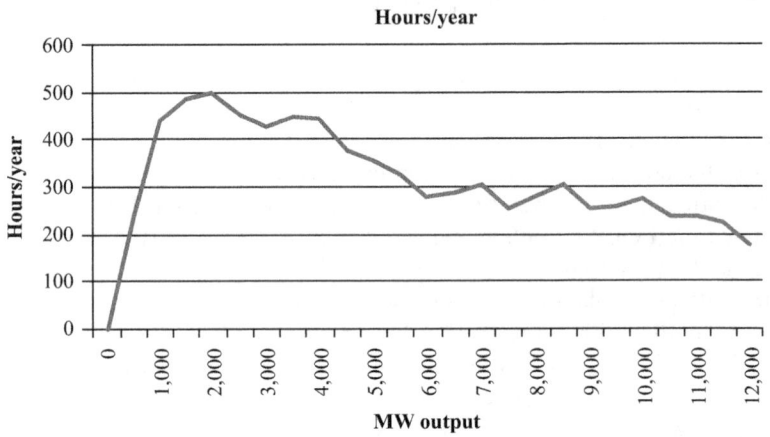

Figure 15.3   Wind output probability function

capacity. Each 500 MW block is included in the generation availability to establish the new consumer hours lost. This is multiplied by the probability of that wind output occurring to establish the number of hours of shortfall that could occur. The process is repeated for successive higher wind outputs until there is no increase in shortfall periods given the associated low probability of occurrence. **This probabilistic analysis indicates a wind contribution to system capacity of between 12% and 15% of installed capacity.** This is of a similar order to the results of practical assessment undertaken in Ireland. Comments by some analysts have suggested higher levels of contribution of up to 20%. The figures will vary with the expected load factor of the installation and could be lower in some less windy onshore sites.

In a longer term planning situation, the mix of generation needs to be matched to their likely operating role. This would include base load generation like nuclear and CHP schemes; mid-merit generation tracking daily demand patterns and peaking generation to cover peaks. The priority despatch afforded to renewable sources has a major impact on the role of more marginal generation as the installed capacity increases. Because of the shorter time taken in installing wind and solar farms they have disrupted normal operating roles before the plant mix can be adjusted to accommodate its impact. Some CCGT operators have converted their plant to operate open cycle to provide the flexibility needed. As the system loads are expected to increase with EV take-up and heat pumps, now is the time to plan to provide the necessary generation and demand-side flexibility to maintain security standards.

The other complication in assessing security is the potential contribution to security from embedded distributed generation. For generation installations based on landfill, waste and CHP capacity contributions of around 50% of DNC (declared net capacity) are estimated, whereas a CCGT installation with a controllable fuel supply a figure of 70% could be assumed.

15.9   Conclusions

The focus for the development of future strategy is to realise the three objectives of affordable energy, security of supply and a sustainable low level of emissions. The latter has proved the driving force leading to the introduction of tranches of subsidised renewable generation that has had a marked impact on other generators and network requirements and energy charges. A consequential effect has been a growth in small-scale embedded generation in preference to centralised generation. This is resulting in the distribution networks becoming more active and to stretch their capacity to accommodate more generation. In part the generation transition is the result of current tariff structures that need to be reviewed to emphasise the flexibility that networks offer to generators and users. In sharp contrast, the priority despatch afforded to renewable sources has reduced the utilisation of conventional generation operating at the margin to the point where it becomes financially non-viable. The uncertainty has led to a dearth in new investment and to premature plant closures, resulting in unprecedented low plant margins.

The introduction of smart meters was aimed at fostering efficiency in the utilisation of energy but is now perceived as facilitating a much wider end-user engagement. Not all countries see the same scope and are limiting roll out to larger end users. Less than 50% of UK consumers offered the meters chose to accept. Some commentators hold the view that customer demand control will significantly reduce the need for generation capacity. The United Kingdom has an energy requirement of around 330 TW h/year with a peak demand approaching 60 GW. It is axiomatic that if the demand were completely flat the capacity requirement would be only 37.7 GW. The mechanisms to realise this utopian position or the costs are less obvious. The growth of new demands for EV charging and heat pumps offer a limited scope if end users are prepared to allow third parties to control their use. Equally generation flexibility will be required and its value needs to be fully rewarded. In contrast to intermittent renewable generation that creates the imbalance problem gas generation with CCS offers flexibility coupled with low emissions to the atmosphere.

The transitions taking place are leading to a massive increase in the number of market players on the generation and supply side as well as in network service provision. In the current market structure, there is no agency with responsibility for taking a longer term holistic view of the evolving industry and the attendant risks and opportunities. It is glibly assumed that the market will solve all with governments and regulators currently responding to events rather than taking a lead. There is an argument at a time of rapid change for a single buyer/seller market model acting on behalf of all users to procure generation and manage the risks. The analysis in this chapter illustrates the scale of the impact of assumptions on key parameters on outturn costs and emissions. The SB (Single Buyer) would also be better placed to react to sponsor and exploit new developments in generation and storage and climate changes.

As well as affordability and containing emissions the third requirement is to maintain standards of security. This requires establishing the right level of generation capacity to realise a LOLP set typically at around 4–8 h/year. An added complication in the analysis is estimating the contribution that can be expected from intermittent renewable sources and embedded generation. Opinions vary but probabilistic analysis shows values of just 12% to 15% of installed capacity for wind and nothing for solar generation when peaks occur in the early part of dark evenings in the Northern hemisphere. There is an EU policy of extending interconnection, but its potential contribution may be limited when weather patterns span several neighbouring countries. The potential contribution from embedded generation will be influenced by local demand conditions and typical availability figures can be used as discussed in Section 14.4. The use of capacity markets is promoted as a way of meeting requirements with the onus on the grid operator to assess requirements. To realise the reduction in capacity needs, it should be suppliers who contract for capacity in the knowledge of their demand management capability.

Question 15.1 The developing system will require an increasing amount of flexibility to accommodate

- The intermittency of renewable generation sources;
- The output from embedded generation that is not subject to central dispatch;
- An increase in domestic demand form EVs and heat pumps that may have limited diversity.

Calculate the value of the provision of flexibility by demand management in terms of the avoided emission costs resulting from a 1% reduction in wind energy curtailment with CO_2 emissions valued at £30/t. Assume that the wind load factor is 25% and replacement energy is based on OCGT (open cycle gas turbine) generation with emissions of 0.54 tCO_2/MW h.

Hence, calculate the value of avoiding curtailment of 30 GW of wind based on the 'more inflexible generation' curtailment line shown in the graph.

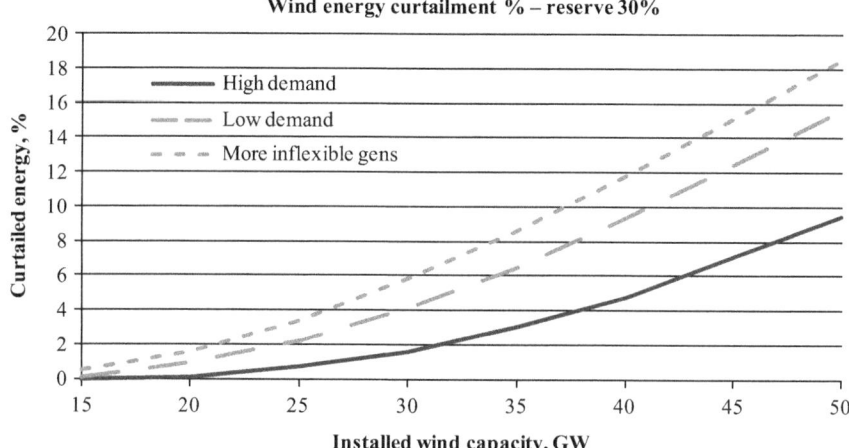

Chapter 16

Scenario and risk evaluation

16.1 Objective

This chapter describes an approach to determine the optimal mix of electricity generation that provides both low generation costs and low levels of emissions under a set of basic assumptions. A two-stage approach is used with a simplified linear programming (LP) model used initially to inform the solution area. Subsequently, a more detailed analysis is used based on a generation dispatch model, including emission calculations, that provides a detailed half-hour simulation of system operation for a selected year. The model provides an estimate of emissions taking account of varying proportions of wind and solar generation including carbon capture and storage (CCS) generation and calculates the total costs of generation. The model uses actual recorded system demand profiles and wind generation output collated by transmission zone in the United Kingdom. The profiles for additional wind farms are based on those for existing generation located in the same zone. The output of solar PV is based on a set of radiation data for a geographical location in the middle of the country. The model takes account of the impact of managing the intermittency of renewable generation on conventional plant and the need for curtailment at times of low system load.

The analysis in this study was based on two Future Energy Scenarios referred to as 'Green' with 56 GW of wind generation and 15 GW of solar and a 'Slow' with 34 GW of wind and 6 GW of solar. A simplified LP model was used to inform the optimal solution area and showed minimum total costs and emissions with a significant tranche of CCS plant resulting in the creation of a 'fast CCS' scenario. The use of storage was also evaluated both to operate to smooth the wind generation intermittency and also to operate buying and selling energy against market price. The analysis assumes that CCS technology could be developed within the study timescale and be commercially available at the higher end of predicted costs. This cost has been used in the analysis rather than a negotiated strike price.

The objective of the analysis is to illustrate how to calculate the overall carbon intensity of the grid and overall generation costs with different mixes of renewable and conventional generation and balancing plant. A model was developed with the following features:

- To provide whole system generation costs including fuel, carbon, capital and operating costs as well as subsides based on a Contract for Differences (CfDs) and use it to identify a cost minimum;

- To provide estimates of the associated CO_2 emissions taking account of the impact of intermittency on generation reserve, ramping and start-ups and fuel utilisation;
- To model and illustrate the impact of the need to curtail the output of high levels of renewable generation at times of lower system load;
- To provide realistic profiles for wind and solar generation and model CCS generation in the merit order whilst accounting for the impact of CCS on its efficiency and added transportation and storage costs;
- To enable studies for a range of plant mix scenarios varying the amount of, wind, nuclear and CCS generation;
- To estimate the benefit from introducing storage to manage intermittency for comparison with the costs of new storage facilities.

The studies focussed on the year 2030 with the option to adjust generation capacities to analyse the costs and emissions resulting from other plant mixes.

The data used in the analysis is based on a combination of

- Data related to national demand profiles and the profile of wind generation by transmission zone;
- A forward view of generation build-out taking account of known data and how balancing services are provisioned;
- Data on generator performance related to the additional energy used in part load operation, ramping and start up and shut down derived from suppliers.

16.2   Methodology

The initial studies were based on scenarios with one based on a continuation of existing policy and the other focussed on reducing emissions with an expansion of wind and solar generation. It was expected that these would not necessarily identify the minimum cost solution. A simplified LP formulation was used to identify a likely optimum solution and to adjust the scenarios to assess the impact on the total costs and emissions. The simplified LP formulation included

- An objective function – to minimise total annual generation fixed capital and operating costs and variable costs including fuel and CO_2;
- A simplified green field study with a number of units of each technology type as the key variables ≥ 0;
- A main constraint was to meet the capacity requirement with margin for security with wind credited with 15% of capacity as firm;
- Demand was represented by a normal probability function giving the number of hours during a year that demand falls within discrete bands;
- Generation scheduled to meet demand for each period in variable cost order;
- The option to run studies with and without the inclusion of the curtailment of wind generation based on the wind capacity and nuclear capacity.

The LP was designed to include the effect of curtailment of wind energy at times of low system demand to meet stability needs and accommodate inflexible generation.

The wind and solar generation output was based on average values for the year with other generation set to meet each demand band based on a defined merit order. The wind generation was subsidised to ensure its inclusion in the optimisation. The results of these studies were used to inform the choice of the optimal plant mix.

The cost minimum and emissions were then verified through sensitivity studies using the more detailed model. Sensitivity studies were based on variations in key variables including the CO_2 price, gas price, CCS capital costs and CCS CO_2 capture rate. The model also provided output on the costs of managing wind and solar generation intermittency including holding reserve, ramping and starting and stopping conventional generation to track renewable output variation. The studies also recorded the required curtailment of wind generation during periods of light load when a proportion of conventional generation has to be kept in service to provide fast reserve and maintain system inertia along with inflexible generation. The inflexible generation included base load nuclear, CHP and other renewable sources that have to be kept in service. The two previously defined scenarios were used to assess the total generation costs and emissions resulting from a 'Green' and 'Slow' strategy. A new scenario based on the LP solutions included more CCS generation to identify the minimum cost. Additionally, sensitivity studies around the new minimum cost scenario were used to both verify its optimality and the changes in the costs of managing intermittency.

16.3   Total costs

The total costs of production were based on the fuel utilisation and variable operating costs and the fixed annual capital and operating costs for conventional generation. The renewable generation costs were based on current CfD feed-in tariffs (FIT) and energy produced excluding any additional charges for energy curtailed to meet system constraints. The fuel utilisation was calculated for gas and coal-fired generation to establish the emissions and was used to calculate fuel costs. This was calculated including CO_2 costs for conventional generation and CCS generation although the CCS generation was included without CO_2 in the merit order to ensure its full utilisation. The saving attributed to the captured and stored CO_2 was calculated retrospectively to enable the capture ratio to be varied. The model was extended to calculate the fixed costs depending on the unit size and the energy production by unit covering both full load and reserve mode utilisation. A typical result is shown in Table 16.1 for the 'Green' scenario. The savings from deployment of CCS are shown at the bottom of the table and are partly offset by the transport and storage costs. For generation subject to a CfD, the capital costs are included in the CfD payment.

To enable different plant mixes to be evaluated, a module was developed to scale the capacity by technology type. The original data was based on the current data set plus known planned developments including wind installations. The capacities are adjusted to match the new proposed capacity mix and reordered according to variable cost to establish the new merit order. The data for other

Table 16.1 Example of annual study costs

Technology	MW 2030	Fixed costs, £m From historic	Energy, TW th	Fuel/var Cost, £m/year	FIT, £/MW h	FIT costs, £m/year
AGR	1,200		8.6		92.5	797
PWR	11,510		82.7		92.5	7,648
BIT	1,987	315	0.4			
CCGT	29,469	937	15.5	8,932		
OCGT	5,579	742	0.3			
CHP	4,982	391	17.3			
OIL	782	62	0.003	0		
Gas CCS	2,000	2,024	76.6			
Coal CCS	2,588	1,509	41.7	3,904		
Interconnectors	7,600					
Offshore wind	35,956		26.8		155	4,197
Onshore wind	20,985		15.8		95	1,511
Solar PV	15,813		7.0		120	840
Biomass	6,736	990	31.6	697		
ROR	2,176		5.1		160	820
PMST	3,356		7.8			
Marine	854	Intermittency		894		
Total, MW	153,573	6,969	334	14,428		15,812
		CCS	Saving	−3,958	Trans/store	997
				Total all, £m/year		34,248
		Total emissions net of CCS		22,216	ktCO₂/year	

quarters of the year is based on the new merit order taking account of maintenance outages.

To evaluate storage options, a model was developed to process basic demand and wind data to minimise the standard deviation of the net output. It was found that the optimum smoothing was realised when the point of zero transfer into storage was set at the average output from the wind source being smoothed. The smoothing process was applied to the basic wind data and the costs of managing intermittency reassessed for comparison with operation without storage. Alternative assessments were also made using an algorithm based on estimated half-hour market prices to assess the potential in a market environment. The process was repeated for different priced storage systems to establish the cost to realise financial viability.

16.4 Initial LP data assumptions

The objective function of the LP was to minimise the cost of capital, fuel, carbon and operating costs including a subsidy for wind (to assess on cost) or a CfD by varying the number of units of each plant type. The list below illustrates the range of data requirements necessary to support studies of the impact of generation mix on costs and emissions.

- The original base case CO_2 price is £110/t with gas at 78 p/therm.
- For the purpose of the LP modelling, wind generation included subsidy (a ROC) of typically £38/MW h onshore and £76/MW h offshore to enable it to feature in the optimisation.
- CCS plant was included based on 'first' costs with high CO_2 transmission/storage cost in the range £20–£32/t.[1]
- Efficiency degradation for CCS plant is included as a variable set at 14% for gas and 15% for coal.[2]
- Nuclear is based on a capital cost of £4,375/kW with equivalent energy cost of £86.1/MW h or a CfD of £92.5.
- Conventional gas plant is included as well as new OCGTs or older CCGT plant operating in open cycle.
- Includes modelling of the need for wind curtailment based on a formulation derived from GEM studies.
- The demand used in the LP dispatch was formulated as a load duration curve based on a normal probability function with the option to reduce the demand by the average PV output.
- The installed capacity was written down by an availability factor for each plant type.

16.5   LP studies

The initial results of the LP optimisation are shown in Figure 16.1(a), where there is no wind subsidy, the wind capacity stays at the level of that already committed. The result shows the optimum capacity where the introduction of CCS was constrained to around 2-GW capacity for both gas and coal; the result shown in Figure 16.1(b) is where the CCS technology is advanced more quickly with its application unconstrained. It can be seen in this case that there is a significant take up of coal CCS and gas CCS, contributing 26% of capacity. This displaces some nuclear generation at lower levels of utilisation below some 95% because of the higher nuclear fixed capital costs. It is also more economic than conventional CCGTs due to CO_2 costs at above medium levels of utilisation. In practice, some of the existing CCGTs may be retrofitted with CCS with others operating in open cycle peaking mode providing quick response to balance wind intermittency. This result is based on relatively high values of capital costs for CCS plant compared to government estimates for 'first' plant. Gas CCS is included at £1,715/kW with coal CCS at £2,538/kW with capture transportation and storage costs at £20/t for gas and £30/t

---

[1]The LP used £20/t for gas and £30/t for coal while the main model used £19.6/t.
[2]Based on author's industrial experience.

Optimal capacity mix – slow CSS

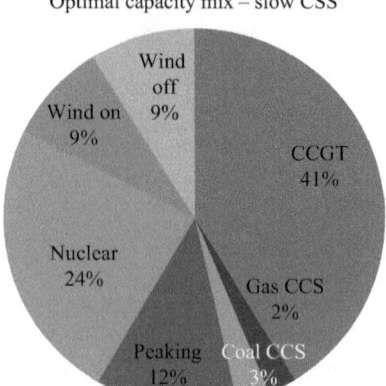

Figure 16.1(a)　Initial LP solution with CCS constrained to 4 GW

Optimal capacity mix – fast CSS

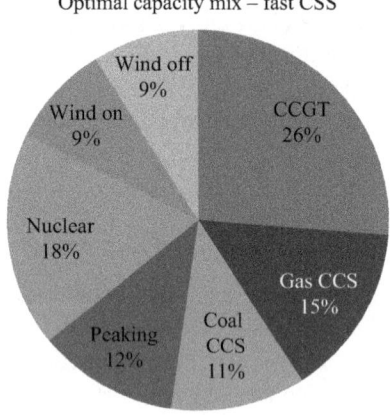

Figure 16.1(b)　Initial LP solution with CCS unconstrained

for coal. Additionally, it is assumed that there is degradation in efficiency of 14% for gas generation and 15% for coal to support the capture process.

The introduction of a minimal subsidy for wind generation based on the ROC concept causes more wind to be included in the optimum solution (when the subsidy is not included in the total cost objective function). Table 16.2 shows the results for a set of scenarios using the LP formulation and the original data assumptions. The total costs are of a similar order to the main model cost studies evidencing that the simple LP model provides a useful test bed. The cheapest solution (including the added subsidy of £38/MW h and £76/MW h for onshore and offshore, respectively) is with fast CCS enabled and the lowest wind subsidy that just causes new wind generation to be built. The LP also included an option to cost wind energy at the current CfD/ FIT values (onshore £95/MW h and offshore £155/MW h). Based on these higher

Table 16.2   Comparison of LP scenario results

Scenario	CCGT	Gas CCS	Coal CCS	OCGT	Nuclear	Wind on	Wind off	Cost, £m/year
LP wind sub. fast CCS	14,935	9,952	5,519	12,375	9,950	3,225	24,000	31,510
LP wind sub. slow CCS	29,153	1,670	1,958	11,134	10,855	3,225	23,671	31,828
LP based on FIT fast CCS	17,922	11,276	6,743	8,250	12,345	5,892	6,177	33,867
LP based on FIT slow CCS	28,494	1,670	1,958	8,199	16,534	5,892	6,177	34,170

costs for wind, the wind build is limited to the level already committed but the overall costs are still higher than the ROC subsidy scenarios. In both fast CCS scenarios, the studies suggest that some 16–17 GW of CCS capacity should be included in the optimal mix with coal slightly cheaper at higher levels of utilisation. The total costs of the two studies including CfD/FIT are higher because the CfDs result in higher costs than the assumed minimum ROC type subsidy.

16.6   LP sensitivities

The LP solution favouring more CCS generation was tested with a range of input data assumptions to evaluate risks. Figure 16.2 shows the variation in the optimal take up of wind generation when the subsidy level is varied, from £28/MWh to £48/MWh for onshore and twice that for off shore i.e. £56/MWh to £96/MWh (the 'x' axis shows on shore with off shore twice the values). The graph illustrates the interaction of wind and nuclear resulting from the need to curtail wind output at light load to accommodate output from conventional inflexible generation including nuclear. A higher subsidy results in more wind capacity being built at the expense of nuclear. To meet the overall capacity and plant margin constraint, less nuclear is required to displace wind capacity because of its higher general availability.

The optimality of the solution was also tested by varying the plant mix manually and checking the impact on overall annual costs. The base case optimal solution was adjusted by reducing the capacity of one plant type whilst proportionately increasing that of another type. It can be seen in Figure 16.3 that relatively small changes have a significant impact on total costs confirming the base case optimality.

The impact of key variables on the total cost of the base case solution was also evaluated as illustrated in Figure 16.4. For each variable, the cost of five equal incremental cases about the base case is calculated (without re-optimisation), see Table 16.3.

These variations show a similar impact on total annual costs of up to ±8%. The subsidy payments are not included directly in the LP formulation but reduce the cost of the wind generation. The result is that the total generation cost rises as

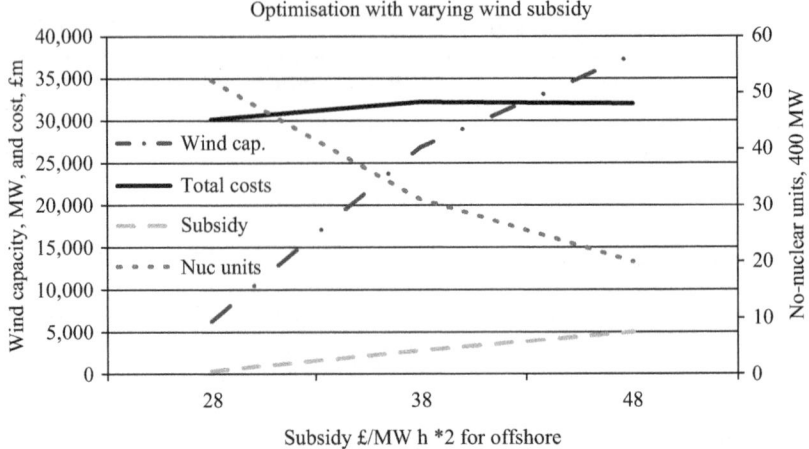

Figure 16.2   Wind/nuclear interaction

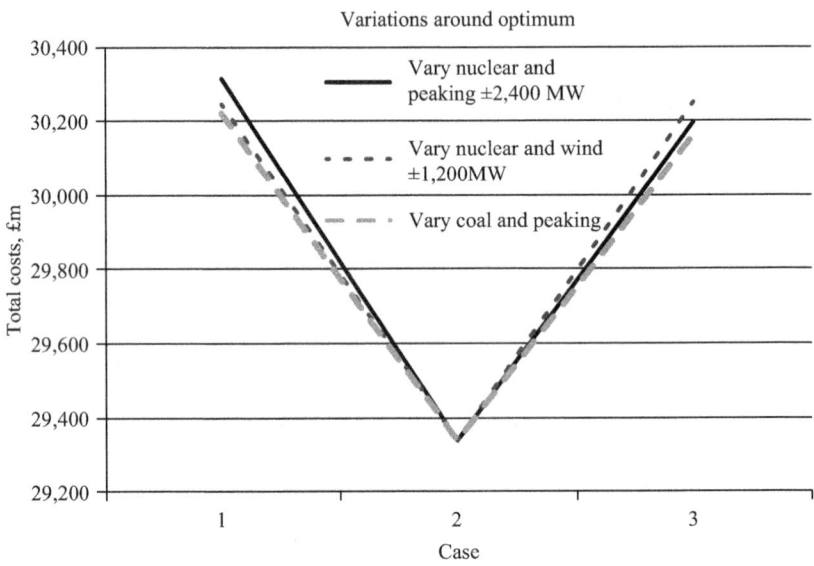

Figure 16.3   Impact of manual plant mix changes

the wind subsidy is decreased as more of the costs are included in the total cost calculation resulting in a variation from the base case by ±10%.

Based on this result, a new plant mix was fed into the more detailed model with a higher proportion of CCS generation, as indicated by the LP studies, to confirm costs and establish emission levels for comparison with the original scenarios.

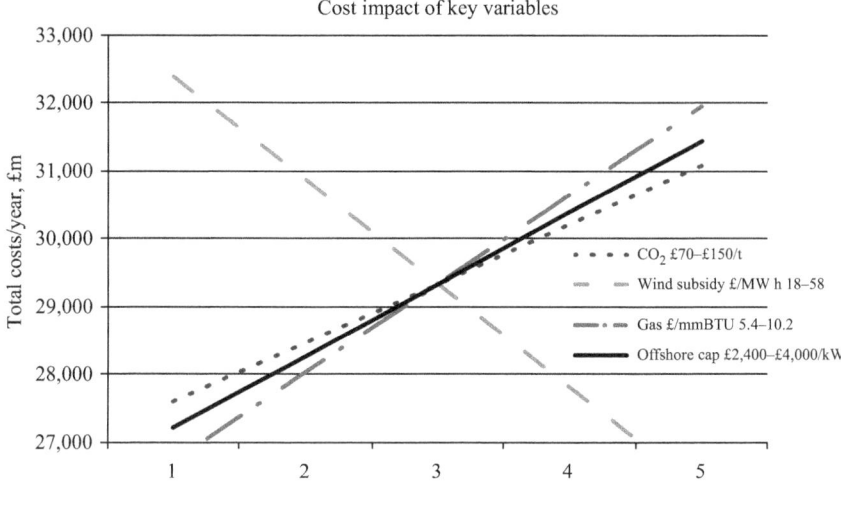

Figure 16.4   Cost sensitivity

Table 16.3   LP sensitivity studies

Key input variables	Scenario					Sensitivity
	1	2	Base 3	4	5	Range (%)
CO_2 price, £/t	70	90	110	130	150	±36
Offshore CAPEX, £/kW	2,400	2,800	3,200	3,600	4,000	±25
Gas price, £/MBTU	5.4	6.6	7.8	9	10.2	±30
Wind subsidy, £/MW h	18	28	38	48	58	±52

16.7   Data assumptions in detailed model

The more detailed analysis was based on a half hourly simulation for a year with base case data as detailed below. The wind, solar and nuclear generation was assumed to be paid a CfD. The key parameters were

- Gas prices £7/GJ (70 p/therm);
- CO_2 costs £70/t;
- Gas CCS capital £1,236/kW; coal 2,475/kW;
- CCS carbon capture rate 90% for both gas and coal;
- CCS transport and storage £19.6/t for both gas and coal;
- Nuclear capital £3,885/kW; nuclear CfD £92.5/MW h;
- Wind CfD onshore £95/MW h, offshore £155/MW h; PV £120/MW h.

This data was used to check the total costs of three scenarios; the original 'Green'; the 'Slow ' and a new 'fast CCS 1' with a higher proportion of CCS as indicated by the LP solution. Subsequently, a fourth scenario was developed to give

Table 16.4   Scenarios

Scenario Technology	2030 slow, MW	2030 green, MW	Fast CCS 1, MW	Fast CCS 2, MW
AGR	1,200	1,200	1,200	1,200
PWR	8,051	11,510	11,510	11,510
BIT	0	1,987	1,987	1,987
CCGT	38,686	29,469	14,935	11,930
OCGT	5,579	5,579	12,375	12,375
CHP	4,253	4,982	4,982	4,982
OIL	748	782	782	782
CCS gas	0	2,000	9,952	12,957
CCS coal	0	2,588	5,518	5,518
Interconnectors	7,200	7,600	7,600	7,600
Offshore wind	20,832	35,956	12,600	12,600
Onshore wind	13,590	20,985	7,400	7,400
Solar PV	6,106	15,813	6,106	6,106
Biomass	4,806	6,736	4,806	4,806
ROR	1,952	2,176	2,176	2,176
PMST	2,744	3,356	3,356	3,356
Marine	40	854	854	854
Total, MW	115,787	153,573	108,139	108,139
Maximum demand		59,812		
Less wind/PV, MW	68,019	72,365	73,579	73,579
Costs, £bn/year	**34**	**41**	**34**	**34**
Emissions, MtCO$_2$/year	**49**	**22**	**29**	**22**

emissions equivalent to those of the 'Green' scenario. This was realised by converting a further 3 GW of CCGT gas generation to CCS to create the 'fast CCS 2' scenario. The four scenario plant mixes are shown in Table 16.4. The total capacities are different because of higher levels of intermittent generation but the firm capacities are similar providing equivalent levels of security.

The main differences are that the new generation mixes have a higher proportion of CCS generation at over 15 GW in fast CCS 1 and up to 18 GW in fast CCS 2 with proportionately less wind generation at 20 GW as opposed to 56 GW with the 'Green' scenario and 34 GW with the 'Slow' scenario.

16.8   Detailed model results

The basic generation set in the model was adjusted to create three models using a spreadsheet to scale the capacity of each type to match the scenarios as shown in Table 16.5. The last row shows the total annual CO$_2$ emissions having netted off that captured by CCS generation. The ninth column shows the total annual costs associated with financing and operating the generation but excluding any transmission costs. The extra costs associated with intermittency were also included. The earlier LP results are shown for comparison although some data assumptions

Table 16.5   Full model/LP scenario results

Scenario	CCGT	Gas CCS	Coal CCS	OCGT	Nuclear	Wind on	Wind off	Cost, £m/ year	Emiss., MtCO$_2$
LP wind sub. fast CCS	14,935	9,952	5,519	12,375	9,950	3,225	24,000	31,510	
LP wind sub. slow CCS	29,153	1,670	1,958	11,134	10,855	3,225	23,671	31,828	
LP based on FIT fast CCS	17,922	11,276	6,743	8,250	12,345	5,892	6,177	33,867	
LP based on FIT slow CCS	28,494	1,670	1,958	8,199	16,534	5,892	6,177	34,170	
Main model 'slow' MW	48,518	0	0	0	9,251	13,590	20,832	33,618	49
Main model 'green' MW	34,451	0	0	5,579	0	20,985	35,956	40,642	22
Fast CCS 1	14,935	9,952	5,518	12,375	12,710	7,400	12,600	33,953	29
Fast CCS 2	11,930	12,957	5,518	12,375	12,710	7,400	12,600	34,248	22

were different. The LP solution cost based on CfDs and fast CCS development at £33,867 compares closely to the new mix solution of £33,953. This suggests that a simplified LP model can provide a suitable base from which to identify optimality.

It can be seen that with the 'fast CCS 1' scenario, the total annual cost is comparable to the 'Slow' scenario, whilst the emissions are less at about 60%. The new mix also costs £6bn/year cheaper than the 'Green' scenario with emissions just 30% higher with a 90% CCS capture rate or just 15% higher with a 100% CCS capture rate. By increasing the proportion of CCGT generation converted to CCS generation in the 'fast CCS 2' scenario, emissions are reduced to match the 'Gone Green' scenario at 22 Mt/year. This result confirms the LP findings suggesting that advancing CCS development offers the prospect of lower costs and emissions. The main difference between the 'Green' and 'fast CCS' scenarios is that some 35 GW (55 down to 20 GW) of wind is displaced by an extra 10 GW of CCS (5 up to 15 GW) in the new mix 1. The CCS generation provides firm capacity realising equivalent security while reducing the CfD costs. A breakdown of the costs of each scenario is shown in Table 16.6. The very high cost associated with CfD payments for the 'Green' scenario is much more than the higher fuel costs in the other scenarios.

16.9   Model sensitivities

The sensitivity of the model results to variations in key parameters was analysed to test the robustness of the solution, where the changes affect the merit order then

Table 16.6   Scenario cost breakdown and emissions 2030

Cost elements, year Scenario	Fixed costs, £bn/ year	Fuel/ var Cost, £bn/ year	FIT costs, £bn/ year	CCS saving, £bn/ year	CCS tran., £bn/year	Total, £bn/ year	Emissions, MtCO$_2$
2030 slow	4.1	11.9	17.5	0.0	0.0	**33.5**	49
2030 green	5.8	8.1	27.5	−1.0	0.3	**40.7**	22
Fast CCS 1	6.1	14.4	15.8	−3.5	1.0	**33.8**	29
Fast CCS 2	6.9	14.4	15.8	−3.9	1.1	**34.3**	22

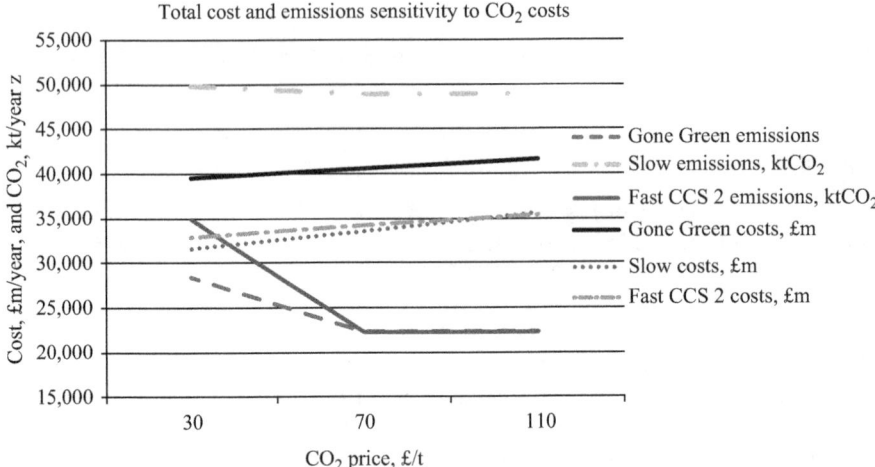

Figure 16.5   Sensitivity to CO$_2$ prices

reordering of all quarters was undertaken, and the generation utilisation calculations were updated. Figure 16.5 shows the variation in the total costs and emissions for different CO$_2$ prices in the range £30–£110/t. The 'Slow' and new mix scenarios show comparable costs rising with CO$_2$ prices while the 'Green' scenario costs are £6bn/year higher. The 'Green' and fast CCS 2 scenarios show emissions falling as CO$_2$ prices rise driving out the use any coal-fired generation above £70/t of CO$_2$. The 'Slow' scenario emissions are relatively stable but some 50% higher.

The base case CCS capture rate was set at 90% of the total emissions. This was varied to establish the impact on costs and emissions as shown in Figure 16.6. The results can equally be interpreted as representing the impact of only a proportion of CCGT generation being converted to CCS as 'rollout' progresses. It can be seen that at higher capture rates the emissions from the 'Gone Green' and 'fast CCS 2' scenarios converge with the fast CCS model emissions falling below those of the 'Gone Green' at CCS capture rates at and above 95%. There is also a slight

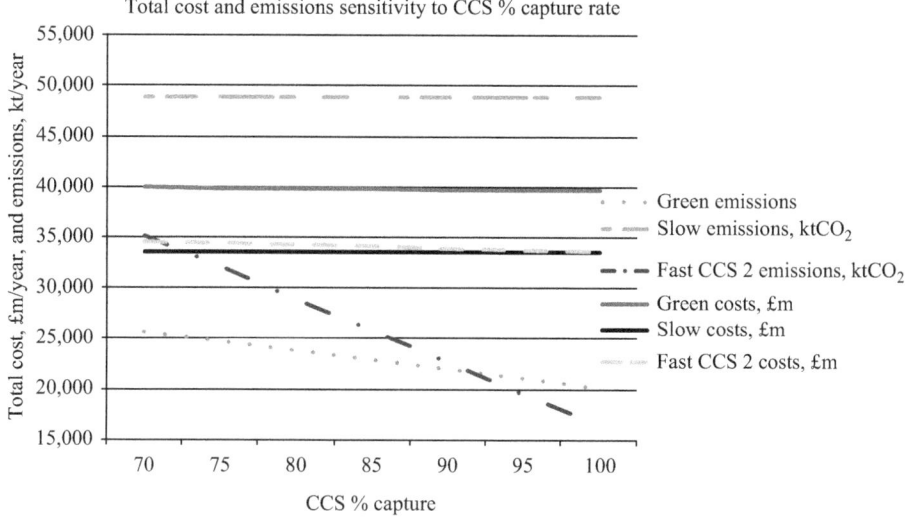

Figure 16.6   Impact of CCS generation CO_2 capture rate

reduction in costs, as the capture rate increases, with less carbon tax being incurred. For the emissions to rise to the level of the 'Slow Progression' scenario, the capture rate would have to drop to around 50% or alternatively only 50% of the CCS ready generation is fitted with CCS.

Figure 16.7 shows the sensitivity of the results to changes in predicted gas prices in the range £4–10/GJ about the base case £7/GJ. The rise in emissions is the result of unabated coal generation coming into merit at higher gas prices. The detailed results show some 12 TW h produced from unabated bituminous coal generation with gas prices of £10/GJ. The 'fast CCS 2' scenario plant mix assumed 9952MW of gas CCS as opposed to 5518MW of coal CCS, restricting the opportunity to shift production from gas to coal at higher gas prices. If the ratio of gas CCS to coal CCS had been chosen to be more even the emissions would have been lower and comparable to the 'green' scenario with a gas price of £10/GJ. The choice of plant mix will depend on the expectation of gas and CO2 prices.

The overall costs of the 'fast CCS 2' scenario remain comparable with the 'Slow' scenario but with much lower emissions. The emissions are comparable with the 'Green' scenario but the costs to end users are around £6bn/year cheaper.

Table 16.7 compares the 'fast CCS 2' scenario with 'Green' through a range of extreme data assumptions. It can be seen that the fast CCS option is always cheaper. The last result combines the extremes of a high WACC (12%), a high CO_2 price (£200/t), a low CfD for offshore wind (£100/MW h) and high CCS capital costs (£2,708/kW) but still results in the 'fast CCS 2' scenario costs being lower than 'Gone Green' by £3bn/year.

The base line capital cost of CCS generation was £1,236/kW for gas-fired generation. The impact of lower and higher capital costs in the range from £1,006

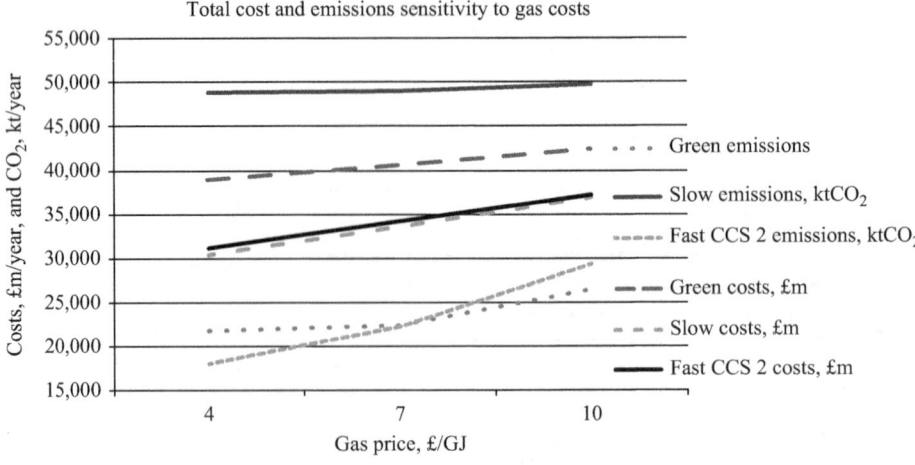

Figure 16.7   Impact of gas prices on costs and emissions

Table 16.7   Data sensitivities

Variable	Units	Fast CCS 2	Green	Difference
Base case costs	Cost, £m/year	34,248	40,641	6,393
Capital at 12% WACC	Cost, £m/year	36,425	42,284	5,859
CO_2, £200/t	Cost, £m/year	37,937	44,012	6,075
Offshore FIT, £120/MW h	Cost, £m/year	33,434	37,973	4,539
Offshore FIT, £100/MW h	Cost, £m/year	32,783	36,402	3,619
CCS cap., £1,422/2,708/kW	Cost, £m/year	34,532	40,719	6,187
12%, CO_2 200, FIT 100, cap	Cost, £m/year	38,491	41,443	2,952
CCS capture 90% base	Emiss., Bt/year	22	22	0

to £1,466/kW is illustrated in Figure 16.8. This shows that the impact on the total costs is minimal for the 'fast CCS 2' scenario. There is no impact on emissions as the SRMC and the merit order does not change.

16.10   Comparison of model and LP

The data changes of the 'fast CCS 2' scenario were inserted back into the LP model to derive a new optimal solution. The resulting optimal capacity mix is shown in Table 16.8 together with the model 'fast CCS 2' generation capacity. It can be seen that they closely agree with just a little more gas CCS and less wind in the LP optimum. This suggests that the new 'fast CCS2' is close to optimum based on the latest data assumptions.

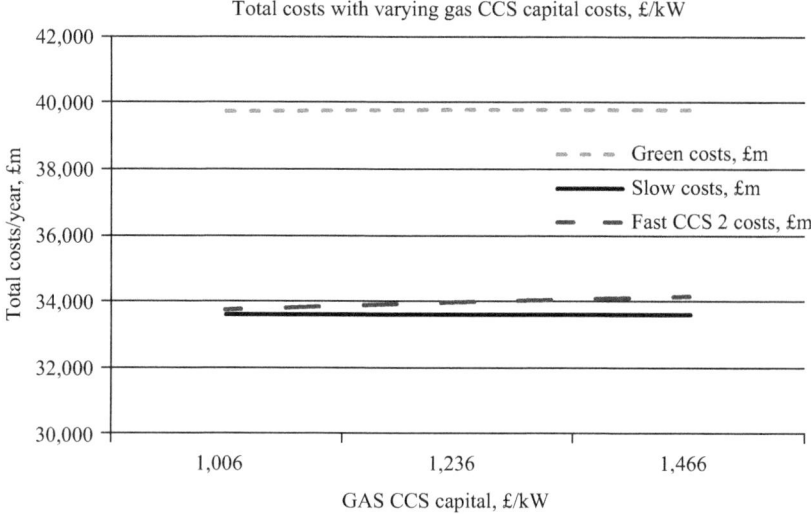

Figure 16.8   Impact of CCS capital costs on total annual costs

Table 16.8   Model mix 2/LP with mix 2 data – % capacity comparison

Type	Model (%)	LP (%)
CCGT	20	22
Gas CCS	13	14
Coal CCS	7	6
OCGT	16	17
Nuclear	21	21
Wind	23	19

16.11   Risk analysis

Given the wide range of variables, there is a need to assess their impact on the total system costs to identify the risks and worst case outcomes. To inform the analysis of the risks, the operation of the detailed model was automated to enable a series of annual runs with key parameters varied for 2030. In this example, the variables were gas prices; CO_2 emission prices and the capture rates of CCS generation. When fuel prices are varied, the merit order has to be re-built each time as part of the automated process. The gas prices were varied through the range £5–£9/GJ from a base of £7/GJ; the CO_2 emission costs from £30 to £140/t around a base of £70/t and CCS CO_2 capture rates from 80% to 100% with a base of 90%. The incremental changes of each parameter lead to 1,331 combinations of the values each giving rise to a total system cost. The values were grouped into bands and ordered as shown in the histogram of Figure 16.9. The diagram also shows the total

Total costs/year, £m (gas 5£–9£/GJ; CO$_2$ £30–£140/t; CCS
capture rate 80%–100% – base £34,239m)

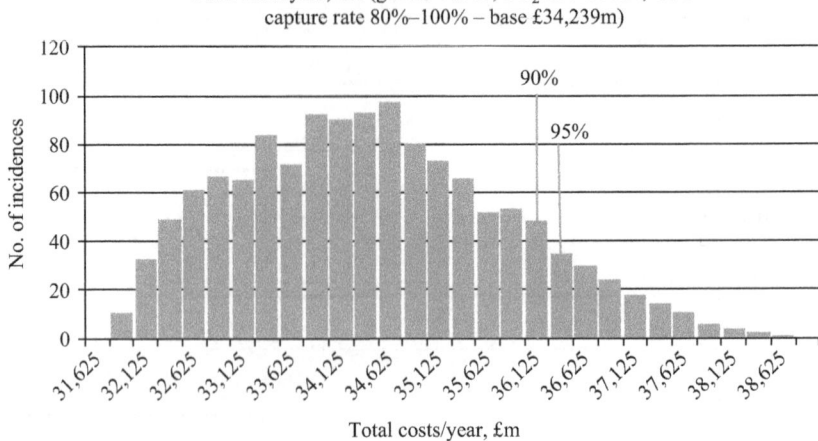

Figure 16.9   Total system cost range

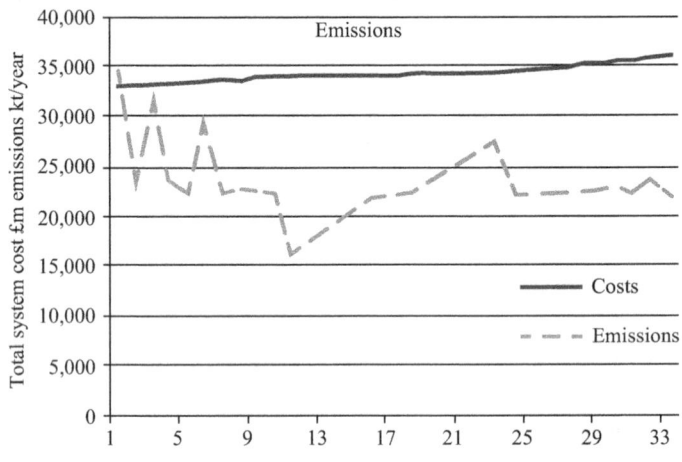

Figure 16.10   Range of emissions across scenario costs

system cost values that could be exceeded in 10% of the cases and 5% of
the scenarios studied. There was a 10% chance that a cost of £36,125m would be
exceeded and a 5% chance that a value of £36,625m would be exceeded with the
central case cost of £34,625m. The total annual estimated emissions were also
captured for each study. The study illustrates the potential impact of a range of
variables all taking an adverse value and would support strategic decision-making
on polices designed to influence the outcome.

The total annual emissions for the same range of total costs are shown in
Figure 16.10. The higher emission values occur with the lower CO$_2$ emission costs
recognising the limits on capture rate with CCS generation.

16.12 Conclusions

This chapter has illustrated an approach to establishing a development strategy that meets emission target reductions at minimum cost. It also demonstrates the analysis of the risks to the outturn due to erroneous data assumptions. In this example, the results suggest that fast development and deployment of generation equipped with CCS offers the prospect of low levels of costs and emissions. It suggests that a lower proportion of wind generation (20–25 GW) coupled with 15 GW or more of CCS generation would have the advantage of both lowering costs and emissions and avoiding the need to frequently curtail wind output to secure system operation.

- The overview LP solution showed minimum total costs with a significant tranche of CCS plant;
- The model base case studies with a 'fast CCS 2' mix confirmed total costs some £7bn/year cheaper than the 'Green' scenario;
- Emissions in the new mix with more CCS are 45% of the 'Slow' scenario, and the same as the 'Green' scenario assuming a 90% CCS capture rate;
- The report assumes that the development of CCS technology could be accelerated and could be made commercially available within the timescale analysed;
- It appears that the use of storage is more financially viable when operating in a market environment buying and selling energy although barely so at current capital costs.

There are a number of other issues that could have an impact on decisions on future technology development:

- With high levels of wind capacity, more energy has to be curtailed during low demand periods and may have to be paid for. This could add £4bn/year to a 'Green' scenario based on CfD strike prices of £155/MW h for offshore wind and £95/MW h for onshore.
- The costs of the extra transmission infrastructure to support remote offshore wind installations are not included;
- The intermittency of renewable energy adds to system operating costs and emissions and the need for reserve plant;
- Generation with CCS is flexible and extra output could be realised at times of peak by limiting the CCS process energy use;
- CO_2 storage in partially depleted oil fields may facilitate enhanced oil recovery.

Those engaged in formulating strategy have a difficult task with so many variables and market participants. There is a need for analytical tools to review the risks and range of potential outcomes and identify actions that avert the worst case scenarios. To support this process, an automated series or studies with varying key parameters was used to identify the probability of high outturn costs.

Question 16.1 Taking account of the plant mix data in Table 16.4, explain the relative changes in total costs from the base case resulting from the parameter changes shown in the table. By what amount would the offshore wind FIT cost have to rise by to make the total scenario costs equal.

Variable	Units	Fast CCS 2	Green	Difference
Base case costs	Cost, £m/year	34,248	40,641	6,393
Capital at 12% WACC	Cost, £m/year	36,425	42,284	5,859
CO_2, £200/t	Cost, £m/year	37,937	44,012	6,075
Offshore FIT, £120/MW h	Cost, £m/year	33,434	37,973	4,539
Offshore FIT, £100/MW h	Cost, £m/year	32,783	36,402	3,619
CCS cap., £1,422/2,708/kW	Cost, £m/year	34,532	40,719	6,187
12%, CO_2 200, FIT 100, cap	Cost, £m/year	38,491	41,443	2,952
CCS capture 90% base	Emiss., Bt/year	22	22	0

Chapter 17

Trading in an uncertain environment

17.1 Trading options

There are a number of areas for trading services that may need to be investigated to fully evaluate the potential of an existing or planned generation installation. As well as physical trades for delivery of services, there is also the option to engage in financial trades including trading out of physical delivery before the event. Trades take place through a range of timescales from power purchase agreements (PPAs) lasting several years, bilateral contracts for supplies for several months up to a year, down to trading close to the event through power exchanges. The volumes tend to be smaller closer to the event with fine adjustments to positions to cover for actual outturn circumstances. The range of options includes:

- Long-term PPAs;
- Capacity market payments;
- Bilateral contracts with local distribution companies or suppliers to provide base load or peaking energy;
- Bilateral contracts with local large energy users, on a monthly basis, six monthly or annual;
- Cross-border trades on a monthly or annual basis;
- Contracts with the TSO to supply ancillary services;
- Spot trades through exchanges for day ahead and within day supply;
- Fuel arbitrage contracting between electricity and gas.

Long-term PPAs are not generally favoured because circumstances can quickly change making the contract look like a bad deal. Recent events that have introduced major changes have been the focus on emission reduction with charges for emissions affecting coal and gas competitive positions. The advancement of shale gas has transformed prices in the gas market in some countries. The change in oil production by OPEC (Organisation of Petroleum Exporting Countries) members has resulted in a collapse in oil prices. PPAs are sometimes necessary to finance major hydro and nuclear developments to ensure return on investment. The PPAs may embrace capacity payments related to realised availability but distinct capacity payments are not usually featured in developed markets but are expected to be embraced in normal trading prices. They may be introduced to help encourage new entry when capacity margins fall below normally acceptable levels to ensure security standards are met.

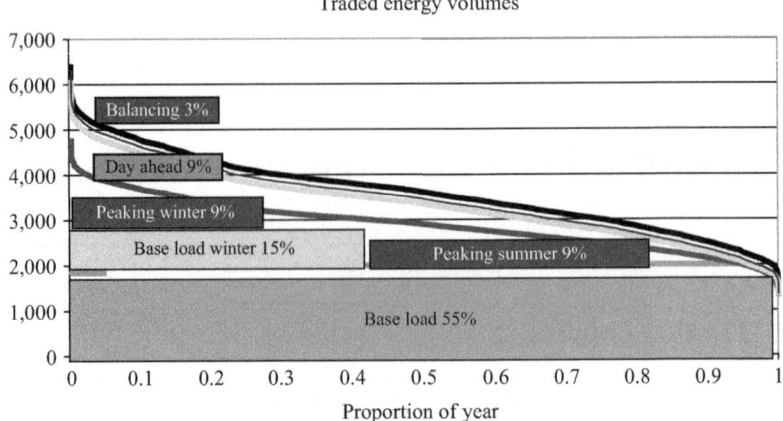

Figure 17.1   Traded energy volumes

Bilateral contracts for base load and peaking energy with distribution or supply companies will often be structured by generators that also own supply companies or through competitive pricing. Larger organisation will often procure energy through auctions for 6 monthly periods with adjustments monthly. Cross-border trades are linked with reservations of interconnector capacity. They are typically contracted for a year or 6 months with options for shorter term adjustments. Finally, day ahead and within day trading is used up to the point of 'gate closure' to balance positions using exchanges. Subsequently, the grid company contracts from generation or the demand side to balance supply and demand in the event. Typical trading blocks are shown in Figure 17.1 for a supplier to contract for delivery for a year to meet the expected demand shown as a load duration curve. It can be seen how the traded volumes match the expected load duration curve varying from continuous base load, daytime periods and peaking periods with reducing volumes close to the margin.

There have been a number of changes that complicate the trading process and predicting needs. Renewable generation receives priority in dispatch and its unpredictable nature makes the net demand to be contracted from conventional generation difficult to define until close to the event. The growth in local embedded generation, that may be used to offset demand, impacts on the grid demand and is not explicitly monitored. There is also the prospect of much more consumer engagement in demand management that will be difficult to predict.

17.2   Trading process

This chapter examines the process of evaluating the future trading options for a power station complex and its potential worth in support of investment decisions. It advances a structured process to review and evaluate the risks. It discusses the

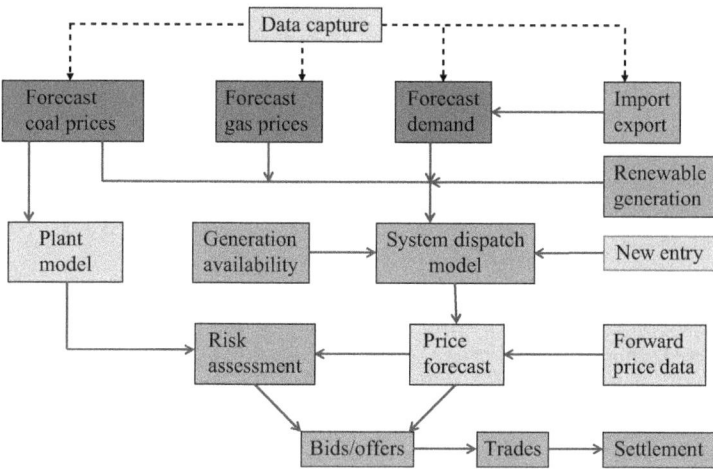

Figure 17.2   Trading process

ongoing process of trading taking account of key variables. The trading process is illustrated in Figure 17.2.

A similar process is used for longer term investment appraisal as well as for shorter term daily or monthly trades. The analysis for investment will include more uncertainty through longer time frames and require more detailed risk assessment. In this example, lignite- and gas-fired generation compete at the margin and market fuel prices need to be predicted. In some circumstances, it may be preferable to exercise arbitrage through the gas market and sell rather than generate. The demand forecasts can be developed based on historical data corrected for weather variations from normal. The demand level needs to be adjusted to take account of expected or planned imports/exports and capacity allocations. This in turn requires some analysis of adjacent market prices and marginal plant and any planned major changes. The demand also has to take account of the expected output from renewable generation like wind and solar that is impressed on the system. Given the development of embedded generation, some modelling of its potential output will reduce the grid system demand. This can be based on details of typical plant load factors and records of any new entry. Generation availability has to be predicted in the short-term based on recent history with longer term figures based on statistics of availabilities for different plant types.

17.3   Data forecasts

Predicting future grid demand is becoming more difficult because records are being distorted by the development of generation embedded within distribution systems that have generally been passive. The output of the many small units has not been specifically recorded and in the case of renewable generation is very intermittent. Longer term forecasts are often based on models linked to historic records of GDP with adjustments to take account of efficiency improvements. **Imports and exports are also less predictable where adjacent countries are equally exposed to increasing volumes of intermittent generation.** Day-ahead predictions need to take account of expected wind and solar conditions. Longer term simulations need to replicate typical patterns of renewable output using recent records. Where new installations are planned, records from existing plant in the same location area can be used to generate representative output profiles. The aspects that need to be considered include:

- Obtaining records of annual/daily demand profile with weather data;
- Historic records of demand and associated GDP;
- Expected future GDP;
- The impact of any energy efficiency schemes;
- Details of any sector developments like electric vehicles;
- Expected renewable energy output with typical profiles;
- Estimates of cross-border imports/exports;
- Estimates of system losses.

Fuel price estimates are based on records of historic exchange prices, reported spot and future contract prices and price volatility. Models are typically based on mean reversion logic where short-term prices follow current trends with prices gradually reverting to the longer term mean trends. The random element is often represented by volatility with a Brownian motion to create a range of potential values as shown in Figure 17.3. These are used to establish the probability of different outturns.

The fuel prices are influenced by demand changes. The collapse in oil prices is evidence to the impact of oversupply. Coal prices are less volatile with a more stable relationship between demand and prices. Because of the costs of transport, power stations are often built close to mines with longer term supply contracts that

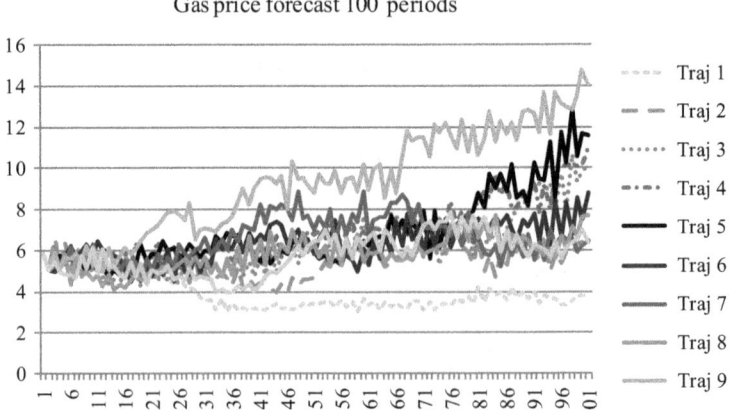

Figure 17.3   Range of gas price forecasts

are less influenced by world markets. The gas market has been disrupted by the development of shale gas. These two factors have a major impact on trading and are often hedged with energy prices linked to prevailing fuel prices. The stations will also hedge with fuel supply contracts for a year or more ahead. In some market models used in the Middle East, a central buyer will procure fuel and supply it to contracted stations to benefit from economies of scale in the same way as state utilities.

17.4   System dispatch

A dispatch simulation is often used to identify the marginal plant and predict prices to support trading. The simulation dispatches available generation to meet the predicted demand for each half hour of the period being analysed. The dispatch is usually based on variable short-run marginal costs with additions to cover fixed costs that depend on the level of competition. This can be estimated based on the prevailing plant margin through the period. The model requires detailed generation data including:

- Installed generation capacity categorised by fuel type;
- Estimates of generation efficiency based on similar plant types, age and size;
- A prediction of generation availability and any new entry;
- Generation start-up costs or fuel burn and variable operating costs;
- Data related to ongoing fixed costs to estimate full market prices.

The modelling should ideally be on a half hour basis to analyse daily utilisation and take account of intermittency and start-up costs. Modelling can be based on a load duration curve to estimate annual base load and peaking prices. The model will also show the expected use of a particular plant through each period by comparison of potential bids against the market price forecast. The plant model needs to take account of start-up costs and minimum on and off times and ramp rates. This model would be used to calculate the full cost of bids through different contract periods that typically cover 4, 2, 1 and 0.5 h. The prices tend to be higher for the shorter periods to take account of start-up costs and part load efficiency degradation. Most trading takes place during the day time period with participants adjusting their position to take account of their outturn compared to their longer term contracted position.

In this example, the requirement was to estimate future revenues for a lignite-fired generator. The data used in this model is shown in Figure 17.4 for a country in Eastern Europe. The dispatch model includes data for all the generators operating in the market and enables the utilisation of each plant to be estimated for future years.

The result of the dispatch process based on variable costs is shown diagrammatically in Figure 17.5. It indicates the types of technology that are used in merit order to meet each demand level and their utilisation during a year of operation. It shows the marginal variable prices associated with each demand level

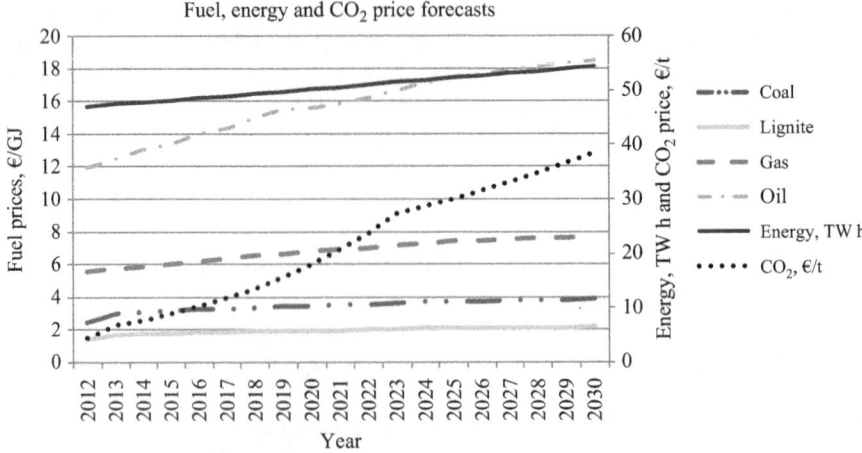

Figure 17.4   Fuel and energy forecasts

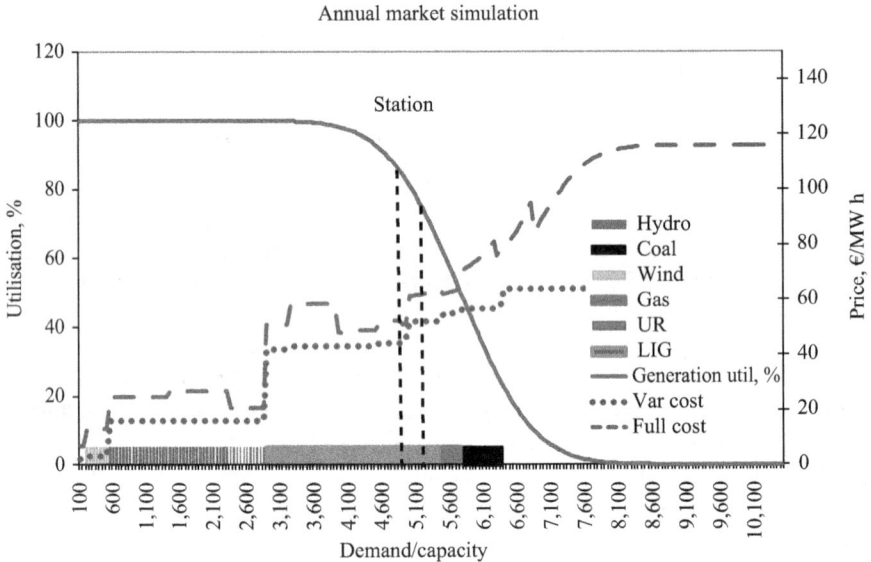

Figure 17.5   Market simulation

and the full prices necessary to also recover fixed costs. As the dispatch is based on variable costs, this curve is monotonically increasing, whereas the inclusion of the additional fixed cost recovery requirement shows discontinuities. The vertical lines relate to the variable cost of the station plant being analysed and illustrate its expected utilisation.

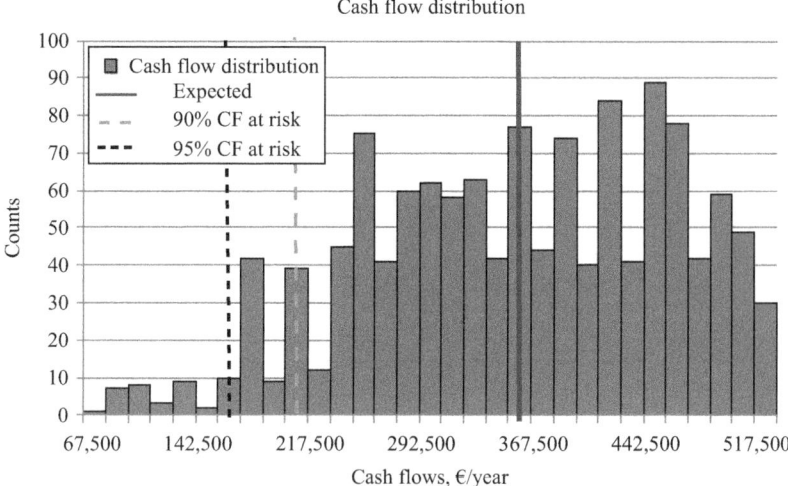

Figure 17.6   Station cash flow for a year

17.5   Value at risk

Risk assessment will often involve establishing a range of possible outturns to identify the probability of the investment resulting in a trading loss. Key variables in this analysis, covering a year, include the demand level and gas and lignite prices that were each assumed to vary through a range of 11 values. The demand was assumed to vary by a cumulative 2% higher and lower than the base case giving a range of ±10% from 45 to 55 TW h. This is increasingly important given the potential impact of impressed renewable generation on demand that undermines the utilisation of conventional plant operating at the margin. Lignite prices are varied by a cumulative 5% from €1.5 to €2.5/GJ and gas prices varying by 5% from around €5 to €8/GJ. The dispatch model was automated to run changing each variable value in turn for 1,330 runs. The generators cash flow for the year was derived from its utilisation and plotted as shown in Figure 17.6 for each data change. The vertical axis shows the number of occasions that the cash flow fell within defined bands. The vertical lines on the distribution show the expected median value and those values that the cash flow falls below for 10% and 5% of the data scenarios. Given a production cost, the associated values at risk can be estimated and the probabilities.

When considering longer term bilateral trades or investment in a new plant, there is a requirement to both predict expected base load and peaking prices and assess their sensitivity to variations in the more dynamic parameters like fuel prices. This will often take the form of establishing the impact of a €1/MW h change in the production price from gas, coal or oil generation on the overall system marginal base load and peaking prices. **A good indication for smaller fuel price changes is provided by identifying the time that each plant type operates at the**

Table 17.1   Time at margin (Percentage)

Year	Gas	Coal	Oil	Hydro
2007	0.37	0.05	0.21	0.37
2008	0.45	0.04	0.22	0.29
2009	0.51	0.03	0.04	0.41
2010	0.59	0.02	0.02	0.37
2011	0.54	0.03	0.11	0.33
2012	0.48	0.17	0.02	0.34
2013	0.34	0.29	0.01	0.36
2014	0.13	0.53	0.02	0.32
2015	0.10	0.54	0.04	0.31

Table 17.2   Annual price predictions

	LRMC, €/MW h	SRMC, €/MW h	Expected price, €/MW h
Base load	61.1	49.6	52.5
Peaking	77.3	54.4	60.3

margin setting the marginal price. Table 17.1 shows the results of an annual dispatch simulation through 9 years analysing the market in Greece. It shows for year 2007 that gas-fired generation is operating at the margin for 37% of the time and a €1/MW h increase in gas generation costs would result in a €0.37/MW h increase in the marginal price for the period studied. As CO_2 prices increase, gas becomes less marginal with coal generation operating more time at the margin and hence having more influence on the marginal prices.

Peaking prices are generally based on the average marginal price through the 12 h of working week days and can refer to a day, a month or a year covering 3,120 h. These will usually be the most expensive periods with the lower base load price referring to the average of the marginal price for the whole year. The averages may be volume weighted by the energy delivered in each half hour. The price estimates will include those based on variable SRMCs and the LRMCs that also include for a return on fixed capital and operating costs. An estimate of the expected price can be obtained from an exponential function of these two prices according to the level of competition in the market as illustrated in Table 17.2.

17.6   Arbitrage spark spread

With the extended use of gas-fired generation and gas markets, there are opportunities to arbitrage between gas and electricity markets. **A gas-fired generator with a take or pay gas contract has the option to either use the gas to generate**

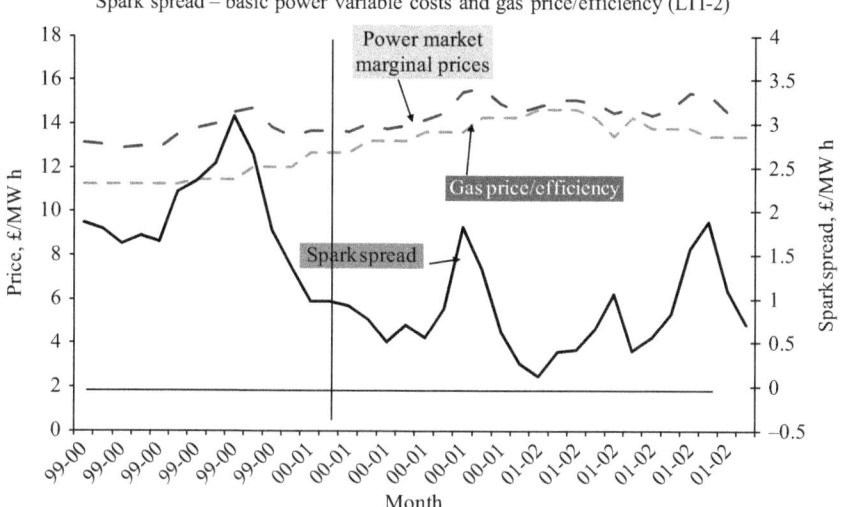

Figure 17.7 Spark spread

electricity or sell the gas back to the market. The decision will be based on the relative worth of the electricity given the prevailing market price vs the spot gas price. This has led to the concept of spark spread to indicate the difference between the gas price expressed in units per MW h based on an assumed efficiency and the actual power price.

An example is shown in Figure 17.7 where it can be seen that the power price is above the converted gas price and the spark spread is positive indicating that it would be preferable to use the gas to produce electricity. There may be other periods when power prices are driven low during low load periods when it is preferable to sell the gas. This form of trading is essentially short-term and seeks to exploit market imperfections. The participating generation needs to be sufficiently flexible to take advantage of premium pricing in the spot electricity market and will usually be a peaking plant.

The option to exercise arbitrage can create problems with the operation of power systems. **If gas prices are driven high due to shortages in winter, it can cause a lot of gas-fired generation to choose to sell gas and be withdrawn from service creating a power shortfall.** This may occur during winter when both gas and power demands are high.

Some of the gas contracts are established with an interruptible clause meaning that supplies can be withdrawn for a period to alleviate meeting a peak gas demand. A similar practise is applied to some contracts with larger electricity consumers. The interruptions are usually limited to a number of hours each year. **This can also lead to power system problems if gas shortages lead to the interruption clause being invoked on a number of generators at the same time.**

17.7   Statistical forecasting

The approach to forecasting based on estimating costs and building a merit order has the disadvantage that it does not explicitly take account of actual market prices. This issue can be partly addressed by 'back-casting'. This involves using the normal model parameters to estimate prices for a historic period and using the result to benchmark the model parameters. There are other approaches that are more explicitly based on the recognition of historic price patterns.

In power markets, the price is very closely linked to the demand level although it tends to be stratified reflecting the different types of plants brought into service on a merit order basis. The graph of Figure 17.8 shows the variation in marginal price with load in Greece. It can be seen that the price falls around three levels based on cheap lignite, gas-fired generation and oil. This is clearly indicated by the data for January with high demand levels, whereas in June, the prices stay lower. The turning points will be influenced by the availability of the different types of generation with the levels varying throughout the year depending on maintenance programmes and fuel prices. The small variations within bands reflect the operation of plant with differing efficiency.

The nature of this type of price function does not lend itself to statistical analysis. If the relationship between demand and price can be established then the prediction of demand presents a more tractable problem. Auto regressive integrated moving average packages have been used successfully to predict demand in the short- to medium-term. The approach identifies the trend and takes account of the impact of weather variation from normal conditions. It relies heavily on a recorded set of similar days when working conditions would have followed a similar pattern.

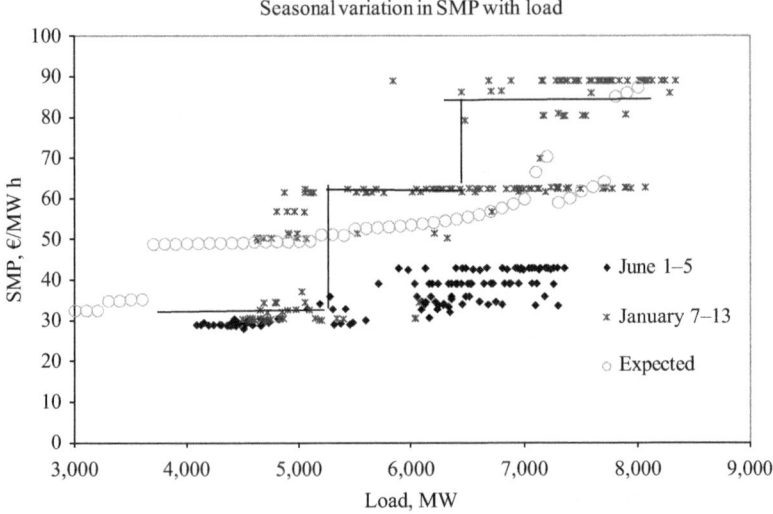

Figure 17.8   SMP variation

Accuracies of around 1.5% can be achieved at the day-ahead stage when weather prediction data is included in the model. The translation of the demand prediction into a price will need to take account of the time of year and generation availability levels.

Other authors (Pilipovic 1997) have suggested that mean reverting models produce the best results in capturing the distribution of energy prices. In the mean reversion to the log of price, the reversion is applied to the log of price rather than the price i.e.

Let $x = \log(\text{Spot price})$

Then,

$$Dx = \text{mean reversion rate (long-term equilibrium of } X - Xt)$$
$$+ \text{volatility} * \text{random stochastic variable.}$$

Another model that has been used includes two factors; the spot price and the long-term equilibrium price. The spot price is assumed to revert towards the equilibrium price level while the equilibrium price level is assumed to have a lognormal distribution.

Given the priority dispatch afforded to renewable generation, there is a requirement to forecast the demand net of renewable output to formulate the price linkage. This places emphasis on predicting the output from wind and solar farms.

17.8   Optimum day-ahead bid

Close to the event, trading is required to enable market participants to balance their generation and demand positions. It is facilitated by power exchanges like the APX (Amsterdam Power Exchange) that operate across Europe and provide trading platforms and market data supporting price discovery. The exchanges focus on short-term day-ahead trading covering short periods of operation and peaks. Generators bidding into the exchanges need to take account of likely running periods, compared to their plant minimum on times, and to factor into their bid the costs associated with start-up.

In this example, a 300-MW generator has a variable operating cost of £20.76/MW h and a start-up cost of £2,250 and is aiming to bid into the market to maximise the profit for the day. The expected market price profile is as shown in Table 17.3 for each half hour of the day ahead. It can be seen that the expected price is often below the generators variable cost and that the bid will have to target the peak periods. The object of the exercise is to establish that bid price above costs that maximises the total profit.

The problem can be formulated as an LP that can be solved using the Microsoft Solver facility. The generator runs for those half hours when the market price is above the bid price. The optimal solution shows a running period of nine half hours with a bid price of £22.67/MW h resulting in a net profit of £324 for the day. The unit produces 1,350 MW h at a cost of £30,276 including the start-up cost. The total revenue is £30,600.

Table 17.3   Optimal dispatch example

Half hour	Market price, £/MW h	Generator load, MW	Gross revenue	Cost of start-up
1	18.7	0	0	0
2	17.7	0	0	0
3	17.3	0	0	0
4	16.0	0	0	0
5	14.0	0	0	0
6	13.3	0	0	0
7	14.0	0	0	0
8	14.0	0	0	0
9	14.0	0	0	0
10	13.3	0	0	0
11	17.3	0	0	0
12	16.2	0	0	0
13	16.7	0	0	0
14	18.6	0	0	0
15	20.6	0	0	0
16	20.3	0	0	0
17	20.7	0	0	0
18	21.8	0	0	0
19	24.5	300	3,400	2,250
20	22.9	300	3,400	0
21	25.8	300	3,400	0
22	23.3	300	3,400	0
23	25.0	300	3,400	0
24	22.7	300	3,400	0
25	24.0	300	3,400	0
26	24.4	300	3,400	0
27	25.7	300	3,400	0
28	20.7	0	0	0
29	20.7	0	0	0
30	20.7	0	0	0
31	20.0	0	0	0
32	20.0	0	0	0
33	22.0	0	0	0
34	22.0	0	0	0
35	22.0	0	0	0
36	20.7	0	0	0
37	20.0	0	0	0
38	19.7	0	0	0
39	18.0	0	0	0
40	17.8	0	0	0
41	17.7	0	0	0
42	17.0	0	0	0
43	17.0	0	0	0
44	17.7	0	0	0
45	17.4	0	0	0
46	18.3	0	0	0
47	19.3	0	0	0
48	19.3	0	0	0

This simple example illustrates the issues facing generators operating at the margin. It highlights the importance of having an estimate of market prices for the following day. **In future, this will be confounded by the impact of intermittent renewable generation coupled with demand side response.** The operation of larger generators is limited by dynamic constraints like minimum on and off times and run up and run down rates. They do not have the flexibility to switch on and off in the short-term. They are able to part load but at the expense of higher operating costs. There will be an expanding market for flexibility services based on storage or demand control. The development of interconnection may enable hydro-based balancing services to be traded.

17.9 Part load operation

Generators operate less efficiently at part load because of the heat losses. These constitute a higher proportion of the costs than when running at full load. A coal-fired unit with a DNC (Declared Net Capacity) of 625 MW has a no-load cost of £896/h with an incremental cost of £12.8/MW h. The total cost per hour is given by:

Output MWs * incremental cost + no-load cost

The effective unit cost is derived by dividing the total cost/h by the output MWs as shown in Table 17.4. The effective efficiency is derived by comparison with the full load efficiency of 38% in this example.

It can be seen that the effective efficiency falls rapidly once the part load drops below 50% as the unit price increases. If the unit is also engaged in regulating to track demand then more energy will be used affecting the economics of operation. Bids into balancing markets will take account of both these effects. Any bid into the market will be constrained by the generator dynamic characteristics shown in Table 17.5.

Table 17.4 Part load efficiency

Load, %	Total cost, £/h	Effective, £/MW h	Effective eff.
0	896		
10	1,697	27.15	19.9
20	2,497	19.98	27.1
30	3,298	17.59	30.8
40	4,099	16.39	33.0
50	4,899	15.68	34.5
60	5,700	15.20	35.6
70	6,500	14.86	36.4
80	7,301	14.60	37.1
90	8,102	14.40	37.6
100	8,902	14.24	38.0

Table 17.5   Generator characteristics

DNC	625	No-load cost, £/h	896
		Incremental cost, £/MW h	12.81
		Start-up cost, £	1,800
Dynamic data	Run up rate 6.0 MW/min based on a DNC of 625 MW Is equivalent to 1.05%/min – this assumes unit is hot Run down rates are 20 MW/min Synchronising generation 60 MW Minimum stable generation 300 MW Start-up cost £1,800 Efficiency 38%		

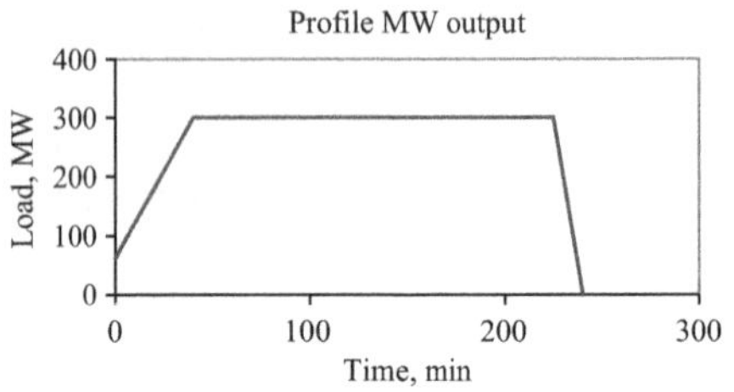

Time, min	MW	MW h
0	60	
40	300	120
225	300	925
240	0	37.5
Total MW h		1,082.5
Total cost, £		19,251
Unit cost, £/MW h		17.78

Figure 17.9   Generator loading profile

If the unit has a minimum on time of 4 h, then the minimum energy supplied can be established and the effective costs estimated as shown in Figure 17.9. The unit synchronises at 60 MW and runs up at 6 MW/min reaching the minimum stable generation of 30 MW after 40 min. The average load through the run-up is 180 MW producing 120 MW h. The unit stays at 300 MW up to 225 min from synchronisation generating 925 MW h when it starts to rundown at 20 MW/min over a further 15 min generating 37.5 MW h. The total generation is 1,082.5 MW h at a cost of £19,251. This cost includes both the start-up cost and no-load heat costs for 4 h and results in a unit cost of £17.78/MW h compared to the basic incremental cost of £12.8/MW h. This example illustrates the impact on costs of intermittent part load operation on generators and trading in markets dominated by intermittent renewable generation.

17.10   Conclusions

Generators are struggling in many areas to maintain financial viability faced with loosing utilisation to renewable generation with priority dispatch. At the same time, a number of changes are taking place that complicate predicting future demands and prices and undermine the energy trading process. The unpredictable nature of renewable generation makes the prediction of the net demand to be contracted from conventional generation difficult to define until close to the event. The growth in local embedded generation, that may be used to offset demand, impacts on the grid demand and is not explicitly monitored. This distorts the historic demand profiles that have traditionally been used as a basis for the prediction of future demand and prices. There is also the prospect of much more consumer engagement in demand management that will further add to the difficulty of predicting demand and prices.

In assessing investments, there is a need for longer term demand and price forecasts that are often based on models linked to historic records of GDP but these now need to be adjusted to take account of policies and measures to improve the efficiency of energy use. Imports and exports are also less predictable where neighbouring countries are equally exposed to increasing volumes of intermittent generation that disrupt planned transfers. Where there are strong links in place, it is necessary to model the adjacent systems to establish comparative price profiles.

In predicting utilisation, it is necessary to place the proposed development in merit order along with competing stations to estimate the utilisation. This uses an annual dispatch model for selected years based on a load duration curve showing the utilisation of stations according to their variable price. Given the costs, estimates can be made of expected revenues and profits. It is necessary to establish the risks by running a number of studies varying key parameters to establish a histogram of possible profit outturns. Decisions are often based on an assessment of the value at risk with a 90% or 95% probability. The fuel price risk is often based on the relationship between system marginal prices and each fuel type. The dispatch process is used to establish the proportion of the year that each fuel type operates at the margin. This then provides the basis for estimating the impact of a fuel price change on the average system marginal price. This approach is satisfactory for relatively small changes but larger changes require reordering the merit order.

Some stations also have the option to exercise arbitrage. A gas-fired generator with a take or pay gas contract has the option to either use the gas to generate electricity or sell the gas back to the market if prices are high. A resulting complication is that if gas prices are driven high due to shortages in winter, it can cause a lot of gas-fired generation to choose to sell gas and be withdrawn from service creating a power shortage. Some generators buy fuel on a contract that enables short-term interruptions of supply. This can also lead to power system problems if gas shortages lead to the interruption clause being invoked on a number of generators at the same time.

A number of analysts hold the view that market prices can be predicted on a purely statistical basis. In practise, prices are often stratified where they take different levels depending on the fuel type operating at the margin. The nature of this type of price function does not lend itself to statistical analysis and a more structured approach is required.

In developed markets, trading is undertaken just based on prices. Generators are expected to include other costs due to start-up and part loading within their price bid. This can be a complicated process in short-term daily trading when the run time is not known. To account for this, trades will often be linked to specific run periods of typically around 4 h.

Question 17.1 Based on the generator data shown below, calculate the effective overall unit cost/MW h of running up at maximum rate to the minimum stable generation of 300 MW and after a period, running down at maximum rate to zero with an overall connection time of 3 h from synchronising to shut down.

DNC	625	No-load cost, £/h	896
		Incremental cost, £/MW h	12.81
		Start-up cost, £	1,800
Dynamic data	Run up rate 6.0 MW/min based on a DNC of 625 MW		
	Is equivalent to 1.05%/min – this assumes unit is hot		
	Run down rates are 20 MW/min		
	Synchronising generation 60 MW		
	Minimum stable generation 300 MW		
	Start-up cost £1,800		
	Efficiency 38%		

Chapter 18
Smart flexible grids

18.1 Introduction

The advent of a growing proportion of intermittent renewable technologies into the plant mix is resulting in the need for significant increase in the flexibility of the power system to adjust to the uncontrolled variations in generation output. There is also a need to curtail renewable output during light load periods due to the need to manage the security of the power system using conventional generation or as a result of system constraints. There are also changes incident upon the planned reductions in emissions from heat production and transport that are likely to be met with a growth in demand from electric vehicles and heat pumps. Table 18.1 shows that for the United Kingdom, a modest **10% electrification in these sectors could result in the system load increasing by 130 TW h/year (60 + 70 TW h)**, depending on conversion efficiency. This is equivalent to adding 37% to future energy demand. The increased loads would generally be connected to the distribution system, and smart grids development has been heralded as the answer to manage these developments and associated problems. **The distribution networks have traditionally been designed based on relatively low after diversity domestic consumer demands of around 1.0 to 1.5 kW/consumer.** This reflects the fact that loads like washing machines, driers and dishwashers exhibit natural diversity. This would not be the case with electric vehicles or heat pumps. Heat pumps would commonly react to ambient temperature. Electric vehicles would generally be charged outside working hours with each vehicle charging at 3 kW or more. Time of use tariffs can be used to focus demand into periods when other loads are low as overnight and will be facilitated by the smart meter rollout, but the scope is limited.

UK government agencies have claimed that the advancement of a smart, flexible energy system would result in 'gross benefits to consumers of £3–8bn/year being realised in 2030' '**saving GB consumers £17–40bn cumulative by 2050**'. They define 'smart to mean enabled by information technology to integrate the action of connected users in order to efficiently deliver secure sustainable and economic electricity supplies'.

The forecast savings of £14–19bn, over the 20-year period, assume 6–9 GW less low carbon generation and 3–9 GW less peaking capacity. This would assume the current 330 TW h demand (equivalent to a flat 37.6 GW) met by an average of

Table 18.1 Energy use transition

	Electricity	Transport	Heating	Totals
Total energy, TW h	350	600	700	1,650
Target reduction, %	34	10	10	15
Renewable, TW h	119	60	70	249

Table 18.2 Capital cost saving from smoothing demand

WACC 7% 20 years	Units	Gas	Peaking
Unit size	MW	400	120
Capital cost	£m/year	28.3	3.78
Capacity reduction low	GW	6	3
Capacity reduction high	GW	8	9
20 years capital cost saving low	£bn	8.5	1.9
20 years capital cost saving high	£bn	11.3	5.7

around 40 GW rather than 50 GW at peak. These estimates appear to be based on the assumption that daily demand can be virtually levelled by the use of storage and demand-side management. If achievable, this would reduce generation capacity costs over 20 years, as shown calculated in Table 18.2, with a low of £10.4bn and a high of £17bn plus operational savings.

These estimates are comparable, but it needs to be shown how the demand profile can be managed to realise these savings in practice. There will also be costs involved in the provision of storage and contracting users to accept demand control that will offset the potential benefits, and these need to be quantified. The cost of smart metering alone, when completed, has been estimated at be around £9 billion.

18.2 Policy and regulation

A paper published for comment by the UK regulator (Ofgem) and the government Business, Energy and Industrial Strategy (BEIS) in 2017 stated '**The age of exclusive control by big energy companies and central government is over**; we must maximise the ability of consumers to play an active role in managing their energy needs. With a smart system we can go further and faster in breaking down barriers to competition – allowing the widest possible range of innovative products and services to prove themselves in the market place'. A smart energy system is defined as 'one which uses information technology to intelligently integrate the actions of users connected to it, in order to efficiently deliver secure, sustainable and economic electricity supplies'. It is less clear how policy and regulation will make this happen to realise the expected benefits.

The concepts advanced are built around more engaged and empowered consumers exploiting smart meters, domestic solar panels and even storage through some form of market mechanisms. The developing countries have practised demand management for years. Faced with inadequate generation capacity, their approach has been to run what generation was available at full load and to selectively switch off demand to match its capacity. The new concept is to provide consumers, or their agents, with the information and encourage their participation through price signals and/or specific incentives. It is recognised that new business models and structures may be required to realise the benefit of the transformation. The establishment of effective markets that enable realisation of the full benefit of flexibility and enable competition is advanced as a cornerstone.

The need for some form of coordination of the many new players is recognised to realise the most cost-effective solution in the short and long term. Several commentators have made reference for the need for a holistic approach but not how it may be realised. Specific regulatory actions would include

- enabling the full benefits of storage and DSR (demand side response) to be available to participants through improved price signals and use of system charging arrangements;
- recognising the need for the distribution network operators (DNOs) to provide active management of its system operation as it moves from being essentially passive to being active;
- enabling and clarifying the role of aggregators acting on behalf of groups of consumers;
- ensuring open access to the balancing market by aggregators and other potential DSR sites based on transparent specifications of requirements.

It is recognised that tariffs and price signals are needed that can engender the required level of response. There are risks in all users reacting through common price signals and overshooting the requirement and leaving suppliers with unsold contracted energy. The likely customer take-up is questionable when wholesale price differentials are small. Surveys by Smart Energy UK have shown just 30% would be interested, particularly if the operation is automated perhaps through an aggregator interfacing to smart appliances. Smart meters will enable smart tariffs based on half hour metering, but some consumers may be deterred by the complexity. The charges for use of the system will need to change to reflect the changes in its use to ensure all users, connected to the system and receiving benefit, contribute to the recovery of the costs of its provision. Ultimately, a more variable dynamic ToU (Time of Use) tariff may be necessary if the complexity can be managed. This may be necessary to manage the unconstrained use of distributed generation.

18.3 How can smart grids help?

Smart grids are aimed at exploiting modern technological developments in monitoring and control to better exploit network and generation assets and meet the trilemma requirements of security of supply, low emissions and affordable energy supply. There are a number of ways that this may be realised involving management of demand and generation sources including

- Using storage to level demand to avoid expensive peaks and reduce both network and generation capacity requirements;
- Applying control of demand to move it from high-demand periods to low-demand periods either through explicit incentives or through time of use tariffs;
- To apply thermal modelling to network assets to exploit short-term overload capability of circuits and transformers;
- To manage local generation sources to reduce supergrid peaks during periods of high wholesale prices;
- To manage demand to balance intermittent generation output to avoid the need to retain and keep cycling spare backup generation capacity;
- To control distribution generation and network assets following faults to restructure and restore the system by shifting loads to other network areas;
- To manage demand to reduce distribution imports from the transmission network, and hence the use of system charges.

These applications are generally most effective when they are coordinated and may be realised through market price incentives or by aggregators operating on behalf of consumer groups. **It needs to be recognised that incentives will be necessary to encourage participation that will offset some of the benefit.**

It would be possible to use storage to take up excess generation from wind and solar generation as may occur during light load periods. In addition to the cost of the facility the energy lost as a result of the low-storage cycle efficiency needs to be accounted as well.

Demand-side management could be applied at domestic, commercial and industrial levels. Small end users are unlikely to participate directly but rather through an aggregator acting on their behalf. Smart meters monitoring and control facilities will be required as well. A payment will be required to encourage users to participate.

Trading across interconnectors where the counter party is prepared to accept varying transfers is practised. This assumes that the counterparty has generation or demand flexibility they are prepared to trade such as hydro.

Regulating on conventional generation like OCGTs and CCGTs is the option currently applied and has prompted some generators to convert their CCGTs to operate open cycle to improve flexibility. This duty affects the operating costs of the unit and their financial viability at reduced overall utilisation.

There is also an option to use excess electricity to produce gas that can be stored and used in the gas network or converted back to electricity.

It needs to be shown that these options for providing system flexibility are technically viable and how they could be financed, managed and coordinated in a market environment.

18.4 The storage option to level demand

There are costs in realising savings in generation capacity through the use of storage in moving demand from high-load to low-load periods to level demand. One suggestion is that consumers may be guided to move demand by time of day tariffs, but the potential is limited without investing in storage capacity. References refer to the falling cost of storage based on Li-ion of 14% from 2007 to 2014 but recognise that falls are expected to slow in future. Tesla has developed a domestic battery system called the 'Powerwall' with capacities of 7–10 kW h. The unit costs in 2016 were around $3,000 plus DC to AC inverter lifetime costs and installation costs of $2,000 giving a total cost of around $5,000 (£5,600 quoted in the United Kingdom). The unit is capable of up to 5,000 cycles under warranty that amounts to less than 1 cycle/day over 15 years. Assuming that all the capacity is used, the cost of the facility/kW h can be calculated as shown in Table 18.3. The capital cost alone of 13c/kW h is similar to the basic unit cost in the USA of about 12.5c/kW h, and the process could only begin to be viable economically if the stored energy was free e.g from wind self generation. In addition, the efficiency is estimated at around 90% or, including inverter losses at 85%. This means that only 85% of the stored energy is returned, and that price differentials between toughs and peaks would have to make up the difference ignoring the cost of the facility. Both the battery and inverter costs would have to fall significantly for this option to be widely adopted.

Current grid installations are usually justified financially on the basis of multiple revenue streams including peak lopping, frequency and voltage control and balancing intermittent generation. These schemes are generally small scale with limited capacity. The revenue is realised through a premium for the speed of response and flexibility. The potential returns for flexibility in SO timescales are influenced by the size of the need that may be limited by successful balancing by BRPs (Balance Responsible Parties) and the level of competition. **In Germany, although the requirement for balancing has not changed significantly, the balancing market prices have collapsed.** This is due to a lot of generators, which have suffered from their utilisation being displaced by wind generation, all competing for the supply of balancing services against a limited requirement.

Larger scale storage connected at higher voltage like pumped storage or compressed air systems are possible contenders if they can be financed by

Table 18.3 Domestic storage

Powerwall	Units	20 years life
Capacity	kW h	10
Battery	$	3,000
Inverter + install	$	2,000
Annual WACC 7%	$	471
Utilisation/day	kW h	10
Capital unit cost	Cents/kW h	13

wholesale price arbitrage. The application of storage schemes needs to recognise the importance of cycle efficiency that may be as low as 70% and less where transmission across the network incurs losses. These effects severely limit the opportunity for arbitrage and undermine the financial case. The market size is limited as prices converge through the application of schemes. The cycle efficiency of pumped storage is around 70%, compressed air systems have an efficiency of around 50%. The cycle efficiency is low because first, the air that heats up during compression must be cooled down again to the ambient temperature before it can be stored in a cavern; second, the cold air must be reheated after discharge from the storage facility since it cools strongly when expanding in a turbine for power generation. RWE (Rheinisch-Westfälisches Elektrizitätswerk) in Germany have a project called ADELE (adiabatic compressed-air energy storage for electricity supply) evaluating adiabatic compressed air storage capturing the compression heat and releasing it back during discharge to generation to improve efficiency to around 70%. This means that the price differential has to be at least 70% to break even and more to cover the capital costs and network losses. **In Section 12.4, it was shown that the storage capital cost would have to around £800/kW for 5 h of storage or £160/kW h to break even** based on forecast market prices in 2020 with 25 GW of wind generation. **These costs are higher than the capital costs for new generation.** Any emissions from producing the stored energy would be higher because of the energy loss during the storage cycle.

18.5   Smart domestic customers

Developments have taken place to enable customers to proactively manage their energy demand. In the United Kingdom, British Gas has developed a Hive Home Hub designed to facilitate demand control remotely from a user's phone or tablet. It includes a hub connected to the user's broadband that in turn links to controllable socket outlets and lights and heating thermostats as illustrated in Figure 18.1.

It would be a short step to enable the controls to react to time of use energy prices. A key element would be the control of the supply to EVs and domestic heat pumps to manage the timing of the increased demand. It is envisaged that this demand could be scheduled in a similar way to generation. The scheduling algorithms would have an objective function to fully charge participating users EVs and

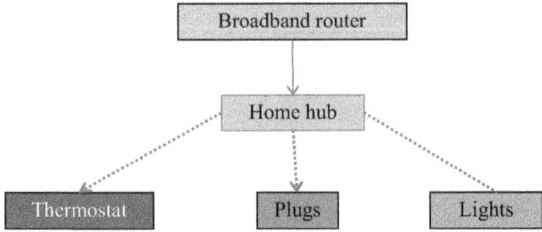

Figure 18.1   Domestic remote demand control

maintain temperatures within bands at minimum cost. It would be necessary to take account of any network constraints and existing loads and avoid all loads reacting to low prices simultaneously. The demand would need to be scheduled like generation. The control signals could be based at local MV (medium voltage) substations where overall loading can be monitored. The costs of the domestic control infrastructure are modest with the hub at £80, controllable plugs at £39, lights from £19 and sensors at £29 (British Gas 2017).

The EC recommended ten functions that should be provided as a minimum with smart meter installations including

1. Consumer
   (a) Provide readings directly to the consumer or any designated party;
   (b) Update readings frequently enough to use energy saving schemes;
2. Metering operator
   (c) Allow remote reading by the operator;
   (d) Provide two-way communication for maintenance and control;
   (e) Allow frequent enough reading for networking planning;
3. Commercial aspects of supply
   (f) Support advanced tariff systems;
   (g) Remote on/off control supply and/or flow or power limitation;
4. Security – data protection
   (h) Provide secure data communications;
   (i) Fraud prevention and detection;
5. Distributed generation
   (j) Provide import/export and reactive metering.

The installations should have sufficient granularity to support time of use tariffs and include two-way communication. However, EC Recommendation 2012/148/UE stated that smart meter installation was not mandatory but should be based on a cost–benefit analysis. Italy started installing smart meters in 2001, and now 35 million consumers have working systems. An early application was to limit supplies to 0.5 kW in the event of bill non-payment and restore supplies following payment. **Other countries like Germany see limited benefit and propose that installation is restricted to large energy users.** The availability of metering data enables forecasting and billing to be managed more frequently improving utility cash flows.

18.6 Demand with EVs

As more vehicles are converted to run on electricity, their expected charging profiles will tend to smooth the national demand as a lot of the load will occur outside normal working hours and mostly overnight. Figure 18.2 shows a UK basic national demand profile over 2 days during winter with the expected charging pattern of 10 million EVs based on ToU tariffs. Each EV is assumed to have a maximum charge of 3 kW with a full capacity of 50 kW h and a charging diversity factor of 0.54.

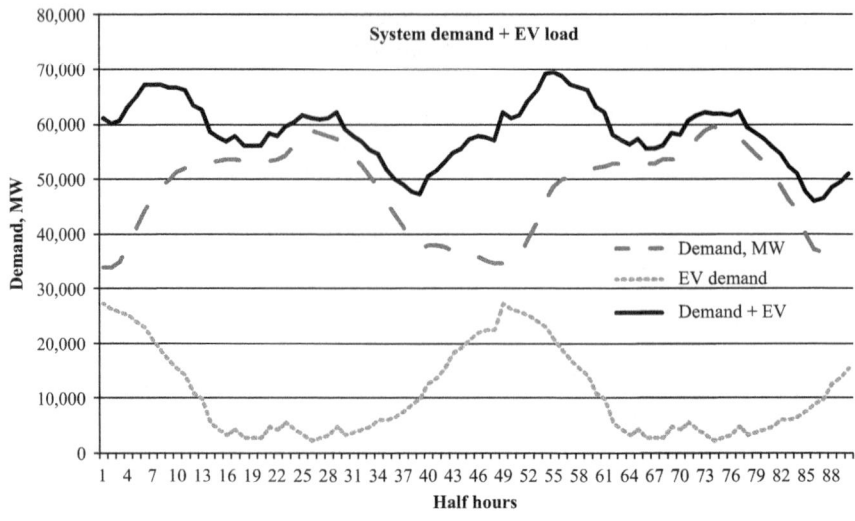

Figure 18.2   Demand + EV load

It can be seen that the combined profile including EVs exhibits a smaller range of variation of 20 GW compared to the basic demand shape of 26 GW. Also the increase in the peak demand of around 10 GW is less than the EV peak demand of around 28 GW. There is assumed to be a restriction on access to EVs during the working day when most will be in use or parked at a place of work.

18.7   Longer term national demand profiles

If the use of heat pumps expands, it introduces a load that can be shifted to help smooth the demand profile exploiting the ability to store heat in hot water tanks and the building fabric. The domestic annual consumption of a heat pump was shown in Chapter 13 to be about 3,000 kW h/year producing 5,000 kW h of hot water and 7,000 kW h of heating energy. It is assumed in this analysis that 15% of households have heat pumps with 4.3 million installations by 2040. Given the increasing pressure on air pollution in cities, there are predicted to be 15 million electric vehicles on the road by 2040. Figure 18.3 shows the expected basic winter demand profile with the addition of EV and heat pump demand using the profiles developed in Chapter 13.

It can be seen that there is a large daily variation in demand varying from a low of 50 GW up to a peak of around 95 GW with a capacity requirement in excess of 100 GW. If new heating systems are designed with associated thermal storage capability in insulated water tanks or in solids like concrete, there will be scope to shift the demand profile. Figure 18.4 is based on the same information as in Figure 18.3 but with the heat pump demand advanced by 2.5 h.

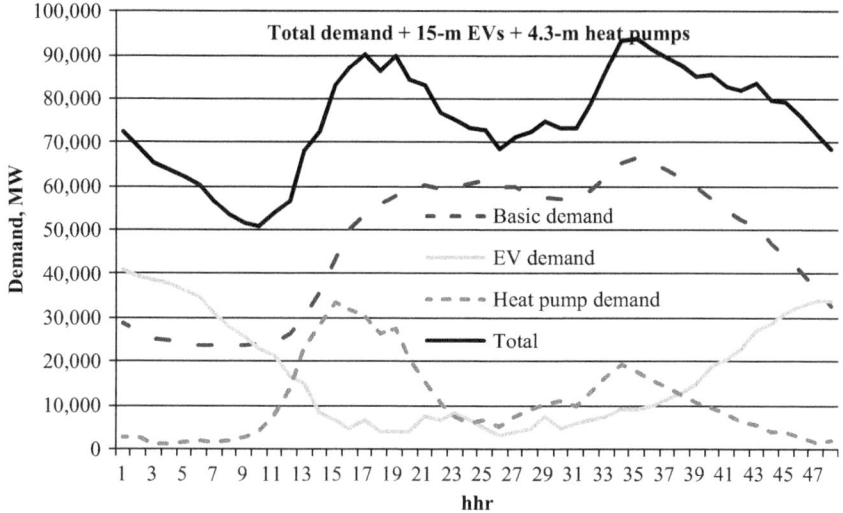

Figure 18.3   Basic demand profiles 2040

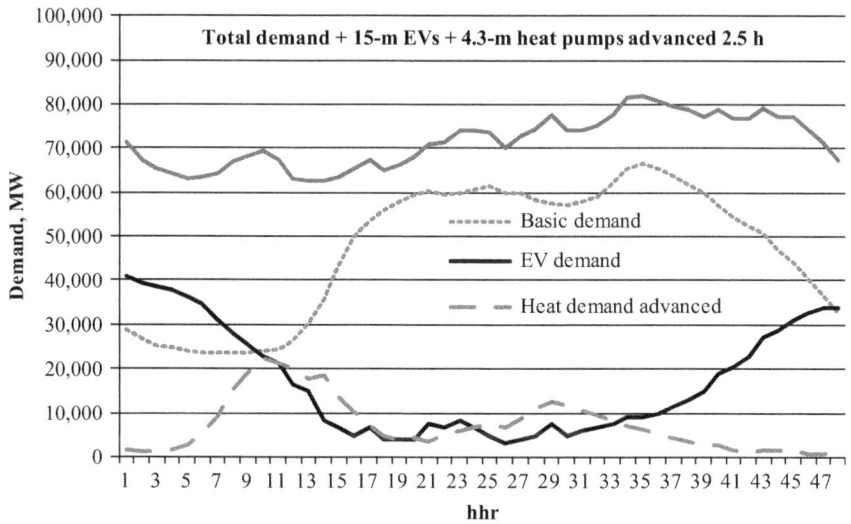

Figure 18.4   Total demand profile with heat pump demand advanced

It can be seen that the adjusted profile is much flatter with a peak of around 80 GW as opposed to about 95 GW in Figure 18.3. Heat-pump-based systems offer the potential to smooth demand profiles and reduce generation capacity needs when they are combined with storage that may be based on phase change materials. There are two aspects associated with demand profiling; at the distribution level, the object would be to contain the requirements for network reinforcement; at national level, the focus would be on flattening the total system demand profile to reduce generation capacity needs. The optimum scheduling of demand would also need to take account of variable renewable generation output that will impact the optimal solution.

18.8 Interconnection trading

There is an option to contract for flexible transfers across interconnectors. It is sometimes used where the counterparty has an abundance of flexible hydro capacity coupled with storage capacity. It is not usual where the adjacent country also has a proportion of intermittent generation that may have limited diversity with the counterparty and similar requirements for flexibility. The exception is emergency transfers to provide support during disturbances. There is a history of high level complaints from the governments of Germany's neighbours about unplanned transfers into their networks resulting from high outputs from their wind generation. It is also important to recognise the losses that will be incurred in transfers across and through networks.

Assumptions are often made about the support that can be expected from interconnectors to provide balancing and emergency support. This is quite different to opportunity trading taking place in the short term. **Firm support requires the availability of capacity that would have to be contracted before the event to secure availability.**

18.9 Flexibility using gas-fired generation

The conventional approach to tracking demand is by dispatching flexible generation. This operating regime impacts on the financial viability of the installation as the unit operates less efficiently at part load and burns more fuel in ramping up and down. The lower level of utilisation also means more of the fixed capital costs have to be recovered from fewer unit sales. **As the unit slips down the merit order, incurring less utilisation, it may become non-viable financially resulting in premature closure unless a premium is paid.** There is also the question of additional wear and tear on the plant resulting from the thermal cycling. Open cycle gas turbines are more flexible but have lower operating efficiencies incurring more fuel burn and emissions.

The graph in Figure 18.5 shows the degradation of efficiency of an actual CCGT plant operating at percentages of full load. The graph can be approximated to a fixed no load heat of 9% with the rest being a function of load i.e.

Part load efficiency $= 100 * \text{load} \% / (9 + 0.91 * \text{load} \%)$

For example, the efficiency and cost of a gas-fired unit when operating at 40% of full load can be calculated. Assuming a full load efficiency $= 54.2\%$ and full load cost of £22.5/MW h at 40% load, the efficiency is reduced to $= 100*40/(9 + 0.91*40) =$ 88% of that at full load i.e. 88% of $54.2 = 47.7\%$ resulting in a part load unit cost of £25.5/MW h $(22.5 * 54.2/47.7)$ as opposed to £22.5/MW h at full load.

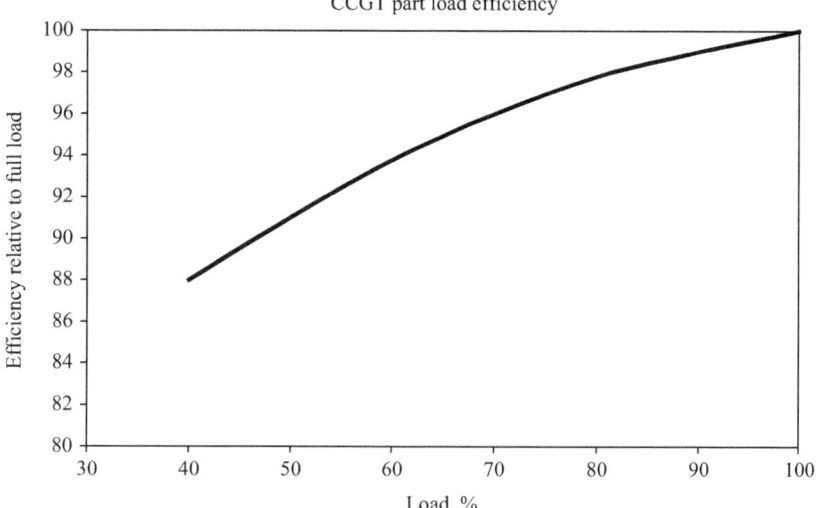

Figure 18.5   Part load efficiency

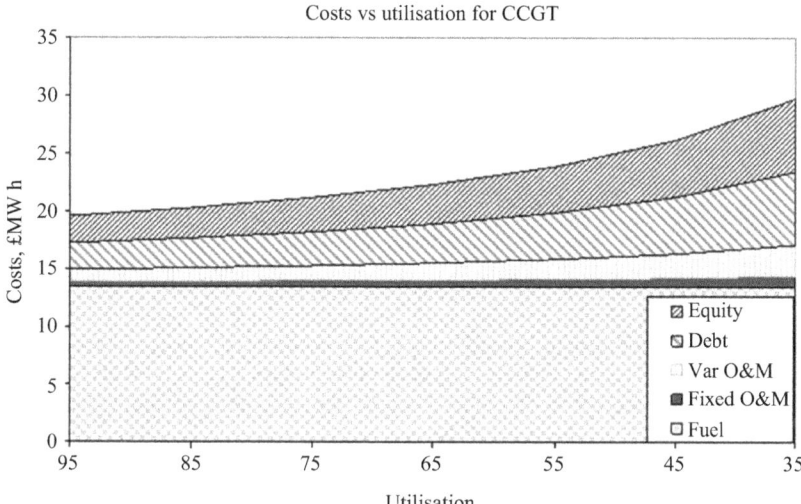

Figure 18.6   Impact on cost of lowering utilisation

The full generation costs including for recovery of the fixed capital and operating costs are also significantly affected by the expected generation utilisation. The lower the utilisation, the fewer generation units produced over which to recover the fixed costs. The graph of Figure 18.6 shows how the effective unit price increases at lower levels of utilisation for a typical gas plant. The proportion of fixed costs added to each unit produced has to increase at the lower levels of utilisation. Accounting for these factors can undermine the competitive position of the plant and lead to premature closure.

18.10   Power to gas

There is significant capacity to store energy in the extensive system of pipes used by the gas supply industry. The UK national gas grid has a 'Linepack' capacity of 4.5 TW h plus 41 TW h of explicit storage capacity with the capability to recover 10% a day. **This capacity dwarfs the UK daily electrical energy consumption of around 1 TW h a day** and has the potential to solve the management of the variations in demand and generation due to intermittent renewable sources. One process involves splitting water into hydrogen and oxygen using electricity and injecting the hydrogen into the gas grid. However, the overall cycle efficiency is shown in Section 13.6 to be low at around 38.5% taking account of the process of producing the gas and then using it to generate electricity. However, it would be useful in absorbing excess electricity from renewable generation that would otherwise have to be curtailed at times of low system load when the energy would effectively be free.

18.11   Demand scheduling and dispatch

The principle of generation scheduling to track consumer demand is well established. Equally, demand-side scheduling is also practised in the developing countries where there is inadequate generation capacity. The United Kingdom used a crude form of demand control during a miner's strike when capacity was short. Generation scheduling is based on minimising the production cost by dispatching generation in a merit order based on variable operating costs. The option to schedule demand reduction is equivalent to scheduling generation. There will be a cost based on the agreement to participate that will be equivalent to a capacity charge and a variable cost based on the duration of the call-off period. If these costs are less than the provision of additional generation capacity, it would be cheaper to contract with the demand side. The demand side includes additional constraints that complicate their use; there will be a requirement to ensure that any EVs are fully charged overnight, and the heating demand interruption should not result in household temperatures falling below a minimum threshold.

There is a need to establish how the process could be managed and the respective roles of interested parties. **To realise the benefit of reduced generation capacity, the demand side would need to forward contract into a capacity market based on a time of day demand.** This would mean being available to reduce load on demand for defined periods. The payments would contribute to the cost of establishing the control facility and any associated storage. In scheduling timescales, down to the day ahead, suppliers would schedule the demand side to flatten their profile along with contracted generation. The energy tariffs offered should reflect their use during off-peak periods and provide additional incentive through reduced prices. The alternative approach would be for aggregators to operate directly in the market to secure the cheapest supplies on behalf of their contracted users. This would enable the benefit of periods of high wind output and low prices to be realised. The supply company would contract for generation supplies and compete on tariffs to supply heat pump and EV demands that are prepared to accept remote control of the timing of their energy use. The appliance remote control would be managed using a system like that illustrated in Figure 18.1.

18.12 Smart grid control and markets

To fully exploit the potential of the distribution systems to facilitate customer engagement will require the development of a system-management capability that parallels that of the supergrid. **A SCADA system will be required that includes a monitoring and control capability with coupling to transmission control and potentially links to end-user internal systems. There will also be a need for a local market platform** as illustrated in Figure 18.7 that shows the data links that would be required to establish a minimum cost for supply. It includes the option to use supplies from local embedded generation or take supplies from the grid depending on prices. Each energy source will have a price/volume sensitivity, or profile and control would also embrace the use of any storage options. The system could also have some potential to smooth demand to reduce peaks as well as accommodate the intermittency of renewable sources. It could be designed to

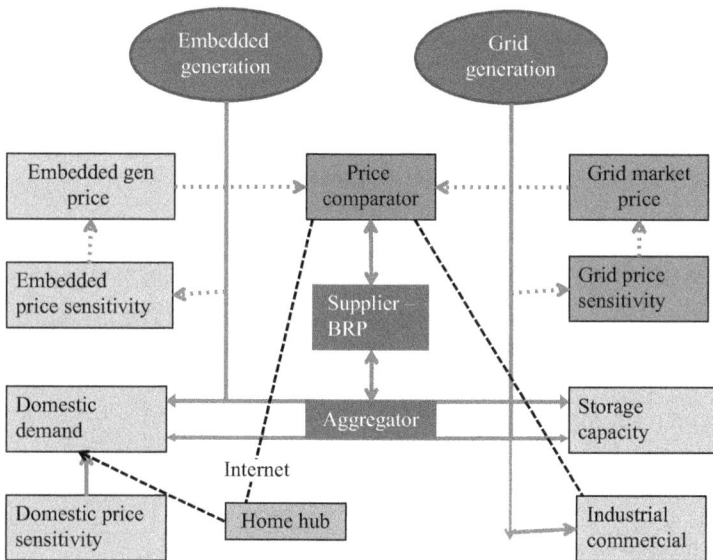

Figure 18.7   Distribution market infrastructures

embrace control of end-user demand at the industrial and commercial level including some domestic demand. The demand control for smaller domestic would probably be managed by an aggregator exploiting the internet to communicate. The aggregator would interface to the supplier/BRP exercising overall management of energy procurement and supply.

It was shown in Section 12.7 that the likely volume of domestic controllable demand is limited. Loads that could potentially be controlled, at least for short periods, are space heating and cold appliances amounting to around 3 GW. Not all consumers would want to participate at all times and take-up could be limited to 1–2 GW of peak lopping by domestic consumers. This is small in relation to the current system diurnal variations of tens of GWs and the duration for which supplies could be interrupted is limited. It would not reduce the system generation capacity requirement by 9–17 GW, as suggested in Section 18.1. New loads like electric vehicles and heat pumps will increase the potential but may be limited by the diversity in demand periods that are acceptable to consumers. Some studies have attempted to gauge if users will be prepared to relinquish control of their electricity usage to outside parties. Customer switching is currently motivated by offering improved tariffs. It will need to be shown that a better tariff will be available by offering to be flexible in time of use. The differences in user-demand patterns would make aggregation of managing demand to exploit time of use tariffs very difficult, and the process will need to be automated as illustrated in Section 18.6 with the option for users to disengage.

In addition to optimising energy purchase, there may be benefits realised in minimising future network capacity requirements that should result in lower network charges to consumers as well. There would be a requirement to fund payments to customers to encourage their participation in demand control schemes that defer the need for local network reinforcement. There will be a cost to establish the SCADA system and control centre and market platforms that should be included in the tariffs that reflects the potential user benefit.

18.13   Smart grid projects in Europe

Since 2002, some 950 projects related to the development of smart grids have been reported across 50 countries in Europe with an investment approaching €5bn (Smart Grids Project outlook 2017 European Commission). The projects have been categorised into six areas with the percentage distribution of EU funding shown including

1. Smart network management (SNM) – 26%
2. Demand-side management (DSM) – 24%
3. Integration of generation and storage (DG&S) – 26%
4. Electric mobility (E-mobility) – 10%
5. Integration of large-scale renewables (L-RES) – 6%
6. Other – 8%

The 26% of applications in smart network management focus on improved monitoring and control facilities at control centres to make the networks more observable. Improved protection and automatic post-fault reconfiguration schemes are being trialled. Dynamic line rating techniques are proposed to revise capacity based on the latest loading conditions. An alternative approach to traditional grid investment and operational practice is being explored.

The 24% of schemes in demand-side applications focus on encouraging users to shift consumption between periods through the application of real-time pricing to make better use of the available network capacity. Other applications focus on improving the efficiency of use. The overall emphasis is on users taking more control of their energy use.

Some 26% of the projects focus on the integration of distributed generation and storage projects including the development of network planning and analysis capability to increase operational flexibility. It reviews how storage can be exploited to manage generation and demand. It implies that the DSOs exercise a degree of control over generation that is connected to their network.

The E-mobility projects include the development of charging arrangements for electric vehicles and hybrids and there integration into network operation. It includes the analysis of charging strategies with the involvement of manufacturers.

Projects aimed at integrating large-scale RES consider both transmission and HV distribution connection. They include offshore wind integration, forecasting and the contribution to system service requirements. An objective is to increase the ability of the networks to accept more RES capacity.

The project funding is split approximately 50% private, 25% EC and 25% national with the United Kingdom, Germany and France being leading contributors. The driving force is generally the DSOs with support from TSOs, utilities, universities and industry. The distribution of funding amongst the project categories is shown in Figure 18.8.

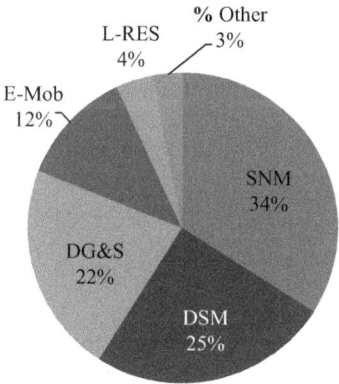

Figure 18.8   Distribution of smart grid funding by category

18.14   Sponsors

The implications to the market of an increased proportion of distributed generation are that liquidity in the wholesale market will fall. **The proportion of large-scale supergrid connected generation has already fallen by 10% to 15% in the last 5 years.** The stations that are built will largely be sponsored by large global energy players better able to manage the risk like EdF. To counter this transmission, interconnection will be developed enabling an increase in cross-border trading and the closer realisation of a pan European market.

The development of distributed generation will mostly be sponsored by local consortia bringing together users and suppliers establishing multilateral deals embracing electricity and heat supply coupled with demand management. Examples already exist where local councils in the United Kingdom, like Woking, have taken an active role in energy management. This has embraced CHP schemes and solar panels applied to their property stock. They have also avoided local distribution charges by directly wiring between some of their town centre sites.

Distribution control will need to become more active with the distributed generation operated in conjunction with storage and demand management so as to appear as a virtual power plant. CHP will be exploited providing heat supplies to local industry and premises. The distribution system will need to be operated as an integrated system to provide the least cost supplies and maintain security. This will require a local market platform to enable trading to establish least cost supplies whilst recognising network constraints. Storage systems with the capability to partially smooth daily demand profiles are likely to be developed improving the overall load factor from an average of 55% up to 65%.

There is a need to distinguish between the different timescales in the provision of balancing services and the SO role. It would help to distinguish between planning; operations in the wholesale market exercising price arbitrage and the essential total system balancing required following gate closure. The roles of the participants are likely to be different in each timescale. BRPs will seek to balance their positions while DSOs will manage any active network constraints prior to gate closure. There could be a case for market splitting as practised in Nordpool when constraints become active. Following gate closure, the TSO must assume balancing control to manage frequency for the system as a whole.

There is also a need to establish what information potential developers need to identify opportunities in the emerging sector. Investors need to be able to evaluate the potential returns and scale of the market and risks. The grid companies have a role to play in informing the market and fostering investment to enable benefits to be realised.

The potential returns for flexibility in SO timescales are influenced by the size of the need that may be limited by successful balancing by BRPs and the level of competition. In Germany, although the requirement for balancing has not changed significantly, the balancing market prices have collapsed. This is due to a lot of generators, which have suffered from their utilisation being displaced by wind generation, all competing for the supply of balancing services against a limited requirement.

Fundamental changes to distribution tariffs will be needed to support the costs of providing and operating a more active network with embedded generation and behind the meter generation. An analogy can be drawn between the uncoordinated use of the road network and its funding. Current use of system tariffs and payments associated with mitigating triad peaks are distorting the market in the provision of generation.

18.15   Changing roles

Several commentators have referenced the need for a holistic approach to managing developments but without explaining how it will be realised. **To manage significant change, there is a need to clearly define respective roles and interface arrangements through a normal business process definition.** This is essential to realise a workable holistic approach given the changes taking place across the energy sector and the increasing number of participants. This involves identifying the key functions that have to be undertaken and the key parties involved in managing them. The responsibilities of the key parties need to be defined and agreed. This stage would review where functions are most efficiently implemented without duplication and overlap. Subsequent steps involve breaking down the processes in more detail and identifying the data exchanges necessary to manage them. The next step would be to translate the process definition into specifications of the IT infrastructure necessary to support it. The major changes being imposed on the processes of managing the system in operational timescales include:

Predicting supergrid demand – The normal process has been for the SO to predict demand for various periods ahead based on algorithms that make use of historic data and expected weather conditions. This is now made more difficult by an increasing proportion of embedded generation that is not subject to central dispatch with the result that its future output will not be known. This includes the output from local wind generators and solar panels as well as other embedded generators of <50 MW and resulted in 216 supergrid points recorded as exporting back to the grid in 2014/15. The SO needs to know what the expected generation and demand flows on the network will be to analyse the security of the network and identify where constraints may become active. This can be realised by an interface to a local market that will supply price information to enable consumer demand-side response to be judged by the DSO who will exchange data with the TSO through a close interaction.

Managing distribution security – It will be incumbent on suppliers and BRPs to provide preliminary demand estimates to DSOs based on their latest traded positions. They are in turn exposed to short-term trading based on outturn positions and variations in renewable generation output. Their position will be influenced by the action taken by aggregators and active large users in reaction to market prices.

The distribution companies will need information on the expected flows on their network to analyse security and identify any active constraints in particular to take account of expanding EV and heat pump loads. In essence, the distribution companies will begin to take on the issues of a grid company with the bulk supply inputs taking on the role of interconnectors. There will be a requirement for a DSO supported by a SCADA (substation control and data acquisition) system capable of monitoring the imports from the grid, key generation inputs and load centres on the local system to manage network security. It is expected that they will need to contract for services to manage the process including shifting demand away from peaks and using storage.

Local market – A need is envisaged for a local market to support the interaction between suppliers and providers of local generation and storage. It is expected that the trading arrangements would mirror the current contracting process adopted by larger users like local councils. Some suppliers would contract 6 months to a year ahead for bulk supplies through a tender process. In the short term, a platform would be required to enable users to adjust their position based on their latest forecasts and take advantage of local renewable generation output. Some participants may choose to own storage and generation to increase their flexibility. Aggregators would play a role for smaller and domestic consumers by trading on their behalf with supplies. **This would embrace trading their customer's flexibility either to suppliers or directly by taking advantage of time of day market price variations.** In the case of EVs, this could include cycling energy through the EVs batteries. The local market infrastructure would need a data capture system to support settlement.

Interface to grid – The TSO could not be expected to monitor and predict the actions of all the embedded generation, storage systems and consumer DSR. An inter-area operating arrangement would devolve the system-management problem to more practical proportions. The defined distribution areas would provide the grid operators with their expected future transfers from and to the grid for each period ahead. The grid operator would use this information together with that from interconnectors and its own projections of supergrid connected generation output to analyse its network and advise on potential constraints. At gate closure, the BRPs would advise on their contracted positions and the TSO would manage the balancing process. To realise a national optimal solution, there would need to be options to trade between areas through the extension of existing facilities.

Planning process – The problems in planning timescales will be most acute on the distribution networks with limited control over what's connected to the system particularly at the domestic level. This will include solar panels and EVs as well as behind the meter small-scale generation. There is already a lot of bulk supply points exporting to the grid during periods of high wind and solar generation output coupled with low demand. The DSOs will need to establish processes to manage the connection of generation, DSR and storage that records all the data necessary to predict their impact on their network. The charging arrangements for use of the system will need to reflect the potential impact on the operating costs and longer term reinforcement needs. The charges for storage will in particular need to reflect their overall benefit in mitigating peaks to delay reinforcement.

18.16   Conclusions

The distribution networks are the focus of changes taking place in the energy sector to enable end-user engagement in realising affordable, secure and sustainable supplies. As their use increases, the capacity limits of the distribution networks will be reached and new tariff structures will be needed with more value placed on flexibility and pricing access to capacity rather than just energy. In contrast, the overall uncertainty in the sector has led to a dearth in investment in large-scale generation leading to unprecedented low plant margins. The proportion of large-scale supergrid connected generation that has already fallen by between 10% and 15% in the last 5 years. The requirement for increased flexibility to meet the increasing demands and manage intermittency could, in part, be met by gas-fired generation equipped with CCS without increasing emissions if viable tariffs are established that recognise the value of flexibility. There is considered to be a world-wide market for this technology that has been estimated at $31bn by 2021 offering export potential for early developers.

In current market structures, there is no agency forward looking with responsibility for taking a holistic view of the developing industry and assessing risks, and it is glibly assumed that the market will solve all. Given the growing number of interested parties, governments should consider the establishment of a single buyer/seller market model to fulfil this role. There are also a number of external potential developments that could drive development paths including health concerns resulting from emissions; technology developments in storage, CCS and fusion generation and changes in global warming. These factors need to be considered in risk assessments by a lead organisation.

Security standards are usually defined as a loss of load probability and are typically set at around a 4–8 h/year in the developed economies. The assessment of security is now complicated by the presence of a tranche of intermittent generation like wind or solar necessitating a probabilistic approach. Analysis indicates a wind contribution to system capacity of between 12% and 15% of installed capacity and of a similar order to the analysis undertaken in Ireland. An enhanced distribution SCADA system will be required to manage security including a monitoring and control capability with coupling to transmission control and potentially links to end-user internal systems. Since 2002, some 950 projects related to the development of smart grids have been reported across 50 countries in Europe with an investment approaching €5bn. These systems embrace network and demand-side management and storage and their operation needs to be embraced in overall control to realise a holistic approach. Fundamental changes to distribution tariffs will be needed to support the costs of providing and operating a more active network with embedded generation and behind the meter generation.

There will also be a need for a local market platform to facilitate coordinated operation. If the claims of reduced capacity through demand-side management are to be realised, the demand side would need to contract into a capacity market that recognises changes from the user normal time of day demand. It is expected that the development of distributed generation will mostly be sponsored by local consortia. They will bring together users and suppliers establishing multilateral deals embracing electricity and heat supply coupled with demand management and storage. This will increase the scope required in local market arrangements to engender price discovery and competition. Aggregators would trade flexibility on behalf of their customer base either to suppliers or directly by taking advantage of time-of-day market-price variations.

To manage the ensuing significant change there is a need to clearly define respective roles and interface arrangements. A starting point would normally be to establish a new business process definition.

Question 18.1 The table shows potential saving in generation capacity from demand profiling. Calculate the saving in capacity in £bn/year based on gas-fired CCGT and peaking generation. If an average consumer can exercise demand control of 1.5 kW calculate the number of consumers that are required to participate and the gross benefit to each consumer per year.

WACC 7% 20 years	Units	Gas	Peaking
Unit size	MW	400	120
Capital cost	£m/year	28.3	3.78
Capacity reduction	GW	9.00	6.00

Chapter 19

Optimum development strategy

19.1 Transformation

This chapter reviews the impact of the transformation taking place across the power industry on the traditional market players. It identifies some of the key adverse consequences and suggests measures that may be necessary to address any market imbalances.

Generation implications – Concerns over global warming led governments to intervene in energy markets to introduce the expected hidden environmental costs. This resulted in subsidies to new renewable generation sources that were principally based on wind and solar that exhibited intermittency. This generation was afforded priority dispatch that meant the conventional generation was exposed to an increasing requirement to balance renewable output that could often be changing counter to the system demand requirements. The regulating duty increased emissions from conventional generation. As the proportion of intermittent generation increased, the utilisation of conventional marginal generation reduced to the point where it became non-viable financially leading to premature closures and very low plant margins. This resulted in security concerns, because conventional plant was still needed for when renewable output was low, and prompted the introduction of capacity payments.

Priority dispatch for renewable generation should be removed and subsidies phased out.

Network implications – The small modular nature of renewable wind and solar generation has meant a lot has been connected at distribution voltages. Subsidies applied to other generation like waste-to-energy schemes and efficient local CHP schemes have also added to the capacity of embedded generation. The result of capacity auctions has been to encourage an increasing proportion of embedded diesel generation often with high levels of emissions. The growth of this embedded generation has in part been due to the structure of transmission network tariffs. They usually include a capacity element based on recorded peak demands that can be reduced by running embedded generation. The trend has resulted in distribution networks moving from being essentially passive to becoming active networks for which they were not designed. The proportion of generation connected to the transmission network has decreased by 10% to 15% while that connected at the distribution level has proportionally increased to the point where they are becoming stressed. There is also talk of defection from the grid with Siemens demonstrating a temporary isolation of a small distribution network based on renewable generation with storage and limited synchronous generation.

The network tariff structures should be reviewed to provide a level playing field for all generation irrespective of connection voltage. Embedded generators should bear a proportion of system costs reflecting the security afforded and access to market.

End-user implications – Governments have introduced policies to encourage energy efficiency through more efficient boilers and insulation. Smart meters are being installed to increase awareness of energy costs and facilitate remote meter reading and time of use tariffs. Customers are being encouraged to engage in demand control unilaterally or through participation in managed schemes. They have experienced high energy prices that in part result for the pass through of ongoing high subsidies for renewable generation. The decarbonisation of heat and transport will increase demand and necessitate demand side control facilities.

A standard intelligent home hub should be developed enabling users or their aggregators to control their demand in response to TOU market price signals or to support peak management.

System operation implications – The introduction of intermittent generation has complicated the process of demand prediction because of its unpredictability. The non-synchronous nature of some renewable sources creates frequency control and protection problems. The growth of embedded generation is resulting in reversals of normal energy flows that impact on the transmission network and voltage control. Intermittent renewable generation makes a limited contribution to firm capacity requirements and undermines plant margin assessment.

A distribution system operation function needs to be established with enhanced facilities and links to the TSO.

19.2   Plant mix development

There is currently no overall control of the plant mix that matches the plant type to the required operating regime. The priority dispatch to renewable generation has dramatically changed the requirement for flexible generation. This has prompted some CCGT owners to convert part of their fleet to operate in open-cycle mode to enhance their load following capability. There is a limit to the proportion of intermittent non-synchronous generation that can be accommodated on the system in operation without the need for curtailment.

The non-synchronous generators should not receive compensation when their output is curtailed for system operation reasons or network constraints.

There is an increasing need for generation flexibility to track variations in intermittent renewable generation. Options being considered to meet this need include demand side control and storage. There is no certainty that end users will be willing to participate, and the available capacity may be limited. The problem with the storage option is the cycle efficiency that may be around 70% or less undermining the business case. The capital costs are currently high and comparable with new generation costs. There is scope to use a limited amount of storage to take excess renewable energy that would otherwise have to be curtailed. There is a general requirement to maintain a proportion of gas-fired generation in the plant mix to meet a proportion of the system requirements for flexibility. This could be fitted with CCS where required to meet emission targets. It is envisaged that these three options should compete on price and performance in a flexibility market. The SO would specify the system requirements and contract for services for periods consistent with the timescale for provision.

A flexibility market should be established to meet system requirements for regulating capability.

Recognising the continuing growth in embedded generation, its operation needs to be integrated into the wider SO process to enable the management of the networks and any security constraints. Recognising its location that is closer to conurbations, it should be subject to costs associated with emissions in the same way as larger scale generation. It should be required to report its expected availability to enable its potential contribution to security to be assessed. In that these generators benefit from system services to maintain frequency and voltage control, they should contribute to their costs accordingly.

Embedded generators should be subject to the same charges as transmission connected generators related to emissions and the provision of system services.

The requirements for the provision of additional generation that is expected to operate baseload should be open to competition. This should include large and small-scale nuclear as well as gas and coal generation equipped for CCS. In a single-buyer (SB) market model, a central agency would be responsible for defining all generation needs by type taking account of the expected load following duty.

There is a need to tailor the plant mix to match the expected duty in tracking the future demand profile.

There is a risk that generation is operated suboptimally with each company dispatching generation to maintain their own agreed transfer. In some circumstances, some generators will be ramping up while others are ramping down. In a SB market model, all generation would be centrally dispatched to track the national demand and maintain agreed interconnector transfers. Generating companies may choose how they meet their target output. The planned output of embedded generation should be aggregated by the DSOs.

All grid connected generation should be centrally dispatched with embedded generators reporting their planned output.

There are several factors that influence the full costs of high proportions of intermittent renewable generation like wind and solar in the plant mix. There are also some technical constraints that limit the proportion in operation on the system at any one time. **The rate of unbridled renewable development has not allowed sufficient time for other generators to adjust their plant portfolio to match the changing requirements.** The added cost elements of intermittent generation include

- The extension of the transmission network to connect the remote wind sites to load centres;
- The need to retain conventional generation to replace the renewable sources when there is little wind or solar radiation;
- The need to curtail renewable output at times of low system load to maintain stable system operation.

These costs are not charged to renewable developers but borne by consumers. The deep network reinforcement costs are socialised. The impact on conventional generation utilisation has caused many stations to prematurely close prompting governments to introduce capacity payments to retain backup plant. The renewable generation has received priority dispatch with compensation if their output is curtailed although this is under review.

There are also technical constraints that impact on the proportion of intermittent generation in operation. Wind generators are non-synchronous in that they do not contribute to system inertia resulting in the system frequency being more volatile following incidents. This can lead to rate of frequency changes that cause generation to trip under rate of change of frequency protection. They also do not offer support to system voltage control and can create local disturbances. The conventional generation is dispatched to counteract the renewable output variations. This results in frequent starts and stops and the need to ramp output up and down adding to the costs of operation and increasing emissions.

These issues become more significant as the proportion of renewable generation capacity is increased above around 40% in normal circumstances. Consideration is being given to increasing system flexibility to increase the ability to be able to use the renewable energy. This issue needs addressing urgently as governments commit to more wind farm schemes as prices fall. The United Kingdom committed to a strike price of £140/MW h for the Hornsea 1 Dong wind farm of 1.2 GW and, following a competitive tender, have committed to a further 1.4-GW Hornsea 2 wind farm at a much lower strike price of £57.5/MW h. This represents a commitment to take 2.6 GW of output at any time irrespective of system need from just one wind farm. The government claims a saving to consumers of £528 m/year from the competitive tender process being the difference in price from the initial strike price of £150/MW h for earlier wind farms and the £57.5/MW h for this project. Assuming a load factor of around 46%, the annual reduction in subsidy is calculated as

$$1,400 * (150 - 58) * (0.46 * 8,760) = £528\text{m/year}.$$

However, given a typical market price for energy of £40/MW h, the subsidy cost/ year for the new farm is still over £103 m/year and for the earlier site £4,835 m/year.

The commitment to priority dispatch of intermittent renewable sources should be urgently reviewed together with the technical limitations on the manageable proportion of wind capacity in system operation.

19.3 Interaction with transport and heat

The process to decarbonise transport and heat has barely begun and represents a much larger challenge in terms of the scale of energy involved and could result in doubling the electricity demand. Some 20% of the total emissions result from domestic energy use for space and water heating, and more than 50% households have vehicles using petrol or diesel. The decarbonisation of this sector will have a profound impact on the distribution network requirements and the need for gen- eration capacity. The extent will depend on the chosen decarbonisation route but is expected to involve electric vehicles and heat pumps. In order to contain their impact, there is scope to profile the demands to minimise the additional network and generation capacity requirements. This option could influence the design of schemes with higher battery capacities and larger heat stores to increase flexibility. Consumers should be encouraged to take up this option in the expectation that their load profiles could be managed by aggregators and suppliers or automatically. Consumers would benefit with favourable flexibility tariffs that match those for generation. The controls would be directed to specific appliances through a broadband router and home hub (see section 18.5).

The nature of EV utilisation and heating demand is that they will not exhibit the same level of diversity as many existing loads. The application of load control would also need to embrace the management of peak demands on distribution transformers.

Tariffs should be developed for consumers to provide demand flexibility that could be applied to EVs and heat pump loads.

The gas grid has massive potential to store energy and could be used to store excess renewable energy. The limiting problem is the overall cycle efficiency of around 38% of producing gas that is then used to generate electricity. However, it could be a viable option to use excess renewable energy that would otherwise have to be curtailed.

Consideration should be given to exploiting the gas grids capacity to store surplus electrical energy from renewable sources.

19.4 Market mechanisms

Given the transition taking place in the sector, the increasing number of indepen- dent players and frequent government intervention, the current market mechanism is not considered to be likely to result in an optimum outturn. Future energy prices and security of supply will be largely determined by today's investment decisions. The pace of change and the interactions between renewable sources, conventional generation, heat and transport and their impact on networks are too complex to be resolved by simple market mechanisms. Whilst the industry recognises the need for a holistic coordinated approach, the complexity and number of players makes realising this ambition unlikely. The current market arrangements have been largely flawed by policy and government intervention. There is a need at this critical stage for a central coordinating authority. It is considered that this will be best realised by establishing a SB/seller market model. The SB would contract for generation through competitive tenders including import options and sell to suppliers. The suppliers would advise their capacity and energy needs taking account of their ability to profile demand based on the use of demand side control and any embedded generation. Capacity needs would be contracted several years ahead consistent with new entry timescales. Energy profiles would be developed and contracted nearer the event. The market prices would reflect the contracted position with generators and importers and be based on a dispatch simulation to meet the overall demand profile. The supply side would pay for energy based on the demand and marginal half hour price plus uplift to cover system management. It is considered that this approach would enable a holistic approach whilst retaining competition in both generation and supply.

A Single Buyer/Seller market model should be established to manage the optimal procurement of energy for all users with equitable tariffs.

It is considered that a SB would be able to underwrite the development and financing of large-scale projects like nuclear and hydro schemes in much the same way as the strike price for energy from Hinkley Point nuclear station in the United Kingdom was underwritten by the government. Suppliers would contract for energy and capacity with the SB in various timescales. Longer term contracts would be for baseload energy requirements. Nearer the event, the requirements for the working weekdays would be defined through the winter and subsequently the summer. The requirements would be refined during the weeks and days before the event. The suppliers would contract taking account of any embedded generation under their control. Independent embedded generators would contract direct with the SB or through an aggregator. A local regional market could be used to enable competition between embedded generators, or they may choose to cooperate with a local demand to establish profiles.

A Single Buyer/Seller market model should be considered as a viable option to realise an optimum holistic outcome from the current transition taking place in the sector.

The aggregator would manage and profile the demand of their customer base and contract for supplies from embedded or grid generation depending on respective prices as illustrated in Figure 19.1. The transaction process could be supported by local as well as national markets. The importance of local markets is their

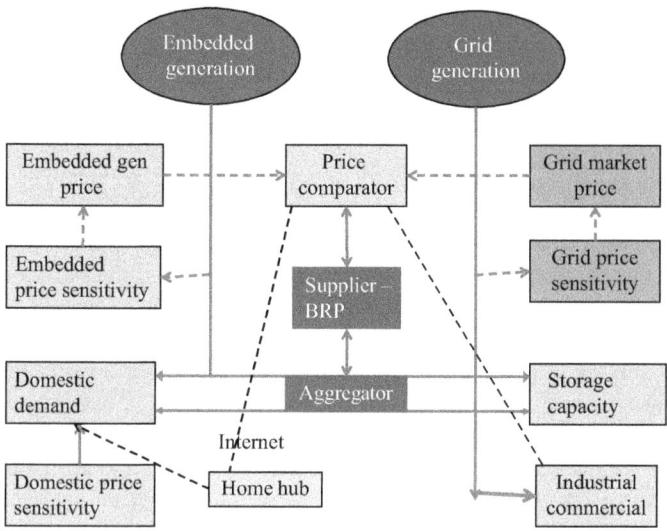

Figure 19.1   Wholesale market

influence on local distribution network constraints that are expected to become more active as loads increase. The bulk of baseload demand would be contracted 6–12 months ahead to get the best price and restrict capacity needs. In the shorter term, the aggregator would trim demand using storage or exchange trades to balance their position. In some circumstances, there is the option to exercise control over previously contracted supplies to end users. This could apply to manage emergencies or sudden increases in renewable generation output.

19.5   Flexibility

There is obvious benefit in smoothing the national demand profile to remove peaks and reduce requirements for generation and network capacity. The realisation of a system with profiled demand depends on establishing flexibility on the demand side to shift demand between periods. In defining the demand flexibility, an analogy can be drawn with generation and attributed with a similar set of characteristics. Including

- An upper and lower limit on demand;
- An indication of availability;
- Advanced notice times equivalent to generation start-up times;
- Capacity available;
- Period of availability (equivalent to energy requirement);
- Run-up and run-down rates.

It is envisaged that these constraints on call-off would impact on payments for services that would be reduced in proportion to the availability. The flexibility service will embrace demand reduction to support peak lopping but equally increasing demand during periods of high renewable output that would otherwise have to be curtailed. There is also benefit in reducing the regulating duty on conventional generation.

It is assumed that domestic consumers would not be interested in engaging directly in monitoring or reporting detailed data at this level. The candidate demands would be electric vehicles and heat pumps or electric heating that may not naturally exhibit much diversity but have inherent storage capability. It is envisaged that an intelligent home hub would monitor user installations and the history of energy use against models that can estimate the scope for reducing demand and the period. The model intelligence could be supported by house and ambient temperature data. In the case of connected electric vehicles, the charge status would be monitored against the full target and normal available charge period to establish the potential to interrupt charging.

The data for each end user would vary from day to day and within the day but is expected to exhibit repeated patterns. It is envisaged that an aggregator would collect data from their customer base and bid the combined total availability into the market as illustrated in Figure 19.2. Based on experience, it should be possible to establish firm capacity commitments over peak periods. To realise the full benefit of reducing generation and system capacity needs, it would be necessary to make longer term commitments on capacity requirements that would be profiled through the day. The aggregator would define and control the basic profile to complement other demands and smooth their overall requirement so that it can be met with generation operating baseload. To manage customer churn, there should be an option to trade flexibility in shorter timescales to enable participating companies to adjust their position. The flexibility trading could be supported along with existing generation markets where the demand flexibility would offset generation needs or take up spare capacity like renewable output. This market would also support the process of price discovery for flexibility from the various options.

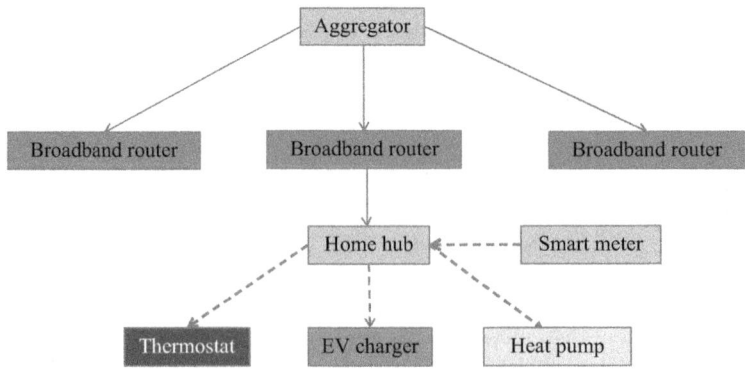

Figure 19.2   Flexibility management

There would be a financial benefit to participating users in avoiding buying energy at peak period premium prices and taking advantage of cheap renewable energy. There is also a wider benefit to all users in containing peak prices and capacity needs. For this benefit to be realised by the participating user, the market mechanism should be allowed to reflect generation capacity costs. **The capacity market should be two sided with both generators and suppliers making commitments to the known proportion of their capacity availability and needs.** The supplier or aggregator would take advantage of their demand flexibility to reduce their contracted peak capacity with generators. In turn, the supplier or aggregator should reflect the benefit in a payment to end users for participation as they will be able to support their customer base with less contracted generation capacity. This payment would be based on the capacity of the demand under control.

Other sources of flexibility should be progressed to establish the cheapest overall cost. There is a strong case to retain a proportion of gas-fired generation in the mix that can be used to support the need for flexibility with its rapid response capability. This could be fitted retrospectively with CCS, if needed, to help meet emission targets. Interconnection to countries with excess hydro would also offer opportunities to contract for flexibility services. An optimum mix of technologies should be established that meets system flexibility requirements at least cost.

19.6   Trading process

In order to realise savings in generation and network capacity, it will be necessary to fix commitments a few years in advance that reflect the capability to profile demand. At this lead time, there will be uncertainties and requirements will be estimated. The capacity market process would include

Capacity market – 4 years ahead of event.

- The supply side would provide estimates of its peak demands and the profiles based on data from end users and aggregators. It would seek to exploit user flexibility to establish the lowest level of required peak capacity. It would have the option to include a definition of its flexibility in the offer or retain it for direct use at a later stage in trading.
- The end user would have the option to agree to have its load profile adjusted for a participation fee. This would reflect the savings in capacity payments that the suppliers collectively have to pay.
- The generation side would participate in an auction to provide capacity to match the total supplier peak demand. It would evaluate the potential demand flexibility and reflect its value in its price submission with a ladder of capacity/ prices.
- The SO and SB would review the provision to exploit the demand flexibility to establish the least capacity price. The SO would include requirements for reserve and security and to cater for any network constraints. Where advantage is taken of the flexibility of a supplier, it would be reflected in its proportion of the capacity charge taking account of the saving realised by the flexibility.

- The capacity provision would be reviewed each year by generators and suppliers to take account of changes due to delays in construction of new generation or customer churn. Market participants would be able to trade capacity commitments to adjust their position.

Suppliers need to provide indicative profiles of their flexibility in the same way that generators have restrictions on their operating profile. These are necessary to enable generators to establish a set of commitments with the best overall profile to match its capability. The contracts with suppliers could include an estimate of the flexibility to take more or less demand through a typical daily period. **As the overall demand profile will be much flatter, it will be less predictable when peaks will occur.** The SO will be responsible for issuing capacity warnings when suppliers will be obliged to use their flexibility to limit their demand to their contracted capacity level. The restriction would not apply during periods of high wind output when the flexibility would be exploited to increase demand.

Energy market futures process – it is expected that most of the energy requirements will be contracted during the period 6 months to day ahead

- Suppliers would invite tenders to meet their energy commitment.
- The suppliers would establish their latest estimates of demand for the period ahead taking account of changes in the end-user base and their loads. They would seek to trade capacity commitments to adjust their final position.
- The generators would take account of their plant availability and any outages and trade to adjust their position.
- The SO may need to contract with generation or/and demand to manage any active network constraints.

Energy trading on the day – smaller volumes will be traded in timescales of 4, 1 and 0.5 h ahead to trim positions

- The SO should provide forecasts of expected renewable generation for the day and expected demand to enable suppliers/aggregators to plan their demand control.
- A generator contracted to supply may negotiate with the demand side to reduce demand if the cost is less than its generation costs.
- The demand side may choose to adjust its demand form one period to another to take advantage of cheaper prices that may be the result of high levels of renewable output during some periods.
- The SO may need to trade to exploit generation and demand flexibility to manage network constraints.
- The SO will provide advanced warning of any periods when capacity will be limited requiring suppliers to contain their demand to their contracted level.

Given the range of uncertainties resulting from renewable intermittency and end-user choice, it is envisaged that contracts with end users would include an interruption clause where in critical situations contracted demands like EV charging and heat pumps could be temporally interrupted.

19.7   Network development

The electrification of heat and transport via EVs and heat pumps will lead to a significant increase in distribution network loading. The connection of this load is currently not specifically regulated or controlled. The rate of development will depend on the take-up and the extent to which the loads can be profiled to avoid high peaks but is likely to extend beyond normal system development capacity. It will be necessary to establish plans and programmes in advance of need to split existing networks to add new distribution transformer installations. This in turn will lead to some restructuring of 11-kV networks.

The tariff structures for use of system are in need of review to better reflect the shift in utilisation. Embedded generators should pay a fair contribution to the provision of system services for frequency and voltage control and having the option to revert to supplies from the grid.

Since 2002, some 950 projects related to the development of smart grids have been reported across 50 countries in Europe with an investment approaching €5bn. Some 26% cover smart network management and focus on improved monitoring and control facilities at control centres to facilitate full use of the capacity. A further 24% of schemes cover demand side applications and focus on encouraging users to shift consumption between periods through the application of real-time pricing to make better use of the available network capacity. Some 26% of the projects focus on the integration of distributed generation and storage projects including the development of network planning and analysis capability to increase operational flexibility. The project funding is usually split approximately 50% private, 25% EC and 25% national.

Recognising the constraints on the pace of network development, there is a need for projects taking a longer term view. Given the scale of the problem of decarbonising heat and transport, the analysis needs to be based on scenarios with high penetrations of EVs and heat pumps amongst other developments. The objectives should be aimed at integrating large-scale RES with both transmission and HV (high voltage) distribution connection. They include offshore wind integration, forecasting and the development of system services. The objective would be to increase the ability of the networks to accept more RES capacity.

19.8   Governance

The transitions taking place in the energy sector are having profound effects but with no overall strategic direction. There are hundreds of new players who are not constrained to operate in a coordinated way but the power system is expected to connect all system users and react instantaneously to any changes. **A fundamental question is how the activities of the disparate players in the new world can be guided and coordinated to meet the user's requirements in an optimum manner.** There is a view that government agencies can coordinate and promote development as illustrated in Figure 19.3 but they tend to operate reactively rather than

Figure 19.3   Development business process

take a lead. The current concept is that research will be sponsored jointly by government and industry to identify the needs and innovative solutions. Data and models can be made available and used to analyse options to identify and evaluate opportunities. Workshops with all interested parties would be used to promote and gain support for the concepts. The expectation is that subsequently industry, government and utilities would collaborate and establish jointly funded development projects to prove concepts.

The business process proposed fails to address the fundamental question of identifying the integrated system of networks, systems and loads that can sustainably meet the requirements to decarbonise heat, transport and electricity at manageable costs. It fails to show how an optimum overall strategy will be identified to provide a basis against which specific developments can be pursued in a competitive environment. Various scenarios have been advanced based on assumptions about the rate of progression to decarbonise energy and the prevailing economic conditions. They also take a view on the extent to which consumers exercise power in a market-driven environment with minimal government intervention as opposed to a scenario directed through policy intervention. **The current situation is one attempting to maintain free market operation whilst distorting investment with subsidies and regulation.** The perceived regulatory risk is discouraging investment and innovation with no clear strategic direction.

The magnitude of the changes that the industry will have to accommodate is unlikely to be managed by singular intervention and will need an overall strategic plan. Given the scale of the development required, the plan needs to project to the middle of the century. The distribution networks will need to be reinforced to manage the extra loads resulting from decarbonisation of transport and heat. The loadings will have limited diversity and either more substations will be required or it may prove economic to use higher intermediate voltages. Given that most of the distribution network in towns is underground and that the requirements will change across the whole network at the same time a 20–30-year development plan is likely to be necessary. Unlike the transmission network, there is less control over when and how demand and generation is added to the distribution networks. User developments will need to be tracked from the planning stage and network plans adapted. Demand side response will need to be exploited to manage the timing of reinforcements and avoid congestion in the workload.

19.9   Open system models

One approach to coordinating developments to ensure they are compatible is to provide a system-wide model that users can access to evaluate their proposed developments e.g. the 2050 Calculator model developed in the United Kingdom. The concept is that users can enter their own data and run the model to check the outturn. The problem is the overall complexity and the availability of robust data like future fuel prices that is not provided. There is also no mechanism to determine what other developers are planning that may have an impact. The models do not embrace system dynamics and operational constraints particularly managing the intermittency of renewable sources. The markets are not modelled, and the market size for a particular system function is not included. It is difficult to model distribution networks with little control over demand side developments. Whilst these general models can provide some insight invariably, any specific development proposal will need to be modelled in much more detail. The intention of developments of this type is to

- Provide a framework that supports system integration;
- Develop a system modelling capability;
- Establish sets of data to support modelling and analysis;
- Enable rapid prototype testing.

These objectives are consistent with the requirements for analysis of proposed developments. In practice, consultants are engaged who have the knowledge of data and likely developments and a suite of modelling tools. They also have a wider experience of current international market developments and are able to make projections that span investment timescales. It is nearly always necessary to adapt modelling approaches to address a specific proposal.

Emphasis is often placed on identifying the need for new functions and products to meet the needs of the evolving energy markets. The approach is often product orientated to meet the needs of suppliers. The scale of the developments needed requires a more fundamental review of the overall power system and its energy sources.

Other models have been developed in the United Kingdom by the Energy Technology Institute called the energy systems modelling environment. An overall

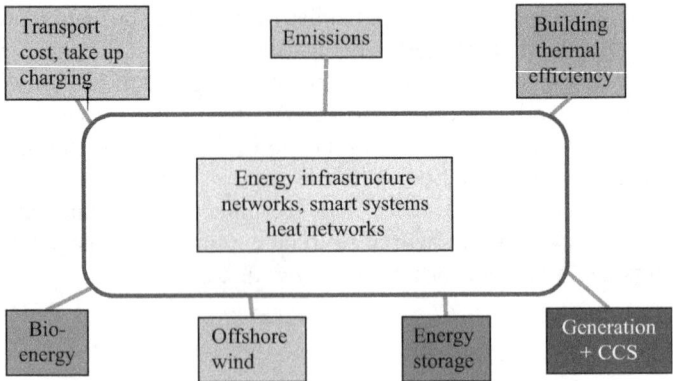

Figure 19.4   ESME (Energy Technology Institute)

schematic illustrating the scope is shown in Figure 19.4. This is designed as a 'bottom-up' model to find the least cost energy system design while meeting emission targets and recognising constraints. User-specified uncertainty is evaluated using a Monte Carlo model. It includes electricity, fuel production, heat and building energy use, transportation and some 250 technologies. Some temporal effects like peak management and spatial factors are included but with limited granularity. The objective function is to minimise the annualised capital costs and operating and resource costs with defined discount rates whilst varying technology capacity and retrofits like CCS, daily production patterns, storage capacity and fuel prices. The constraints include meeting demand, balancing supply and consumption and meeting peak demands while operating within emission targets.

The scope of the model limits the granularity to a few season days every few years and limits its veracity in representing the temporal effects associated with renewable intermittent sources and storage systems. The probabilistic elements are analysed based on user-specified ranges. This type of model is not designed to simulate market operation.

19.10   Market-based models

The overall complexity of all inclusive models can make it difficult to interpret the results and isolate cause and effect. Inevitably, it is difficult to fully model the dynamics of an energy system like electricity that has to balance at each instant in time introducing extra constraints. An alternative modelling technique successfully used by the author is based on a more modular approach with a suite of linked models replicating market operation. A simplified overview model is built to identify the key interactions between the main elements of the system. This supports visualisation of the problem and informs the level of detail that is material to the interacting coupling variables. More detailed models are used to represent plant operation with full granularity that enables the impact of intermittency, storage and system balancing to be fully evaluated. The approach is

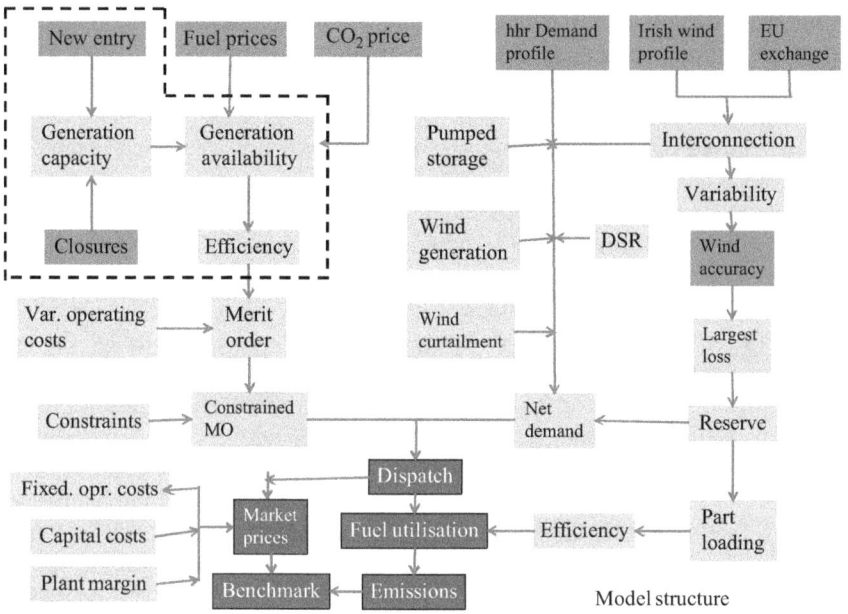

Figure 19.5   Market-based modelling

designed to reflect how the market will operate in investment and operational timescales. The coupling between modules is realised by price linkages reflecting supply and demand.

Longer term modelling includes evaluation of new entry based on comparison of expected market prices and investment and operating costs and the timing of closures as shown in Figure 19.5. On the demand side, GDP growth and price sensitivity support prediction of future demands. The demand model also includes for the response of demand to price. Interconnection modelling is essential to reflect what is happening in adjacent markets. The model also includes the impact of managing wind and solar energy variability on requirements for reserve holding and generation part loading and costs. A dispatch algorithm is used to order the plant to meet demand based on a merit order constrained by any transmission limitations.

The market-based approach to modelling also enables the impact of policy intervention to be modelled and accounted for. The objective is to mimic more closely how the business world operates.

19.11   Functionality required

To manage the developing system, a number of functional requirements can be identified:

Planning – Much has been commented on the savings in capacity that could be realised through demand side control but little on how it might be realised. Decisions on capacity requirements need to be made several years in advance of the event. For savings to be realised, suppliers need to commit to specific levels of demand that they are able to manage in conjunction with their customer base. **The future generation capacity requirements should be set by the suppliers, rather than the system operator, in the knowledge of their demand control capability.** They in turn would structure contracts and appropriate tariffs with their customer base and aggregators. The SO would be engaged in taking account of any active network constraints, security requirements and the requirements for ancillary services. Whereas suppliers usually contract 6 months or a year ahead, they would need to take a longer term view that would be complicated by customer switching. This reinforces the need for a SB/seller market model with a central organisation aggregating the needs of all suppliers.

Demand side scheduling and dispatch – This is required to manage the extra demand on the distribution system resulting from the roll-out of EVs and heat pumps in the domestic sector. The objective will be to phase their control to smooth the overall demand to minimise the requirement for network reinforcement and generation capacity. This role may be performed directly by suppliers or by aggregators and controls would be directed to specific loads exploiting internet and intranet connectivity. End users would be incentivised to participate through beneficial energy tariffs that are likely to include interruption clauses for selected demands. This type of control has been practised for years in the developing countries where remote villages, with limited micro-hydro capacity, use schedules that define when residents could iron washing.

Distribution extended SCADA – This is required to extend the distribution system SCADA recognising the more active network status. It will monitor demand and the output of embedded generation to assess any active network security constraints. An interface to suppliers and aggregators would be necessary in various timeframes to capture data on intended operational plans affecting the load of their consumer base.

Regional market – This will provide the opportunity to exercise trading options within the local area whilst taking account of local network constraints. It will provide visibility of the options and capture data to facilitate settlement. It will support the development of data submissions by BRPs to the grid company.

19.12   Development timescales

The pace of development of some of the key drivers is likely to outstrip the ability of the system infrastructure to accommodate it. With the prices of wind farms and solar energy farms falling, there is acceleration in the expansion of their capacity connected to the system. In parallel, concerns over the impact of diesel and petrol car emissions in cities are resulting in pressure to establish low-emission alternatives. In some cities, diesel and petrol vehicles are likely to be charged for entry or banned, and this will accelerate the introduction of EVs. The decarbonisation of heat will lead to the take-up of more heat pumps to replace older gas boilers. The third driving factor is the growth in embedded generation, partly as a result of policy on network charges linked to the peak demand and high network charges.

All these factors impact directly on the distribution networks with some parts of the network having already reached the limit to absorb more generation. The distribution networks have traditionally been developed assuming low after diversity maximum demand (ADMD) as low as 1 kW based on diverse use of appliances like washing machines and cookers. The new loads will not exhibit this diversity; most EVs will be put on charge on returning from work; heat pumps will operate together in response to ambient temperatures. The choice consumers make is not subject to control other than through intervention in markets.

The distribution networks in conurbations are usually based on underground cables that are not easily changed or ungraded like overhead lines. Substations are designed to serve a set of local consumers based on expected ADMDs. It may be more tractable to split networks adding more transformers connected into the area 11-kV network to increase capacity. At the same time, the control and data acquisition systems will need to be upgraded to support management of the more active networks and the demand side. With normal resource capability limited to maintenance and minor network extensions, managing the system transition is likely to be a major constraint on development. Some networks have already been reported as having reached their limit to accept more embedded generation.

There is a requirement to establish the timeframe for development of the distribution networks and control infrastructure to meet the emerging needs of system user.

19.13 Conclusions

A major concern is the implication of a large number of new participants in the energy sector whose actions are not coordinated or regulated. The approach is to encourage end-user engagement based on the philosophy that this will promote more competition. A lot of the developments in the electrification of transport and heat are focussed on the distribution networks where end users can make unilateral decisions. Whilst there is recognition of the need for a holistic approach, there is no process in place by which it may be realised. In this situation, developments will be impeded by the rate at which networks can be transformed to match the transition in its use. Various models have been developed that are designed to provide insight into the wider impact of user-driven developments. Inevitably their use is limited by the availability of a central data set related to future conditions and regulator intervention. They also have insufficient detail to underwrite specific investments.

A number of actions have been identified in this chapter that are considered necessary support the transition process including:

- Priority dispatch for renewable generation should be removed and subsides phased out.
- The network tariff structures should be reviewed to provide a level playing field for all generation irrespective of connection voltage.
- An intelligent home hub should be developed enabling users to control their demand in response to TOU market price signals.
- A distribution system operation function needs to be established with enhanced facilities and links to the TSO.

- The non-synchronous generators should not receive compensation when their output is curtailed for system operation reasons or network constraints.
- A flexibility market should be established to meet system requirements for regulating capability.
- Embedded generators should be subject to the same charges as transmission-connected generators related to emissions and the provision of system services.
- There is a need to tailor the plant mix to match the expected duty in tracking the future demand profile.
- All grid-connected generation should be centrally dispatched with embedded generators reporting their planned output.
- The commitment to priority dispatch of intermittent renewable sources should be urgently reviewed together with the technical limitations on the manageable proportion of wind capacity in system operation.
- Tariffs should be developed for consumers to provide demand flexibility that could be applied to EVs and heat pump loads.
- Consideration should be given to exploiting the gas grids capacity to store surplus electrical energy from renewable sources.
- A SB/seller market model should be established to manage the optimal procurement of energy for all users with equitable tariffs.
- A SB/seller market model should be considered as a viable option to realise an optimum holistic outcome from the current transition taking place in the sector.
- The capacity market should be two sided with both generators and suppliers making commitments to the known proportion of their future capacity availability and needs.
- Recognising the constraints on the pace of network development, there is a need for more projects taking a longer term view.
- A fundamental question is how the activities of the disparate players in the new world can be guided and coordinated to meet the user's requirements in an optimum manner.
- The future generation capacity requirements should be set by the suppliers, rather than the system operator, in the knowledge of their demand control capability.
- Consideration should be given to sponsoring the development of modular nuclear generation and generation with CCS to provide alternatives that can be deployed to meet global requirements to reduce emissions.

Governments should establish organisations that can support the process of formulating an indicative development strategy.

Question 19.1

1. Draw a comparison between the cost of the provision of flexibility from generation, demand side control and storage.
2. Comment on the impact of intermittent renewable generation on the optimal plant mix that meets system dynamic requirements.

Chapter 20

Key findings

20.1 Drivers of change

In the current environment, governments are faced with a spectrum of objectives including securing future supplies, building up renewable energy to meet emission targets whilst maintaining fuel diversity and containing end-user bills (the so-called trilemma). In attempting to meet these objectives, governments have intervened in the market process by using subsidies to promote development of renewable technology and setting strike prices for energy contracts with clean generators. This has resulted in a market that is neither free nor centrally coordinated and has created a number of problems:

- Subsidies to renewable generators and priority dispatch has undermined the operation of the wholesale market;
- Conventional generators have suffered reduced utilisation coupled with an increase in regulating duty to track variations in wind output undermining their financial viability;
- There is a shift in the operating regimes of conventional generation that will impact on the optimal plant mix;
- Investors see the regulatory bias in favour of renewables as introducing risks that undermine the business case to build new plants leading to unprecedented low margins;
- The costs of subsidies to renewable generation and the costs of extending transmission and distribution to accommodate it has led to high increases in end-user costs;
- System operation has become more complicated due to the need to accept on the system an increasing amount of intermittent renewable generation output, predict demand and manage dispatch and interface with Distribution System Operators (DSOs);
- Use of system charging arrangements on the distribution and transmission networks have distorted competition between grid connected and embedded generation leading to an expansion of distribution-connected generation;
- The distribution networks are becoming more active rather than passive stretching the capacity of the network and leading to the need for reinforcement and the establishment of a DSO;

- Policies to promote competition have resulted in hundreds of new players in the sector that undermines the ability to establish a holistic approach to system development.

Given the scale of the problem with the decarbonisation of heating and transport still pending, actions need to be initiated now to manage the increasing impact on the electricity sector. Regulators need to review the mechanisms used to encourage renewable generation and review the need for subsidies and priority dispatch. The distribution companies need to take up the role of a DSO to manage their active networks. The DSO and TSO need to interact to manage the more active nature of the distribution network and exchange data to support demand prediction and manage security. Mechanisms need to be established to coordinate the interaction of an increasing number of market players.

20.2   Technical challenges

There is no competitive market operating for generation at the distribution level to support competition between local generation sources, suppliers and grid-connected generation.

The impact of large tranches of renewable generation has meant that the net demand profile is much more variable and difficult to predict with in extreme cases distributors exporting energy back to the grid. The traditional approach of using historic records is invalidated, and the problem may need to be devolved to DSOs.

It is more difficult for potential investors in generation to predict revenues or even establish the data to make an analysis of the optimal plant mix other than much more flexibility will be required.

The renewable generation output will have to be curtailed at times of low system demand to maintain sufficient reserve and regulating capability to secure the system. The need for wind output curtailment increases exponentially when the non-synchronous capacity reaches around 50%.

The potential contribution of wind generation to the secure capacity available at peak times is estimated on a probabilistic basic to be around 10%–15% of the total installed. This means that conventional capacity has to be retained to maintain plant margins. Because of the low utilisation of this generation, it has been necessary to introduce capacity payments to retain the availability.

The intermittent loading of generation providing backup to renewable generation results in a loading regime that increases the normal fuel burn and incurs additional costs in wear and tear. The extra fuel burn results in more emissions offsetting the saving resulting from the use of the renewable sources.

As the power system non-synchronous generation penetration increases, the system inertia falls and the system becomes more volatile increasing the chance of ROCOF relay operation. This could lead to reductions in wind generation output following faults and cascade tripping.

The net impact on emissions of converting to EVs will depend on the emissions from generation used to charge the EV battery being below the target emission levels. The loading on the distribution networks due to EVs may not have a lot of diversity. The rapid growth of domestic EVs could seriously stress a distribution network designed for an after-diversity demand of 1.5 kW per consumer leading to the need for reinforcement as the penetration reaches around 30%.

Network costs have increased to accommodate renewable generation that is often remotely sited. Some commentators have suggested that increasing network charges could promote mass defection from the grid where there are cheaper alternatives.

Indicative of the rate of introduction of EVs, the Norwegian Government plans to ban the sales of petrol and diesel cars by 2025. This is consistent with policies advanced by other countries and would lead to most vehicles being based on electricity or hybrids by 2040.

There is a developing problem due to the impact of variations in the output of large tranches of wind generation that is not balanced close to source, on interconnector flows.

These technical issues constitute constraints on the future development of the industry that will impinge on economic assessments. They are likely to result in the need to reinforce distribution networks and their control capability to accommodate embedded generation and new EV and heating loads. System operation and demand prediction will become more complicated requiring a greater need for TSO/DSO interaction and cooperation.

20.3 Impact on market players

As the capacity of intermittent wind generation is increased, the utilisation of conventional generation decreases, leading to closures and reduced plant margins that impact on system security. Most of the large players in the market have suffered from falling revenues and share prices. The German utilities in particular have experienced a collapse in share prices, falling by 80% with earnings/share halved. This has been driven by ambitious targets for renewable generation with a target for 30% of supply by 2020. In parallel, costs are being added to conventional generation to reduce CO_2 emissions. These developments taken together add significant unit costs to a wide range of conventional generation, and it is inevitable that some will cease to be economically viable and will close. The generating company assessments indicate an industry driven into crisis by shifts in policy that were not signalled and that are having far reaching consequences.

The United Kingdom introduced an emission performance standard set at 450 g/kW h. It was shown that conventional coal generation with emissions of over 600 g/kW h would not comply with the standard and will lead to it being phased out. This will impact on the plant margin at a time when investors are nervous of the perceived risks and are unlikely to pursue replacement projects. Gas-fired

CCGT generation at around 324 g/kW h would comply, but if the expected utilisation is low, the project may not appear profitable.

In contrast, there is a growth in small-scale generation being connected at distribution voltages. This is in part the result of their potential impact on the use of system charges levelled on suppliers. By generating at the time of peaks, the embedded generators can reduce the distribution demand and the supplier capacity charges. The smaller generators also do not pay transmission charges. As a result, some 12 GW of capacity has been transferred from being connected at transmission voltages to the distribution networks. This is stretching the capacity of the distribution networks with some networks reported as being at their limit. Proposals are being considered to revise the network charging arrangements.

As the distribution networks move from being essentially passive to active, their systems for monitoring and control will need to be enhanced. The costs can be expected to add around 10% to distribution tariffs. The capacity of the networks will also need reinforcement particularly to accommodate EV charging and heat pump demand resulting in costs rising in proportion. These costs may shift the balance back in favour of HV-connected generation where the control infrastructure is already in place.

The transmission network is experiencing less utilisation because of the increase in distribution-connected generation that sometimes exports back to the grid. Extra residual costs have to be charged to the demand side to cover transmission costs. At times, the circuit loadings are below the natural loading level and circuits have to be switched out to contain voltage rises. The connection of offshore wind farms has been put out to tender with the result that by 2017 there were 18 transmission operators in the UK sector.

Subsidy schemes to promote renewable generation have committed consumers to very high energy premiums for years to come, and the premium for nuclear energy in the United Kingdom will add significantly to the burden. Subsidies for wind and solar energy in the United Kingdom for 2016 resulted in extra costs to consumers above normal energy rates, of around £100/year. To meet the climate change targets for 2030, the subsidy premium for UK customers would rise to £312/household/year based on the proposed plant mix.

The firm contribution of intermittent generation like wind and solar to the plant margin is questionable. A study based on the probability of generation availability and demand was run both with and without wind generation to realise the same loss of load probability (LOLP). The study analysed the amount of wind generation necessary to restore the plant margin to the same level based on thermal generation; the result showed that a capacity contribution of around 12% of installed wind capacity could be expected from wind generation.

The maintenance of supplies is now more dependent on the actions of an increasingly diverse range of players that are principally motivated by opportunities to make money, as is their job. Whilst the need for a holistic coordinated approach to system development is generally accepted, how it can be realised in the current liberal environment is less clear.

20.4 Impact of renewable generation

Progress on the reduction of emissions is proceeding slowly across the world by the largest emitters. Emissions in China were 26 times those of the United Kingdom in 2015, but initiatives are being made to expand their nuclear fleet. In the USA, the success of fracking for gas has resulted in increases in gas-fired generation at the expense of coal, and it is claimed that their reduction in emissions exceeds that of Europe with all its subsidised wind and solar generation. This is the result of emissions from gas CCGT generation being around 40% of those from coal-fired generation.

There are also questions about the impact on emissions from conventional generation required to complement the variable output from intermittent wind and solar generation. These result from operating in reserve mode part loaded at reduced efficiency; frequently ramping up and down and more unit starts and stops. The impact of these effects has been modelled based on half-hour-recorded demand and renewable generation for a year. The impact is modest with wind capacity up to around 50% of the peak demand but at higher levels increases significantly with more conventional unit starts and stops required. The other contributing factor is the need to more frequently curtail wind generation at times of low load to maintain operational security. This occurs when the wind generation output exceeds the demand less the conventional generation needed to manage stable operation plus other inflexible generation like CHP and some nuclear. The implication is that there is a practical limit to the proportion of installed wind generation before technical constraints start to undermine the economic case.

The assessment of new wind farm installations needs to fully account for the effect on net output of maintenance outages, system transmission losses, wake losses and blade degradation that can reduce the output to 72% of the theoretical maximum based on wind speed data. It is also usual to take a conservative view of the expected load factor by basing the assessment on that value that will probably be exceeded in 90% of cases. There are clear advantages in locating wind farms offshore where wind speeds are higher. The London array of 630 MW is located in waters 25-m deep. The Gemini wind farm off the coast of the Netherlands is located in water 45-m deep. Limits are reached in deeper waters and RWE abandoned plans for a 1.2-GW farm in the Bristol Channel due to difficult seabed conditions. However, a floating wind farm (5 * 6 MW) that operates in deeper waters has been located off the coast of Scotland. If the foundation costs double, then the load factor has to increase from 30% to 37% to maintain the same unit cost. Wind makes a limited contribution to the provision of firm capacity, estimated on a probability basis to be around 10% of installed capacity.

There are other generation options that reduce emissions whilst providing a degree of flexibility to manage system operation. One near-zero carbon emission option is a biomass-fired station burning woodchips. Several 660-MW units have been converted to biomass at the UK Drax ex-coal-fired station. The generation emissions are complemented by the absorption of CO_2 during their growth cycle.

Waste-to-energy schemes are an option that also helps with waste disposal and can provide some flexibility. Solar generation prices are falling but a problem in the Northern Hemisphere is that when the sun is out demand is generally very low and may not be able to cope with a high injection of solar energy. Another option to reduce emissions is to convert all existing coal plant to burn gas with just 40% of the emissions from an equivalent coal-fired station. This alone would reduce world emissions from all sources by 25%. And more than that from all the worlds wind farms.

A financial comparison of low carbon generation options shows that thermal generation with CCS can compete with nuclear and wind on price while also offering flexibility. The analysis shows the impact of different levels of utilisation on relative costs and illustrates the importance of establishing a mixture of technologies that match the expected load duration curve at least cost.

20.5   Embedded generation issues

A rapid growth is occurring in embedded generation or self-generation as labelled by the EU that the traditional industry has been slow to adjust to. A number of large generators are suffering financially, and uncertainty is curtailing investment in new stations. Some distribution networks are reaching saturation in their ability to accommodate more SGs. The distribution networks have become more active rather than passive requiring a more sophisticated monitoring and control infrastructure to manage security and the impact of intermittency. The SGs like wind and solar do not inherently provide inertia making frequency more volatile following system disturbances that can cause local trips. The reduction in supergrid generation reduces the reactive support to manage voltages necessitating switching out some transmission circuits at times of low load. The distribution control framework needs to interact with an increasing number of players and has yet to be established.

One of the reasons for the growth in SGs is related to increases in network charges that in part are the result of the extra funding required to develop the network to accommodate the new renewable generation sources. A contributing factor is the charging arrangements for use of system where the transmission charges to the host supplier are based on recorded peak demands. The embedded generator can reduce these charges by running at the peak times to reduce the import and sharing the benefit with the supplier. The smaller generators also do not usually contribute to transmission costs and may avoid emission charges, other than were they are added to fuel costs.

Some distributed generation schemes involving combined heat and power can operate at efficiencies of up to 85% where there is a local demand for heat. This applies to both large-scale systems of tens of MWs as well as systems with embedded generators of 100 kW. These systems can be very cost effective and enable arbitrage between local and grid supplies.

A comparison of unit costs shows that the smaller generator costs may be 30% to 35% higher than from large-scale units. An analysis of recorded data confirms economies of scale with larger CCGT units being 50% cheaper/kW and 50% more efficient than smaller embedded units. A key differentiating factor, in the financial analysis of small embedded units, is the inclusion of network costs in the avoided electricity price. At the distribution level, the network costs can equal the energy costs in the tariffs to end users. It is not surprising that energy from local generation appears cheaper. However, the local installation will require backup from the network and will rely on the network for frequency and voltage control. They may also export surplus energy to the grid, and it could be argued they should pay for that option. It appears unjust that a neighbour has to contribute to subsidies for the energy from a neighbour's solar panels. Other factors to be considered in relation to SGs are the emissions particularly from local diesel generators that may be located in densely populated areas.

The current popularity of small generating units is partly fuelled by the structure of regulation, and this will likely be changed. The desire for end users to take control of their energy supplies is understandable given the recent rises in costs. But this needs to be founded on accurate and stable information on the developing regulatory environment. The industry needs to rethink the network tariffs that should be applied to local generation embedded within the distribution network to better reflect the longer term costs that will be incurred in managing their introduction onto the power system.

20.6   The nuclear option

The market for nuclear technology is expanding with 65 plants under construction in 14 counties. In countries like China with growing pollution problems, there is an urgent need to reduce dependence on coal-fired generation that could not be quickly met from renewable sources. By 2015, China had 27 nuclear generating units in operation with a capacity of 25 GW with another 25 units under construction equivalent to a third of the world total. The planned 58 GW of nuclear generation operating at 85% load factor will produce energy equivalent to about 250 GW of wind power installations.

An estimate of costs results in a unit price of £88/MW h and consistent with the strike price for energy from the latest UK station of £92.5/MW h. Based on estimates of prices from conventional generation sources of £56.5/MW h, this results in a premium of £36/MW h. Based on the expected output, this would result in an extra cost to consumers over 35 years of £30 billion or around a £1,000/consumer. If gas from 'fracking' keeps gas prices low, then the subsidy could rise to £40 billion.

The development of large-scale nuclear stations appears to be often fraught with problems and delays. Small modular reactors (SMRs) have been developed and used for many years to power nuclear submarines. Based on this design, commercial nuclear stations are expected to be a tenth of the size of a traditional nuclear power station and at least a fifth cheaper. Advantages are that units could be built in a factory and installed underground. They could also be located and used to meet the needs of more remote areas.

SMRs represent a major opportunity for the nuclear industry to contribute to meeting targets for the world transition to a low carbon society. A preferred design and licencing requirements need to be established to pave the way for prototype installations. The smaller units could potentially be designed to offer more flexibility than the larger counterparts and increase their potential market share. Against small distributed nuclear stations is the problem of managing site security. Large-scale nuclear parks are attractive in facilitating the management of site security as opposed to plant being scattered over several sites.

A longer term option is fusion generation, and experimental installations have been operating to support development for many years, and it is estimated that plants could be operating by 2040. It has been predicted that energy prices would be comparable with existing sources with the added advantages of no long-life radioactive waste and more readily available fuel supplies.

The market for nuclear technology is expanding with 65 plants under construction in 14 counties with a lot in China, India and Russia. In 2016, there were some 440 nuclear power reactors operating in 32 countries producing 11% of the world's electricity. In contrast, a few countries like Germany and Japan are choosing to reduce their dependence on nuclear because of safety concerns. However, given the potential increase in demand for carbon-free electricity to supply the growth in electric vehicles and heat pumps, there is an expanding need for nuclear generation. In particular, it will contribute most economically if operated in baseload mode.

20.7   Carbon capture and storage

Given that some 50% of the world's emissions result from the use of gas and coal to provide energy CCS offers a viable alternative to reduce emissions in the short-to-medium term. It will not be easy, financially or politically, to wean countries like China and India away from the use of coal with pit closures, putting miners out of work and closing stations without alternatives being available. Particularly, when there are estimated to be hundreds of years of cheap coal available that could be exploited. There is the option to retrofit CCS to recently built stations and begin to progress decarbonisation in the short term. It will require positive regulation to ensure that the reductions in station efficiency are more than compensated for by financial benefits from an emission reduction scheme. There will also be a need for support in the establishment of the shared CO_2 transport and storage infrastructure to minimise costs. The overall costs, including provision for CO_2 transmission and storage, are estimated at £97.6/MW h for gas and 86.5/MW h for coal stations based on increases in fuel prices of 50% from values in 2017. These costs are of the same order as new contracts for large-scale nuclear and cheaper than offshore wind generation. The economics are improved if the CO_2 is pumped into partially depleted oilfields to realise enhanced oil recovery.

The CCS generation can make a valuable contribution to the generation mix because its output is controllable, and it can be used to help balance the output of intermittent renewable generation. The inclusion of a tranche of CCS generation reduces the overall system capacity needs and does not require backup like intermittent renewable sources. The CCS generation would take advantage of existing generation sites with transmission connection already established as opposed to the need for increasing amounts of transmission to remote wind sites. Modelling overall capital and production costs to meet demand for a system with a peak of 50 GW shows that annual costs are reduced by 17% with a tranche of CCS generation as opposed to one based on high proportions of subsidised intermittent wind generation.

The Global CCS Institute website reports 22 large-scale CCS installations in operation or construction with the capacity to capture 42 $MtCO_2$/year. The technology components are generally available, and one large-scale plant in the USA is capturing 1.4 $MtCO_2$/year. The CO_2 is used for EOR with 130 injection wells and 130 production wells that could be used over the 20-year life of the project.

20.8 Wholesale markets

The performance of wholesale markets has been undermined by the introduction of increasing renewable generation capacity that receives priority in dispatch. Prices have become more volatile and less predictable due to renewable output changes that are not linked to demand. The renewable generation output may be increasing when the demand is falling making prediction of the demand and marginal prices very difficult. The result is to place a premium on flexibility in generation and demand.

As the renewable capacity is increased, the utilisation of marginal plant is reduced to the point where it becomes non-viable and closes. However, it is still required to meet demand when the renewable output falls, and to retain its availability, it has been necessary to introduce capacity markets. At the same time, fossil-fuelled generation is subject to additional charges related to CO_2 emissions causing coal-fired generation to be prematurely closed. In contrast, the network charging arrangements have promoted the development of small embedded generation connected to the distribution system. These market distortions have had an adverse effect on generation companies that have been recognised by regulatory authorities who have proposed a number of changes including

- The removal of priority dispatch for renewable generation;
- The removal of 90% compensation when their output is curtailed;
- The requirement for self-generators to contribute to system costs.

In the current environment, governments are faced with a spectrum of objectives including securing future supplies, building up renewable energy to meet emission targets whilst maintaining fuel diversity and containing end-user bills (the so-called trilemma). **The view of the author is that in a rapidly changing situation the single-buyer (SB) model offers the best approach to meeting these diverse objectives.** The market would act as an interface between the generation side and an active demand side. The process would be augmented by a future capacity market including flexibility ranges with costs. It would realise the benefit from competition in generation and provides a mechanism to manage the evolvement of the plant mix and the capacity margin as well as acting as the counterparty for large-scale capital developments like nuclear or hydro. Rather than the SO defining the system generation capacity requirement, it would be established by the demand and generation side contracting directly. **The demand side would need to contract forward for capacity taking account of its capability to smooth overall demand using DSR and embedded generation.** This is essential to realise the full benefits of demand smoothing in reducing capacity requirements.

The SB model is appropriate where there is a need to attract new entry to bolster the plant margin but at the same time ensure fuel diversity and supply security. It also affords the opportunity to retain tariffs on the basis of average rather than marginal costs to meet wider social objectives. Competition in this environment can be managed through auctions for new capacity as well as for additional tranches of energy import at each annual, monthly and the day-ahead stage. New entry can be fostered by the sale of existing generation clusters backed by power purchase agreements, for a limited period, pending wider market liberalisation. A large residual state generator can be retained to manage balancing and also provide make-up and spill options for smaller independent power producers to balance their commitment to end users.

The SB model is not too difficult or costly a mechanism to establish and can be implemented relatively quickly. Subsequently, the market can be developed further by enabling some direct contracting between generators and end users. This could initially include supply to works on the same site or the definition of a percentage of output capacity that should be traded directly to provide price discovery. The security that this direct trading option affords may appeal to large end users operating continuous processes.

20.9   Balancing markets

The growth of renewable generation is having a significant impact on the requirement for reserve in balancing supply and demand. The SO defines the requirements for these facilities and calls them off to maintain secure operation and restore services. The primary reserve for immediate frequency change containment following loss incidents is the most expensive at around £20/MW/h reflecting the need for very fast response. Reserves available for automatic or fast manual dispatch to arrest frequency fall are around £6/MW/h and manually dispatched reserve to restore normal frequency costs around £3.6/M/h based on less stringent response requirements.

An enquiry by the SO for fast acting primary reserve for frequency deviation containment elicited responses based on lithium-ion batteries leading to 200 MW being contracted with eight suppliers at costs averaging £328/kW. The specification called for full output for just 15 min, and the facilities could not contribute to support wind energy balancing over hours.

The sources of short-term operating reserve (classed as tertiary) include OCGTs 44%, diesel 37%, pumped storage 9%, CCGTs 7%, biomass 4%, hydro 4%, CHP 3% with just 4% from the demand side. The annual costs for the UK 60 GW system were £98m in 2015 with £58m for availability and £40m for utilisation. It is worth noting that the tender round was oversubscribed with a third being unsuccessful. This indicates a competitive market operating with a limited requirement. This has caused prices in Germany to collapse by 50% for secondary and tertiary reserve although the capacity requirement has barely changed. It is envisaged that as the use of EVs grows, they could provide an additional source of reserve when connected to the grid. A million vehicles each with a battery capacity of 50 kW h contains 50 GW h of energy that could provide tertiary reserve with an output of 2 GW that over 1 h would deplete the total charge/vehicle from 50 kW h down marginally to 48 kW h/vehicle.

The requirement to provide reserve to manage intermittent renewable generation output is a function of the forecast accuracy. The RMS forecast error in scheduling timescales is expected to improve from around 17% initially to 13% by mid-decade down to 10% by 2020/21. Based on three standard deviations, the equivalent reserve requirements would be 51%, 39% and 30% of the average wind capacity output. The output depends on the load factor, assuming 30% for 20 GW of wind, the average output would be 6,000 MW less 500 MW catered for by other provisions = 5,500 MW requiring 30% reserve, i.e. 1,650 MW. The same assessment applied to Germany gives results similar to their provisions based on lower load factors of 17.6%.

There is also an increase in reserve provisions to cater for variations in interconnector flows. This is likely the result of variations in renewable generation output that has not been balanced locally. Germany is a known offender in this capacity where variations in the output from its large wind capacity alter planned interconnection flows and disturb the secure operation of the adjacent networks in the Netherlands and Poland resulting in complaints at government level. The high levels of renewable output also impact on market prices. At times, the solar energy output leads to negative energy prices in some southern states of the USA.

20.10 Capacity markets

There is a fundamental limitation in fully liberalised markets in using short-term energy markets to guide the development of the optimum level of generation capacity to meet future security requirements. The simplistic economic approach is that market price rises will encourage new capacity. But, there are increasing levels of government intervention in the market in the form of subsidies, capacity payments and contracting for nuclear generation aimed at meeting wider environmental and security objectives. Understandably, these have the effect of undermining the market and general investor confidence leading to potential capacity shortfalls in future years. The expansion of intermittent energy sources, embedded generation, demand side regulation and interconnection as discussed in this book further add to the complication. There are also questions to be addressed related to the interaction of intermittent sources with less flexible generation like nuclear and CHP that affect the plant mix. At times of light load, wind output may need to be curtailed to accommodate inflexible generation and maintain sufficient conventional generation reserve in service for the system operator to maintain regulating capacity and secure system operation.

There is an urgent need for the development of an overall indicative energy plan that analyses and quantifies all the issues and enables investors to analyse their options and assess their risks. There are examples from other countries of publishing medium-term energy plans providing a background to investment analysis. In the current situation, the subject area is too complex to assume that all will be resolved by market mechanisms, and it is too open to influence from lobby groups with particular vested interests. The mixture of generation development also has significant implications to the future requirements of the transmission and distribution systems. The overall costs may run into hundreds of billion pounds, and a mechanism is needed to encourage development that is close to the optimum to maintain business competitiveness.

There are a number of areas of research that are necessary to support future regulatory policy related to setting security standards:

- Establishing the capacity contribution to system security that can be expected from renewable sources;
- Facilitating demand side participation in energy and capacity markets through the application of smart meter installations;
- Determining the levels of security that should apply at the transmission and distribution levels consistent with consumer needs and overall system security;
- Ensuring access to the networks and markets by new local embedded generation sources and energy storage systems;
- Restructuring the regulatory price review processes to accommodate the changes in system infrastructure and facilitate innovation;
- Determine a regulatory regime appropriate to the application of smart meter technology to ensure consumers receive a share of any benefits.

As distribution networks become more active with embedded renewable generation, CHP schemes and domestic heat pumps, solar energy systems and electric vehicles, modelling capability will need to be developed to establish the collective impact on demand levels and profiles. The regulatory agencies will play an expanding role in building a framework to meet the changing environment whilst protecting consumer interests.

The probability analysis in this chapter indicates that wind generation contribution to firm capacity is equivalent to around 15% of the installed total capacity with load factors of 30% falling to 12% with 25% LF. This was confirmed by a full year half hour dispatch with and without wind to identify the conventional generation displaced. It is also noteworthy that on cold days in the United Kingdom, there is often no wind.

Analysis of plant margins will often assume support from interconnectors that may not be available from adjacent countries that are also experiencing adverse weather. Support from interconnectors can only be guaranteed when firm capacity contracts are in place for both generation and interconnector capacity with the neighbour.

The UK capacity scheme was intended to encourage the build of new efficient combined cycle gas-fired generation but failed to do so with new small-scale diesel and gas generators bidding capacity at lower prices. The outturn price of the first auction at £19.4/kW/year was much lower than the government estimate of £49/kW/year that was based on open cycle gas turbine prices. The net result has been to end up with the lowest forecast plant margin on record. The UK capacity prices compare with the average capacity prices realised in capacity auctions in the PJM auction of £21.4/kW/year in the USA.

The contribution to capacity provided by demand side response amounts to 54% of the total procured by PJM in the USA. This may in part be realised by contracting interruptible air conditioning load. The scope in the United Kingdom is currently limited but may change with the take-up of electric vehicles and heat pumps.

20.11   Interconnection

The expansion of interconnection capacity in Europe is targeted to be 10% of generation capacity rising to 15% by 2030. The objective is to facilitate cross-border trading resulting in prices convergence. It will also provide benefit in enabling reserve sharing and in managing the intermittency of renewable generation through flexibility sharing.

There is a danger in crediting interconnection capacity with a contribution to the plant margin. There is no guarantee of availability of generation or link capacity unless it is backed by firm capacity contracts. Adverse weather may affect wide areas influencing the availability of generation in neighbouring countries at the same time reducing any spare capacity.

The evaluation of the potential benefit of trading across a proposed new link needs to be based on a comparison of expected marginal prices through the life of the link. This in turn requires an analysis for each system of future demand, fuel prices and generation additions and closures. The respective profiles of demand will also be relevant in evaluating potential within day trading opportunities. The process is made more complicated with interconnected networks with associated loop flows. In some instances, where the plant mixture is very different, there may be clear opportunities e.g. exploiting Norwegian Hydro to balance wind intermittency, using spare French nuclear capacity to displace the use of oil in Italy, a link from Iceland to the United Kingdom to export spare geothermal energy.

From an investment perspective, it's the long-term potential for energy trading rather than within day trading that will determine the likely return for sponsors. This can be reviewed by an analysis of historic energy flows and an understanding of their rationale. The process provides an indication of the average annual energy flows around Europe that is useful to support initial investment analysis and expected utilisation of proposed new generation.

Interconnection can also support the wider use of renewable generation. There are technical constraints on the proportion of non-synchronous generation in operation at any one time. These relate to ensuring system reserve capability to cater for changes in renewable output and managing frequency changes following sudden loss of generation. The non-synchronous renewable generation does not contribute to the system inertia that slows the initial frequency change. The rapidly falling frequency may cause other generation to trip initiating a cascade in loss of capacity. In Ireland, a proportion of non-synchronous generation of up to around 50% is considered just viable based on practical operating experience. The availability of additional interconnection to England would enable more renewable generation to be installed in Ireland and Scotland with the excess energy exported.

The EU sponsorship of interconnection development is based on the perceived potential improvement in market coupling. But, progress on realising a fully integrated European-wide single market has been slow and the transformation now taking place in the industry is likely to hamper harmonisation because of the large increase in the number of players that are largely autonomous.

20.12   Demand side management

Demand control has always been used in the power sector primarily to reduce short-duration demand peaks. At the system-wide level, techniques include tripping demand on low frequency relays, direct control of contracted consumers demand, pumped storage schemes and voltage reduction. The applications have generally been restricted to larger consumers where the control infrastructure costs/MW is small compared to the domestic sector. Applications at the distribution level have generally focussed on local peak demand reduction to defer the need for reinforcement. The UK domestic demand represents around 30% of the total demand but with some 95% of consumers indicating the scale of the problem.

In the current situation, the wider application of demand side control is driven by the need to manage the intermittency of increasing levels of renewable generation. In its wake will follow the expected increase in network and generation capacity requirements to support decarbonisation of heat and transport. The roll-out of smart meters offers the potential to engage the smaller consumers in this demand control through time-of-day tariffs or through participation in direct control of selected loads. The current scope would be limited to a few GWs and would not be sufficient to smooth the output of a large tranche of wind farms. As the wind capacity reaches around 50% of the conventional capacity, there are periods of low system demand when wind output has to be curtailed. Detailed modelling showed curtailment average levels of around 3 GW over 1,600 half hours/year costing over £300m/year in lost FIT payments. This could be managed by 10 million consumers increasing their demand by just 1 kW to avoid the need for any compensation payments worth £30/consumer/year. The potential in the domestic sector is expected to increase further as and when electric vehicles and domestic heat pumps are deployed. A fleet of a million vehicles would, depending on the time of day, have a charge demand varying from around 400 MW up to 2,800 MW that could be regulated. There is currently scope to influence the design of these schemes to include sufficient energy capacity to support DSR.

Storage schemes based principally on pumped hydro have featured in power system operation as a mechanism to lop peaks. The wider application is limited by the cycle efficiency that may be as low as 70% necessitating a wide spread in peak/trough prices for arbitrage to be viable. Lithium battery technology is being deployed in small volumes to provide fast response to arrest frequency changes following incidents. These applications are viable in specialist areas where the speed of response is critical but would not, at current price projections, be viable smoothing the output of large tranches of wind generation.

Current tariff design is very varied and does not specifically encourage flexibility. There are wide variations in the proportion of charges related to capacity compared to energy. In Hungary, there was no capacity charge with Estonia at 77%, while the energy part varies from 100% in Italy down to 11% in the Czech Republic. A high-capacity charge encourages consumers to manage their own peaks, but in balancing renewable output, the requirement may be to increase demand to use up excess wind/solar generation. A variable time-of-day tariff is required to encourage demand profiling but consumers will need some revenue certainty to justify their investment in flexibility. This could in part be covered by a participation fee linked to the range of flexibility offered.

20.13   Interaction with gas, heat and transport

Whereas emissions in North America and South and Central America fell by 2% and 2.4% in 2016, those in the Asia-Pacific region increased by 0.9% as did those in the Middle East 1.6% and Africa 1.1%. This resulted in a net global increase of 0.1%. The emissions from the Asia-Pacific region were almost equal to those of the rest of the world reflecting a dependency on coal. US emissions at 1,925 $MtCO_2$/year were less than those of China at 2,943 $MtCO_2$/year. The United Kingdom had a target to reduce emissions from 247 $MtCO_2$/year by 34% and has already exceeded this with levels of 112 $MtCO_2$/year, i.e. about 4% of those of China. Electricity production is responsible for 37% of the emissions in the USA compared to 34% in the United Kingdom. This reflects a continuing reliance on coal for electricity production in the USA.

Some 20% of total emissions result from domestic use for space and water heating. To decarbonise heat, one solution is to use heat pumps that extract heat from the environment increasing the overall efficiency by a factor of three. In operation, the heat pump is cheaper than a conventional boiler but the overall costs are much higher because of the capital costs and the high price of electricity compared to gas. District heating schemes can be viable if there is a local source of waste heat and a dense group of potential users, and they are widely used in Austria, Denmark and Copenhagen.

Electric vehicles provide the opportunity to reduce emissions from transport and also to support balancing the intermittency of renewable sources through demand management and exploiting their storage potential. The reduction in emissions depends on the generation used to charge the EV batteries. Where conventional CCGTs are used, the net reduction compared to a petrol car is about 38% with half the operating costs. The use of local solar panels for charging results in no emissions or extra costs for those that have them installed.

The gas grid has massive potential to store energy and could be used to store excess renewable energy. The limiting problem is the overall cycle efficiency of around 38% of producing gas that is then used to generate electricity. It could be a viable option to use excess renewable energy that would otherwise have to be curtailed.

Combined heat and power schemes can operate at high efficiencies if the waste heat can be utilised. This generally results in small schemes with a local requirement for heat that are relatively more expensive than larger generating units. The cost effectiveness is made more attractive by network charges being avoided when compared to grid prices for electricity.

The requirement to decarbonise heat and transport will have a significant impact on demand for electricity. A 10% reduction in UK heat and transport would require 28.4 TW h/year, to realise 100% reduction would require 284 TW h/year and virtually double the UK electricity demand. However, there is an opportunity to employ a more integrated approach to their management to realise a more optimal provision of energy that exploits the inherent capabilities of each sector. It is considered that the provision of heat storage offers the potential to smooth domestic demand profiles at the lowest cost with little loss of energy. A longer term view is necessary to determine what needs to be in place to facilitate development.

20.14   Network issues

During 2007/08, the EU pressed for ownership of the networks to be unbundled to establish truly independent SOs able to manage operation of the system without any prejudice. Ownership unbundling is seen as the most effective means of avoiding discrimination. Similar initiatives have taken place in the United States where the Federal Energy Regulatory Council orders 2,000 required utilities to hand control of their networks to an independent entity. These initiatives have been coupled with the introduction of competition in the provision of networks as well as engaging the SO in the network development process. The high use of system costs for the provision of networks to system users has become an important factor in determining their choice of local vs remote generation.

Distribution networks have traditionally been based on consumer after diversity maximum demand (ADMD). For the average domestic consumers, the ADMD may be below 2 kW, but this natural diversity may be destroyed by a number of factors – time of use tariffs, local solar panels all exporting and local wind turbines all generating. The management of the network utilisation would have to be fostered by a process of real-time pricing that reflects both energy prices and network limitations.

Use of system charges has traditionally been based on the level of utilisation with both the network and energy charges reducing as it increases to the benefit of larger users. This basis of charging does not cover the requirements of embedded generation. The current tariffs for embedded generators reflect the fact that their operation may reduce network demands at peaks, and hence, the transmission use of system charge and also network losses, and they may even be negative. The introduction of embedded generation also alters the network security assessment and management.

The other key change is the increase in demand that can be expected with the take up of electric vehicles and heat pumps that will have less diversity than more conventional loads. Significant savings could result from managing the increased demand to provide a smoother profile that better exploits both the network and generation assets.

The implication to the distribution networks is that they will move from being essentially passive to requiring active management along the lines of the transmission system. The IREN2 project in the Wildpoldsried area of Germany has demonstrated island operation of a microgrid serving 2,500 consumers. The additional costs incurred will need to be reflected in the use of system charges to embedded generators and system users.

It is suggested that changes in the use of networks will require new functionality that operators need to react to. There are many more participants in the sector that will bring new requirements. Some of the developments that are expected to have an impact are

- A shift from connecting large generation at transmission voltages to connecting more smaller generators at distribution voltages;
- The small embedded generators are not normally dispatched and may be wind- or solar-powered exhibiting intermittency;
- Smart metering offers the opportunity for real-time pricing of energy encouraging changes in the network profile of demand and peak lopping;
- New players in the form of aggregators will wish to establish interactive arrangements for managing load;
- There could be a significant increase in demand resulting from the take-up of electric vehicles and heat pumps with less diversity than normally assumed.

The implication to the distribution networks is that they will move from being essentially passive to requiring active management along the lines of the transmission system. The difference is the degree of control that can be exercised over the many potential active participants that could run into several hundred on one distribution group. This in turn will make it very difficult to plan for network development. End users will not expect to seek permission as when to run their embedded generation or when to charge their electric vehicle or alter their demand. It would also be impractical to manage. **Management of the network utilisation would have to be fostered by a process of real-time pricing that reflects both energy prices and network limitations.** This type of arrangement has been successfully applied at transmission levels with zonal pricing. Time-of-day pricing has also been applied to encourage night-time use of energy to charge heat storage devices. When the network is stretched, prices would be high to discourage use and the use of own generation. At lower demands, prices would be low encouraging energy import and less self-generation. The import/export prices would be determined by conditions on the supergrid. The distribution system would need a monitoring and control infrastructure interfaced to the national control centre as well as reinforcement. The costs should be levied on the system users that are creating the change in requirements including embedded generators and storage operators.

Network planning in this environment will be challenging with difficulty in predicting the many small-scale developments that can be expected. One scenario advanced by NGC for the United Kingdom suggests that by 2030, there could be 3.3-million EVs, 6.6 million heat pumps and 71 GW of solar and wind capacity generating 39% of total electricity. Given this potential rate of change, the distribution companies will need to actively control development and define connection agreements that will meet longer term requirements for the exchange of data with active participants. This in turn will need to be facilitated by a regulatory and governance framework that embraces research and takes a whole system view. There will also be technical issues that need to be analysed and resolved at the planning stage related to voltage control, flicker and harmonics and the network protection taking account of the effect of rate of change of frequency on embedded generation, all the issues that affect transmission system development. This will inevitably lead to a step change in distribution charges.

20.15   Future scenarios

The basis for the development of future strategy is to realise the three objectives – affordable energy, security of supply and a sustainable low level of emissions. The latter has proved the driving force leading to the introduction of tranches of subsidised renewable generation that has had a marked impact on other generators and network requirements and charges. A consequential effect has been a growth in small-scale embedded generation in preference to centralised generation. This is resulting in the distribution networks becoming more active and to stretch their capacity to accommodate more generation. In part, the generation transition is the result of current tariff structures that need to be reviewed to emphasise the flexibility and network capacity offered to generators and users. In sharp contrast, the priority dispatch afforded to renewable sources has reduced the utilisation of conventional generation operating at the margin to the point where it becomes financially non-viable. The uncertainty has led to a dearth in new investment and to premature plant closures resulting in unprecedented low plant margins.

The introduction of smart meters was aimed at fostering efficiency in the utilisation of energy but is now perceived as facilitating a much wider end-user engagement. Not all countries see the same scope and are restricting roll-out to larger end users. Less than 50% of UK consumers offered the meters chose to accept. Some commentators have that view that customer demand control will significantly reduce the need for generation capacity. The United Kingdom has an energy requirement of around 330 TW h/year with a peak demand approaching 60 GW. It is axiomatic that if the demand were completely flat, the capacity requirement would be only 37.7 GW. The mechanisms to realise this utopian position or the costs are less obvious. The growth of new demands for EV charging and heat pumps offers a limited scope if end users are prepared to allow third parties to control their use. Equally, generation flexibility will be required, and its value needs to be fully rewarded. In contrast to intermittent renewable generation that creates the imbalance problem, gas generation with CCS offers flexibility with low emissions.

The transitions taking place are leading to a massive increase in the number of market players on the generation and supply side as well as in network service provision. In the current market structure, there is no agency with responsibility for taking a longer term holistic view of the evolving industry and the attendant risks and opportunities. It is glibly assumed that the market will solve all with governments and regulators currently responding to events rather than taking a lead. There is an argument at a time of rapid change for a SB market model acting on behalf of all users to procure generation and manage the risks. The analysis in this chapter illustrates the scale of the impact of assumptions on key parameters on outturn costs and emissions. The SB would also be better placed to react to sponsor and exploit new developments in generation and storage and climate changes.

As well as affordability and containing emissions, the third requirement is to maintain standards of security. This requires establishing the right level of generation capacity to realise a LOLP set typically at around 4–8 h/year. An added complication in the analysis is estimating the contribution that can be expected from intermittent renewable sources and embedded generation. Opinions vary but probabilistic analysis shows values of just 12% to 15% of installed capacity for wind and nothing for solar generation when peaks occur in the early part of dark evenings in the Northern Hemisphere. There is an EU policy of extending interconnection but its potential contribution may be limited when weather patterns span several neighbouring countries. The potential contribution from embedded generation will be influenced by local demand conditions, and typical availability figures can be used as discussed in Section 14.4. The use of capacity markets is promoted as a way of meeting requirements with the onus on the grid operator to assess requirements. To realise the reduction in capacity needs, it should be suppliers who contract for capacity in the knowledge of their demand management capability.

20.16   Scenario evaluation

This chapter has illustrated an approach to establishing a development strategy that meets emission target reductions at minimum cost. It also demonstrates the analysis of the risks to the outturn due to erroneous data assumptions. In this example, the results suggest that fast development and deployment of generation equipped with CCS offers the prospect of low levels of costs and emissions. It suggests that a lower proportion of wind generation (20–25 GW) coupled with 15 GW or more of CCS generation would have the advantage of both lowering costs and emissions and avoiding the need to frequently curtail wind output to secure system operation.

- The overview LP solution showed minimum total costs with a significant tranche of CCS plant;
- The model base case studies with a 'fast CCS 2' mix confirmed total costs some £7bn/year cheaper than the 'Green' scenario;
- Emissions in the new mix with more CCS are 45% of the 'Slow' scenario and the same as the 'Green' scenario assuming a 90% CCS capture rate;
- The report assumes that the development of CCS technology could be accelerated and could be made commercially available within the timescale analysed;
- It appears that the use of storage is more financially viable when operating in a market environment buying and selling energy although barely so at current capital costs.

There are a number of other issues that could have an impact on decisions on future technology development:

- With high levels of wind capacity, more energy has to be curtailed during low-demand periods and may have to be paid for. This could add £4bn/year to a 'Green' scenario assuming CfD strike prices of £155/MW h for offshore wind and £95/MW h for onshore;
- The costs of the extra transmission infrastructure to support remote offshore wind installations are not included;
- The intermittency of renewable energy adds to system operating costs and emissions and the need for reserve plant;
- Generation with CCS is flexible and extra output could be realised at times of peak by limiting the CCS process energy use;
- CO_2 storage in partially depleted oilfields may facilitate enhanced oil recovery.

Those engaged in formulating strategy have a difficult task with so many variables and market participants. There is a need for analytical tools to review the risks and range of potential outcomes and identify actions that avert the worst case scenarios. To support this process, an automated series or studies with varying key parameters was used to identify the probability of high outturn costs.

20.17 Trading in an uncertain environment

Generators are struggling in many areas to maintain financial viability faced with loosing utilisation to renewable generation with priority dispatch. At the same time, a number of changes are taking place that complicate predicting future demands and prices and undermine the energy trading process. The unpredictable nature of renewable generation makes the prediction of the net demand to be contracted from conventional generation difficult to define until close to the event. The growth in local embedded generation, that may be used to offset demand, impacts on the grid demand and is not explicitly monitored. This distorts the historic demand profiles that have traditionally been used as a basis for the prediction of future demand and prices. There is also the prospect of much more consumer engagement in demand management that will further add to the difficulty of predicting demand and prices.

In assessing investments, there is a need for longer term demand and price forecasts that are often based on models linked to historic records of GDP but these now need to be adjusted to take account of policies and measures to improve the efficiency of energy use. Imports and exports are also less predictable where neighbouring countries are equally exposed to increasing volumes of intermittent generation that disrupt planned transfers. Where there are strong links in place, it is necessary to model the adjacent systems to establish comparative price profiles.

In predicting utilisation, it is necessary to place the proposed development in merit order along with competing stations to estimate the utilisation. This uses an annual dispatch model for selected years based on a load duration curve showing the utilisation of stations according to their variable price. Given the costs, estimates can be made of expected revenues and profits. It is necessary to establish the risks by running a number of studies varying key parameters to establish a histogram of possible profit outturns. Decisions are often based on an assessment of the value at risk with a 90% or 95% probability. The fuel price risk is often based on the relationship between system marginal prices and each fuel type. The dispatch process is used to establish the proportion of the year that each fuel type operates at the margin. This then provides the basis for estimating the impact of a fuel price change on the average system marginal price. This approach is satisfactory for relatively small changes but larger changes require reordering the merit order.

Some stations also have the option to exercise arbitrage. A gas-fired generator with a take or pay gas contract has the option to either use the gas to generate electricity or sell the gas back to the market if prices are high. A resulting complication is that if gas prices are driven high due to shortages in winter, it can cause a lot of gas-fired generation to choose to sell gas and be withdrawn from service creating a power shortage. Some generators buy fuel on a contract that enables short-term interruptions of supply. This can also lead to power system problems if gas shortages lead to the interruption clause being invoked on a number of generators at the same time.

A number of analysts hold the view that market prices can be predicted on a purely statistical basis. In practice, prices are often stratified where they take different levels depending on the fuel type operating at the margin. The nature of this type of price function does not lend itself to statistical analysis and a more structured approach is required.

In developed markets, trading is undertaken just based on prices. Generators are expected to include other costs due to start-up and part loading within their price bid. This can be a complicated process in short-term daily trading when the runtime is not known. To account for this, trades will often be linked to specific run periods of typically around 4 h.

20.18   Smart grids development

The distribution networks are the focus of changes taking place in the energy sector to enable end-user engagement in realising affordable, secure and sustainable supplies. As their use increases, the capacity limits of the distribution networks will be reached and new tariff structures will be needed with more value placed on flexibility and pricing access to capacity rather than just energy. In contrast, the overall uncertainty in the sector has led to a dearth in investment in large-scale generation leading to unprecedented low plant margins. The proportion of large-scale supergrid-connected generation has already fallen by 10% in the last 5 years. The requirement for increased flexibility to meet the increasing demands and manage intermittency could, in part, be met by gas-fired generation equipped with CCS without increasing emissions if viable tariffs are established that recognise the value of flexibility. There is considered to be a world-wide market for CCS technology that has been estimated at $31bn by 2021 offering export potential for early developers.

In current market structures, there is no agency forward looking with responsibility for taking a holistic view of the developing industry and assessing risks, and it is glibly assumed that the market will solve all. Given the growing number of interested parties, governments should consider the establishment of a SB market model to fulfil this role. There are also a number of external potential developments that could drive development paths including health concerns resulting from emission, technology developments in storage, CCS or fusion and changes in global warming. These factors need to be considered in risk assessments by a lead organisation.

Security standards are usually defined as a LOLP and are typically set at around a 4–8 h/year in developed economies. The assessment of security is now complicated by the presence of a tranche of intermittent generation like wind or solar necessitating a probabilistic approach. Analysis indicates a wind contribution to system capacity of between 12% and 15% and of a similar order to the analysis undertaken in Ireland. An enhanced distribution SCADA system will be required to manage security including a monitoring and control capability and coupling to transmission control and potentially links to end-user internal systems. Since 2002, some 950 projects related to the development of smart grids have been reported across 50 countries in Europe with an investment approaching €5bn. These systems embrace network and demand side management and storage and their operation needs to be embraced in overall control to realise a holistic approach. Fundamental changes to distribution tariffs will be needed to support the costs of providing and operating a more active network with embedded generation and behind the meter generation.

There will also be a need for a local market platform to facilitate coordinated operation. If the claims of reduced capacity through demand side management are to be realised, the demand side would need to contract into a capacity market that recognises changes from the user normal time-of-day demand. It is expected that the development of distributed generation will mostly be sponsored by local consortia bringing together users and suppliers establishing multilateral deals embracing electricity and heat supply coupled with demand management and storage. This will increase the scope required in local market arrangements to engender price discovery and competition. Aggregators would trade flexibility on behalf of their customer base either to suppliers or directly by taking advantage of time-of-day market price variations.

To manage the ensuing significant change, there is a need to clearly define respective roles and interface arrangements. A starting point would normally be to establish a business process definition.

20.19   Optimum development strategy

A major concern is the implication of a large number of new participants in the energy sector whose actions are not coordinated or regulated. The approach is to encourage end user engagement based on the philosophy that this will promote more competition. A lot of the developments in the electrification of transport and heat are focussed on the distribution networks where end users can make unilateral decisions. Whilst there is recognition of the need for a holistic approach there is no process in place by which it may be realised. In this situation developments will be impeded by the rate at which networks can be transformed to match the transition in its use. Various models have been developed that are designed to provide insight into the wider impact of user driven developments. Inevitably their use is limited by the availability of a central data set related to future conditions and regulator intervention. They also have insufficient detail to underwrite specific investments.

A number of actions have been identified in this chapter that are considered necessary support the transition process including:

- Priority dispatch for renewable generation should be removed and subsides phased out.
- The network tariff structures should be reviewed to provide a level playing field for all generation irrespective of connection voltage.
- An intelligent home hub should be developed enabling users to control their demand in response to TOU market price signals.
- A distribution system operation function needs to be established with enhanced facilities and links to the TSO.

- The non-synchronous generators should not receive compensation when their output is curtailed for system operation reasons or network constraints.
- A flexibility market should be established to meet system requirements for regulating capability.
- Embedded generators should be subject to the same charges as transmission connected generators related to emissions and the provision of system services.
- There is a need to tailor the plant mix to match the expected duty in tracking the future demand profile.
- All grid connected generation should be centrally dispatched with embedded generators reporting their planned output.
- The commitment to priority dispatch of intermittent renewable sources should be urgently reviewed together with the technical limitations on the manageable proportion of wind capacity in system operation.
- Tariffs should be developed for consumers to provide demand flexibility that could be applied to EVs and heat pump loads.
- Consideration should be given to exploiting the gas grids capacity to store surplus electrical energy from renewable sources.
- A Single Buyer/Seller market model should be established to manage the optimal procurement of energy for all users with equitable tariffs.
- A Single Buyer/Seller market model should be considered as a viable option to realise an optimum holistic outcome from the current transition taking place in the sector.
- The capacity market should be two sided with both generators and suppliers making commitments to the known proportion of their future capacity availability and needs.
- Recognising the constraints on the pace of network development there is a need for more projects taking a longer term view.
- A fundamental question is how the activities of the disparate players in the new world can be guided and coordinated to meet the user's requirements in an optimum manner.
- The future generation capacity requirements should be set by the suppliers, rather than the system operator, in the knowledge of their demand control capability.
- Consideration should be given to sponsoring the development of modular nuclear generation and generation with CCS to provide alternatives that can be deployed to meet global requirements to reduce emissions.

Governments should establish organisations that can support the process of formulating an indicative development strategy.

Appendix

Conversion tables

Table A.1   Energy conversion

To From	TJ	Gcal	Mtoe	MBtu	GW h
TJ	1	238.8	$2.388 * 10^{-5}$	947.8	0.2778
Gcal	$4.187 * 10^{-3}$	1	10^{-7}	3.968	$1.163 * 10^{-3}$
Mtoe	$4.186 * 10^{4}$	10^{7}	1	$3.968 * 10^{7}$	11,630
MBtu	$1.0551 * 10^{-3}$	0.252	$2.52 * 10^{-5}$	1	$2.931 * 10^{-4}$
GW h	3.6	860	$8.6 * 10^{-5}$	3,412	1

Relationship between oil and gas price (6:1)
1 t of oil contains between 6.3 and 7.9 barrels, assuming 6.61 barrels then,
1 barrel contains $41.87/6.61 = 6.33$ GJ
since 1 GJ $= 0.9478$ MBtu
hence, 1 barrel $= 6.33 * 0.9478 = 6$ MBtu
i.e. one barrel of oil costs six times 1 MBtu of gas on an equivalent energy basis.

Relationship between GJ and MW h
1 GJ $= 1$ GW for 1 s so to convert to MW h divide by 3,600 and multiply by 1,000
i.e. 1 GJ $= 0.2778$ MW h

GJ to therms
1 GJ $= 440/46.4$ therms $= 9.48$ therms

kcal/kg to GJ/t
Multiply by 4.18/1,000 e.g. 4,000 kcal/kg is equivalent to 16.7 GJ/t
6,000 kcal/kg is equivalent to 25.05 GJ/t

Mtoe to MW h
1 Mtoe $= 4.1868$ TJ $\times 10^{4}$
1 toe $= 41.8$ GJ
1 toe $= 41.8 * 1,000/3,600$ MW h $= 11.63$ MW h

Barrels of oil to litres
1 barrel of oil $= 35$ imperial gallons or 40 US
Equivalent to 158.9 l

Specific net calorific values

CV oil = 41.87 GJ/t
CV coal (bituminous) = 25 GJ/t (varies with source, see Table A.2)
CV lignite (brown coal) = 10 GJ/t (may be between 5 and 15 GJ/t)
CV natural gas = 40,000 kJ/m³ (varies with source, see Table A.3)
Net heat content of gas = 0.9 ∗ gross heat content

Table A.2   CV coal

Coal	Toe/tonne
China	0.531
The United States	0.634
India	0.441
Australia	0.614
South Africa	0.564
Russia	0.545
Indonesia	0.615
Poland	0.551
Kazakhstan	0.444
Ukraine	0.505

Table A.3   CV natural gas

Natural gas	kJ/m³
Russia	37,578
The United States	38,347
Canada	38,260
Iran	39,536
Algeria	42,000
The United Kingdom	39,790
Norway	40,029
The Netherlands	33,320
Indonesia	40,600
Turkmenistan	37,700

Answers

Question 1.1
Given data shown on the costs of the capital and operating costs of gas- and coal-fired generation establish the total generation costs if the load factor is reduced from 85% to 35% due to priority dispatch being given to renewable generation.

Answer: At 35% load factor, gas costs £49.95/MW h and coal £65.91/MW h.

Question 2.1
Given data in the table below calculate the emissions per kW h from bituminous coal- and gas-fired generation.

Using data from Table 2.2 calculate the average kW h/mi for a Tesla 60 kW h and the emissions/mile when charged from coal- and gas-fired generation for comparison with emissions from a typical petrol car at 200 g CO_2/mi.

Answer:

Fuel	gC/GJ	tC/ mtoe	tCO_2/ GJ	Price/ GJ	Efficiency	Price/ MW h	tCO_2/ MW h	$kgCO_2$/ kW h
Lignite/ brown coal	25,200	1.169	0.0924	0.924	33	10.08	1.008	1.008
Bituminous coal	23,700	1.1	0.0869	0.869	38	8.23	0.8232632	0.82326316
Oil	19,900	0.923	0.072967	0.729667	40	6.57	0.6567	0.6567
Natural gas	13,500	0.626	0.0495	0.495	52	3.43	0.3426923	0.34269231

Tesla 60 kW h uses   0.244898 kW h/mi
Emissions from gas   83.92465 g/mi
Emissions from coal 201.6155 g/mi

Question 3.1
The fossil generation gas and coal fuel costs were based on coal at £4.4/GJ and gas at £7.8/GJ with CO_2 at £70/t giving total costs of £10.48 and £11.26/GJ and are summated for all plant types in the table. Assuming that fuel prices do not rise as high then based on the data in Table 3.1 and assuming that the CO_2 price was £35/t with coal at £3/GJ and gas at £6/GJ, calculate the new gas and coal fuel costs and hence the estimated total fossil costs for comparison with the renewable costs. Assuming that 20 GW of the offshore wind is contracted by competitive tender at

£100/MW h and the same level of curtailment applies, find the revised renewable costs on a pro rata basis and the new total.

Answer:

	Existing CO$_2$ £70/t		New CO$_2$ £35/t	
	Gas, £/GJ	Coal, £/GJ	Gas	Coal
Fuel	7.8	4.4	6.00	3.00
CO$_2$	3.46	6.08	1.73	3.04
Total	11.26	10.48	7.73	6.04
Total costs	5,003.4	1,719.8	3,434.8	991.2
All fuel costs	6,723.2		4,426.0	
+Capital/year	18,142.2		15,845.0	

With 20 GW of offshore wind at £100/MW h
New offshore wind cost = 11,948*(2,000*100+15,956*155)/(35,956*155) = 9,469.6
This gives a new total of £19,763 m/year

Question 4.1
Using data in Table 4.7, calculate the unit cost if the onshore wind load factor is 20% and the offshore 25% and also with an onshore load factor of 30% and offshore of 38%.

Answer:

	Wind on	Wind off
	Unit cost, £/MW h	
Load factor	30	38
Unit cost	**74.8**	**126.5**
Load factor	25	30
Unit cost	**89.6**	**147.2**
Load factor	20	25
Unit cost	**111.7**	**176.2**

Question 5.1
Given data in the table for a microgeneration calculate that price of avoided electricity that would result in a payback within 15 years based on the lower capital costs.
   What would the payback period be if the boiler gas price is doubled?

Answer: The payback period is equal to the project life of 15 years with an electricity price of €55.65/MW h.
   If the gas price is doubled to €50/MW h, the payback period for the lower capital cost unit is 15.3 years.

Question 6.1
What would be the impact on the unit cost of extending the life of the nuclear station to 40 years rather than 30 as shown in the table below. What would be the impact of the load factor extending to 85% rather than 80%.

Answer:

40-year life	83.95	£/MW h
85% LF	83.51	£/MW h

Question 7.1

Using data shown in the table calculate the cost of conventional gas-fired generation assuming an efficiency of 55%. At what CO_2 price would the costs of conventional generation equal that form gas-fired generation with CCS.

Answer:

Basic unit cost no CCS	77.83	£/MW h
Price difference with CCS	19.69	£/MW h
Emissions no CCS	0.32	tCO_2/MW h
CO_2 price to equal with CCS	60.77	£/tCO_2

Question 8.1

Given wind output shown in the table, identify the new marginal unit form dispatch at each load level and the associated marginal price. Hence, find the average marginal price for the day.

Answer: £22.87/MW h.

Hour	Hourly demand, MW	Marginal gen, no wind	Marginal gen. price, £/MW h	Wind output, MW	Marginal gen, with wind	Marginal gen. price, £/MW h
1	1,520	4	23	75	4	23
2	1,470	4	23	135	4	23
3	1,400	4	23	195	3	22
4	1,250	4	23	270	3	22
5	1,300	4	23	330	3	22
6	1,350	4	23	450	3	22
7	1,450	4	23	540	3	22
8	1,600	5	24	630	3	22
9	1,650	5	24	570	3	22
10	1,750	5	24	630	3	22
11	1,800	5	24	600	3	22
12	1,825	5	24	450	4	23
13	1,785	5	24	300	4	23
14	1,800	5	24	390	4	23
15	1,850	6	24.5	300	5	24
16	1,870	6	24.5	210	5	24
17	1,975	6	24.5	150	5	24
18	1,840	6	24.5	240	5	24
19	1,780	5	24	270	4	23
20	1,700	5	24	195	4	23
21	1,670	5	24	165	4	23
22	1,650	5	24	75	5	24
23	1,600	5	24	60	4	23
24	1,580	5	24	30	5	24
						22.875

Question 9.1

Given data in the table calculate the cost in €/MW h of delivered wind energy in catering for the provision of reserve to manage intermittency in each of the 3 years.

Answer: Reserve cost/MW h of delivered wind energy in Germany.

$2008 = 82m/(8{,}760 * 3{,}520) = €2.6/MW\ h$
$2015 = 159m/(8{,}760 * 7{,}040) = €2.5/MW\ h$
$2020 = 217m/(8{,}760 * 10{,}560) = €2.3/MW\ h$

Germany		Year		
		2008	**2015**	**2020**
Reserve for wind		Based on 4 h forecast error		
Wind capacity	MW	20,000	40,000	60,000
Expected average output	MW	3,520	7,040	10,560
Wind LF	%	18	18	18
Forecast RMS error (std)	%	17	13	10
Marginal wind effect 3* std	%	51	39	30
Average MW reserve	MW	1,795	2,746	3,168
Reserve energy used	TW h	2.73	4.18	4.82
Price premium	€/MW h	30	38	45
Total cost	**€m**	**82**	**159**	**217**

Question 10.1

Assuming that the data in the table below can be represented by an average unit size of 61 MW and that the volume-weighted average availability is 43% use equation (10.1) to calculate the availability profile of the embedded generation.

Answer:

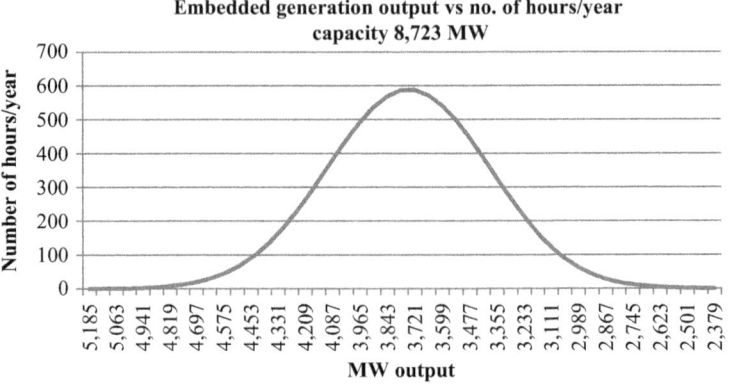

Question 11.1

To evaluate risks, calculate the impact on the price of energy delivered to the United Kingdom from Iceland if the link cost/km is €3,000k rather than €2,500k and also if the link availability reduces utilisation to 75% rather than 85%.

Answer: At a capital cost of €3,000k/km, the UK energy cost would be £69.02/MW h.

At a utilisation of 75%, the UK energy cost would be £66.99/MW h.

Question 12.1

Compare the total annual cost of realising a reduction in peak demand using DSR with the characteristics as illustrated in Table 12.3 with the alternative of installing OCGTs at a capital cost of $400/kW with a 7% interest rate to meet the same level of demand. The DSR set-up cost is assumed to be $182/kW. Calculate the extra fuel costs in using OCGTs instead of CCGTs assuming that CCGT fuel costs are $26.43/MW h compared to OCGT fuel costs of $36.81/MW h with peak lopping for 500 h/year.

Answer: The DSR incentive costs are $39.6/kW/year plus a set-up cost equivalent to $17/kW/year i.e. a total of $56.6/kW/year.

The OCGT capital costs at $400/kW are equivalent to an annual charge of $38/kW/year.

The extra costs of meeting peak demands with less efficient OCGTs for 500 h/year is given by

$$(\$36.81 - \$26.43)/\text{MW h} * 500/\text{year} = \$5,190/\text{year or } \$5.19/\text{kW/year}.$$

Question 13.1

Calculate the impact on global emissions, if Europe/Eurasia replaced the use of coal with gas. Also calculate the impact on emissions if N. America doubled coal use at the expense of gas utilisation.

Answer: The impact on global emissions, if Europe/Eurasia replaced the use of coal with gas at 16,015 TW h emissions $= 32,994$ MtCO$_2$ i.e. a reduction of 2.1%. The impact on emissions if N. America doubled coal use at the expense of gas utilisation – coal at 9,002 TW h and gas at 5,803 TW h emissions $= 34,306$ MtCO$_2$, an increase of 606 MtCO$_2$ or 1.8%.

Question 14.1

Using data in the table calculate the impact on 'le' and 'li' end-user prices if 20% of the peaking energy comes from renewable sources at €90/MW h displacing conventional generation.

Answer: The peak price ratio to baseload prices increases to 1.4 with 'le' prices rising from €61.5/MW h to €66.9/MW h or by 8.8%.

And 'li' prices rising from €49.6/MW h to €51.6/MW h or by 4%.

Question 15.1

The developing system will require an increasing amount of flexibility to accommodate

- the intermittency of renewable generation sources;
- the output from embedded generation that is not subject to central dispatch;
- an increase in domestic demand form EVs and heat pumps that may have limited diversity.

Calculate the value of the provision of flexibility by demand management in terms of the avoided costs resulting from a 1% reduction in wind energy curtailment with CO_2 emissions valued at £30/t. Assume that the wind load factor is 25% and replacement energy is based on OCGT generation with emissions of 0.54 tCO_2/MW h.

Hence, calculate the value of avoiding curtailment of 30 GW of wind based on the more inflexible curtailment estimate.

Answer: Based on a price of £30/t, the cost/MW h = 30 * 0.54 = £16.2/MW h

Based on a load factor of 25%, each MW of wind capacity generates

$$1 * 8,760 * 0.25 = 2,190 \text{ MW h/year.}$$

A curtailment of 1% reduces the energy output by 21.9 MW h/year.

Then, the value of 1% of the avoided annual emissions = 21.90 * 16.2 = £354/year/MW of installed wind capacity.

The curtailment of 30 GW of wind = 6% or 30,000 * 8,760 * 0.25 * 0.06 = 3.942 TW h.

At a cost of £16.2/MW h/year cost = 16.2 * 3.942 = £63.8 M/year.

Question 16.1

Taking account of the plant mix data in Table 16.4, explain the relative changes in total costs from the base case resulting from the parameter changes shown in the table. By what amount would the offshore wind FIT cost have to rise to make the scenario costs equal.

Variable	Units	Fast CCS 2	Green	Difference
Base case costs	Cost, £m/year	34,248	40,641	6,393
Capital at 12% WACC	Cost, £m/year	36,425	42,284	5,859
CO_2, £200/t	Cost, £m/year	37,937	44,012	6,075
Offshore FIT £120/MW h	Cost, £m/year	33,434	37,973	4,539
Offshore FIT £100/MW h	Cost, £m/year	32,783	36,402	3,619
CCS cap., £1,422/2,708/kW	Cost, £m/year	34,532	40,719	6,187
12%, CO_2 200, FIT 100, cap	Cost, £m/year	38,491	41,443	2,952
CCS capture 90% base	Emiss., Bt/year	22	22	0

Answer: The higher WACC impacts more on the CCS scenario more because of the higher capital cost with wind capital included in the FIT.

The higher CO_2 cost impact on CCS is because of its limited capture rate offset by less unabated generation.

The lower wind FIT costs benefit the green scenario.

The higher CCS capital costs impact directly on the CCS option costs.

A £20 drop reduces costs from 37,973 to 36,402 i.e. by £1,571 m/year. To rise, 6,393 would require a rise of 20*6,393/1,571 = £81.4/MW h.

Question 17.1
Based on the generator data shown below, calculate the effective overall unit cost/MW h of running up at maximum rate to the minimum stable generation of 300 MW and after a period running down at maximum rate to zero with an overall connection time of 3 h from synchronising to shut down.

Answer:

Time, min	MW	MW h
0	60	
40	300	120
165	300	625
180	0	37.5
Total, MW h		782.5
Total cost, £		15,408
Unit cost, £/MW h		19.69

Profile MW output

Question 18.1
The table shows potential saving in generation capacity from demand profiling. Calculate the saving in capacity in £bn/year based on gas-fired CCGT and peaking generation. If an average consumer can exercise demand control of 1.5 kW, calculate the number of consumers that are required to participate and the gross benefit to each consumer per year.

Answer:

WACC 7% 20 years	Units	Gas	Peaking
Unit size	MW	400	120
Capital cost	£m/year	28.3	3.78
Capacity reduction	GW	9.00	6.00
Capital cost saving	£bn/year	0.64	0.19
Potential saving	£/kW/year	70.8	31.5
Number of consumers at 1.5 kW	Million	6.00	4.00
Saving/consumer/year	£/year	106.1	47.3

Questions 19

1. Draw a comparison between the cost implication of the provision of flexibility from generation, demand side control and storage.
2. Comment on the impact of intermittent renewable generation on the optimal plant mix that meets system dynamic requirements.

Answers:

1. Generation flexibility helps to meet varying demand rather than influence it;
    Storage facilitates moving energy from one period to another but has an associated efficiency penalty;
    Demand side control exploits the end-user inherent flexibility to store heat and energy in batteries that enables the timing of demand to be adjusted.
2. The requirement is for more generation, demand side flexibility or storage to track the variations in the output from intermittent renewable sources with lower overall utilisation. For generators, this will mean plant with less emphasis on efficiency but with lower capital costs like OCGTs instead of more efficient CCGTs.

Glossary

AAC	Already Allocated Capacity of interconnecting transmission routes
Active Network	one containing active energy sources
ADMD	after diversity maximum demand
AGC	Automatic Generation Control used to effect short-term generation control
APX	Amsterdam Power Exchange supporting trading in NW Europe
ATC	Available Transmission Capacity on interconnecting transmission routes
AVR	Automatic Voltage Regulators used on generation to control voltage
Ancillary Services	that enable the system operator to maintain a stable system
Balancing Market	market used to effect short-term balance of supply and demand
Base Load	describes a generator operating throughout a period of time
Bilateral Contracts	contracts for supply of energy between generators and suppliers
Black Start	capability of generation to start up without external supplies
BRP	a Balance Responsible Party able to make submissions of contracted positions
BST	Bulk Supply Tariff used to supply energy wholesale from generation groups to suppliers or distribution companies that engage in supply
BSUoS	Balancing System Use of System Charges
Capacity Charge	charge made to cover the provision of assets to effect supply
CAPEX	the Capital Expenditure of an organisation
CCGT	Combined Cycle Gas Turbine
CCS	Carbon Capture and storage from generation processes and its transportation

CDCA	Central Data Collection Agent responsible for collecting tariff metering data
CEER	Council of European Energy Regulators
CEGB	UK Central Electricity Generating Board now defunct
CfD	Contract for Differences
CGN	Chinese General Nuclear Power Corporation
CHP	Combined Heat and Power system producing electricity and heat
Contingency Reserve	reserve to cover loss of planned generation
Contracts for Difference	bilateral contract with a set price independent of pool price
DCLF	DC Load Flow model used to analyse the flow of MWs on networks
DNC	Declared net Capacity
DSO	Distribution System Operator
EEX	European Energy Exchange supporting trading across central Europe
Equal Lambda Criteria	occurs when marginal production prices are equal
Embedded Generation (EG)	generation connected to the distribution network
Energy Rate	tariff for supply of energy
EOR	Enhanced Oil Recovery through pumping CO_2 into partially depleted field
EPR	European Pressurised Reactor
ERGEG	Association of European Regulators Group for Electricity and Gas
ETS	Emission Trading Scheme
ETSO	Association of European Transmission System Operators
EU	European Union
Eurostat	European organisation publishing energy-related statistics
EV	Electric Vehicle
Exchange	centres supporting trading in energy products
FIT	Feed in Tariff
FPN	Final Physical Notification of expected transfers through a designated node
Gate Closure	time before event for submissions of all final contracted positions
GDP	Gross Domestic Product
GHG	Green House Gases

GSP	Grid Supply Point from super-grid to lower voltage grid distribution system
Hedging Contracts	to defray the risks associated with contracted positions
HVDC	High Voltage Direct Current
ICRP	Investment Cost-Related Prices for use of transmission systems based on the investment cost of assets used
IPP	Independent Power Producer being a generator separate from any residual state organisation
LDC	Load Duration Curve expressing demand for period as a curve against time
LF	Load Factor
Liberalisation	the process of enabling open access to the power market
Liquidity	a measure of the number of times a physical commodity is traded
Load Factor (LF)	the relation between the peak load and average load expressed as %
LMP	Locational Marginal Price for energy at a reference node
LNG	Liquid Natural Gas created to enable transportation of gas from remote sites
LOLP	Loss of Load Probability is the probability of load exceeding generation
LP	Linear Programme
LRMC	Long Run Marginal Cost
Makeup Cost	the cost in a market of buying extra energy from a reserve generator
Marginal Cost	the cost of an extra increment of energy on the system
MCP	Marginal Cost of Production
Merit Order	a list of available generation in ascending price order
NAP	National Allocation Plan – a plan of how CO_2 allowances are distributed
NGC	National Grid Corporation
NISM	Notification of Inadequate System Margin
Nordpool	the market organisation covering the Scandinavian area
NSG	None Synchronous Generation
NUG	Non-Utility Generator being separate from the residual utility

OCGT	Open Cycle Gas Turbine
OPEC	Oil Producing Exporting Countries
OPEX	Operating Expenditure of organisation
Passive Network	radial without active energy sources
PCC	Post Combustion Capture
Peak Load	the highest load on a system during a defined period
PJM	Pennsylvania Jersey Maryland US utility
Powernext	the power exchange in France covering Northern Europe
PPA	Power Purchase Agreement for the sale of energy and services from a generator
Primary Reserve	reserve available in seconds to support system frequency control
PSP	Pool Selling Price for energy from a pool to suppliers/distribution companies
PV	Photo Voltaic
Quality of Supply	the quality of voltage and frequency and security
RAV	Regulated Asset Value – value of those assets of monopoly subject to regulation
RES	Renewable Energy Source
Reserve	spare generation or interruptible demand held in reserve for system needs
RIIO	Regulation, Incentives, Innovation Output
RMS	root mean square
ROC	Renewable Obligation Certificate
ROCOF	Rate of Change of Frequency
RTO	Regional Transmission Operator – an operator covering several control areas
RPI	Retail Price Index monitoring the change in retail prices to customers
SB	Single Buyer
SCADA	Substation Control and Data Acquisition
Secondary Reserve	spare generation available within minutes to meet system needs
SG	Self Generation
SMP	System Marginal Price being the price of the next increment of generation
SMR	Small Modular Reactor
SNSP	System Non-Synchronous Penetration
SO	System Operator managing the operating of the network in conjunction with generators and users

Spill Price	the price paid to independent generators for energy supplied in excess of their output contracted to customers
SRMC	Short Run marginal Cost
STOR	Short Term Operating Reserve
Super-grid	the high-voltage network facilitating the pooling of large-scale generation
TOU	Time of Use Tariff
Transmission Constraints	limitations on transfers due to system security needs
Triad	the three peak demands in a year separated by more than 10 days
TRM	Transmission Reserve Margin capacity left available for emergency transfers
TSO	Transmission System Operator responsible for network operation and security
TTC	Total Transmission Capacity of interconnecting routes between systems
UCTE	Union for the Coordination and Transmission of Electricity
UoS	Use of System – used to describe charges for use of a network
Unit Commitment	the process of selecting which generators should be committed to run through a period
Uplift	the added cost in a pool of accommodating network constraints and services
Vertically Integrated Company	company owning generation and supply business
VLL	Value of Lost Load being the cost to consumers of failure to supply energy
WACC	Weighted Average Cost of Capital
Wheeling	the process of transferring energy between two systems through a third

Bibliography

APX Amsterdam Power Exchange. www.apx.nl.

Barrie Murray – Electricity Markets – Investment, Performance & Analysis – J. Wiley Chichester 1998. ISBN 0-471-98507-4.

Barrie Murray – Power Markets & Economics – Energy Costs, Trading & Emissions – J. Wiley Chichester 2009. ISBN 978-0-470-77966-8.

Dragana Pilipovic – Energy Risk – McGraw Hill 1997. ISBN 0-7863-1231-9.

DUKES – Digest of UK Statistics 2015.

EEX European Energy Exchange. www.eex.de.

ERG European Regulators Group. www.erg.eu.int.

ETSO European Transmission System Operators. www.etso-net.org.

European Commission – EU-Smart Grids Project Outlook 2017.

FERC Federal Energy Regulatory Commission. www.ferc.fed.us.

Loi Lei Lai – Power System Restructuring and Deregulation – J. Wiley. ISBN 0 47149500 X.

SOF – National Grid System Operation Framework 2016.

UCTE Union for the Coordination of the Transmission of Electricity. www.ucte.org.

Index